Medical Implications of Biofilms

Human tissues often support large, complex microbial communities growing as biofilms that can cause a variety of infections. As a result of an increased use of implanted medical devices, the incidence of these biofilm-associated diseases is increasing: the non-shedding surfaces of these devices provide ideal substrata for colonisation by biofilm-forming microbes. The consequences of this mode of growth are far-reaching. As microbes in biofilms exhibit increased tolerance towards antimicrobial agents and decreased susceptibility to host defense systems, biofilm-associated diseases are becoming increasingly difficult to treat. Not surprisingly, therefore, interest in biofilms has increased dramatically in recent years. The application of new microscopic and molecular techniques has revolutionised our understanding of biofilm structure, composition, organisation, and activities, resulting in important advances in the prevention and treatment of biofilm-related diseases. The purpose of this book is to bring these advances to the attention of clinicians and medical researchers.

Michael Wilson is currently Professor of Microbiology in the Faculty of Clinical Sciences at University College London. Wilson's research interests include biofilms, antibiotic resistance, and bacterial virulence factors. Wilson has authored more than 200 scientific papers as well as six books, including *Bacterial Disease Mechanisms: An Introduction to Cellular Microbiology* (Cambridge, 2002).

Deirdre Devine is currently a Senior Lecturer in Microbiology at Leeds Dental Institute, University of Leeds. In addition to biofilms, Devine's research interests include antimicrobial peptides in control of resident microflora, oral anaerobes, and environmental regulation of bacterial virulence factors.

Medical Implications of Biofilms

Edited by

MICHAEL WILSON

Eastman Dental Institute
University College London

DEIRDRE DEVINE

Leeds Dental Institute
University of Leeds

CAMBRIDGE UNIVERSITY PRESS
Cambridge, New York, Melbourne, Madrid, Cape Town,
Singapore, São Paulo, Delhi, Tokyo, Mexico City

Cambridge University Press
The Edinburgh Building, Cambridge CB2 8RU, UK

Published in the United States of America by Cambridge University Press, New York

www.cambridge.org
Information on this title: www.cambridge.org/9781107403451

© Cambridge University Press 2003

This publication is in copyright. Subject to statutory exception
and to the provisions of relevant collective licensing agreements,
no reproduction of any part may take place without the written
permission of Cambridge University Press.

First published 2003
First paperback edition 2011

A catalogue record for this publication is available from the British Library

Library of Congress Cataloguing in Publication data
Medical implications of biofilms / [edited by] Michael Wilson, Deirdre Devine.
p. cm.
Includes bibliographical references and index.
ISBN 0-521-81240-2
1. Biofilms. 2. Medical microbiology. I. Wilson, Michael, 1947–
II. Devine, Deirdre, 1961–
QR100.8.B55 M435 2003
616′.01 – dc21 2002034957

ISBN 978-0-521-81240-5 Hardback
ISBN 978-1-107-40345-1 Paperback

Additional resources for this publication at www.cambridge.org/9781107403451

Cambridge University Press has no responsibility for the persistence or
accuracy of URLs for external or third-party internet websites referred to in
this publication, and does not guarantee that any content on such websites is,
or will remain, accurate or appropriate.

Contents

*All figures in this section are available in color for download from www.cambridge.org/9781107403451

Preface

Biofilms are increasingly recognised as the preferred mode of growth of bacteria in a wide range of sites, including the many, and varied, habitats present in man. Until approximately 10 years ago, biofilms were regarded as consisting simply of an accumulation of bacteria and their products on a surface. But recent technical advances have significantly changed our understanding of biofilm organisation and function. Non-destructive methods of examining organisms in their living, hydrated state (for example, confocal laser scanning microscopy) have revealed that they have an ordered structure, often permeated by water channels, which can function as a primitive circulation system. Modern molecular techniques have identified genes that are up- and down-regulated in different regions of a biofilm – this has important implications for assessing the behaviour of organisms in biofilms (for example, virulence potential, susceptibility to antimicrobiol agents, and host defence mechanisms). Furthermore, the discovery of population-dependent gene regulation (quorum sensing) in bacteria has meant revising our concept of bacteria as being independently operating cells. Quorum-regulated gene regulation is, obviously, likely to play an important role in determining the collective properties of bacteria in biofilms.

During the last few years, interest in microbial biofilms has increased dramatically and now encompasses a broad range of disciplines – microbiology, molecular biology, microscopy, medicine, engineering, ecology, and marine biology. This increased specialisation within the field has led to a need for books that focus on particular aspects of biofilms. This book is concerned with the roles played by biofilms in infections of man, including those associated with prosthetic devices (catheters, implants, etc.) and teeth (caries, periodontitis, stomatitis), as well as certain lung, gut, and vaginal infections.

This book consists of four main sections. The opening chapters review key general aspects of the subject – gene expression and quorum sensing in biofilms – as well as the susceptibility of these communities to antimicrobial agents. Succeeding sections then deal with the three types of infection with which biofilms are associated – those of prosthetic devices, teeth, and shedding surfaces.

This book is written by leading researchers in the field and will be of interest to both scientists and clinicians.

List of Contributors

Yuehuei H. An
Orthopaedic Research Laboratory
Medical University of South Carolina
Charleston, South Carolina, USA

Samir Bhagwat
Department of Medicine
University of Rochester
Rochester, New York, USA

Werner Bokranz
Department of Cell Biology
Gesellschaft für Biotechnologische Forschung
Braunschweig, Germany

David J. Bradshaw
Bioscience Department
Quest International
Ashford, Kent, UK

Michael R. W. Brown
Department of Pharmacy and Pharmacology
University of Bath
Bath, UK

Robert A. Burne
Department of Oral Biology
University of Florida
Gainesville, Florida, USA

Yi-Ywan M. Chen
Department of Oral Biology
University of Florida
Gainesville, Florida, USA

Teresa R. de Kievit
Department of Microbiology
University of Manitoba
Winnipeg, Manitoba, Canada

Richard J. Friedman
Orthopaedic Research Laboratory
Medical University of South Carolina
Charleston, South Carolina, USA

Ulrich Gerstel
Microbiology and Tumorbiology Centre
Karolinska Institute
Stockholm, Sweden

Sean P. Gorman
School of Pharmacy
Medical Biology Centre
Queen's University of Belfast
Belfast, UK

Hend A. Hanna
Department of Infectious Diseases
Infection Control and Employee Health
University of Texas M.D. Anderson Cancer Center
Houston, Texas, USA

Christine Heilmann
Institute of Medical Microbiology
University of Münster
Muenster, Germany

Christopher M. Hill
Orthopaedic Research Laboratory
Medical University of South Carolina
Charleston, South Carolina, USA

Barbara H. Iglewski
Department of Microbiology and Immunology
University of Rochester School of Medicine and Dentistry
Rochester, New York, USA

Kimberly K. Jefferson
Channing Laboratory
Department of Medicine
Brigham and Women's Hospital
Harvard Medical School
Boston, Massachusetts, USA

David S. Jones
School of Pharmacy
Medical Biology Centre
Queen's University of Belfast
Belfast, UK

Philippe Lejeune
Laboratoire de Microbiologie et Génétiquè
Institut National des Sciences Appliquées de Lyon
Lyon, France

Yunghua Li
Faculty of Dentistry
University of Toronto
Toronto, Ontario, Canada

C. Y. Loo
Boston University, Goldman School of Dental Medicine
Department of Paediatric Dentistry
Boston, Massachusetts, USA

Heinrich Lünsdorf
Department of Microbiology
Gesellschaft für Biotechnologische Forschung
Braunschweig, Germany

G. T. Macfarlane
MRC Microbiology and Gut Biology Group
University of Dundee
Ninewells Hospital Medical School
Dundee, UK

S. Macfarlane
MRC Microbiology and Gut Biology Group
University of Dundee
Ninewells Hospital Medical School
Dundee, UK

Philip D. Marsh
Department of Oral Microbiology
Leeds Dental Institute
Leeds, UK

Manfred Nimtz
Department of Structural Biology
Gesellschaft für Biotechnologische Forschung
Braunschweig, Germany

Gerald B. Pier
Channing Laboratory
Department of Medicine
Brigham and Women's Hospital
Harvard Medical School
Boston, Massachusetts, USA

Issam Raad
Department of Infectious Diseases
Infection Control and Employee Health
The University of Texas M.D. Anderson Cancer Center
Houston, Texas, USA

Wolfgang Rabsch
National Reference Centre for Salmonellae and
Other Enteric Pathogens
Robert-Koch-Institut
Wernigerode, Germany

Ute Römling
Microbiology and Tumorbiology Centre
Karolinska Institute
Stockholm, Sweden

Anthony W. Smith
Department of Pharmacy and Pharmacology
University of Bath
Bath, UK

Helmuth Tschäpe
National Reference Centre for Salmonellae and
Other Enteric Pathogens
Robert-Koch-Institut
Wernigerode, Germany

Zezhang Wen
Department of Oral Biology
University of Florida
Gainesville, Florida, USA

Xhavit Zogaj
Microbiology and Tumorbiology Centre
Karolinska Institute
Stockholm, Sweden

PART ONE

INTRODUCTORY CHAPTERS

CHAPTER ONE

Biofilm-Dependent Regulation of Gene Expression

Philippe Lejeune

1 INTRODUCTION

Microbial development and biofilm formation on implanted biomaterials and hospital equipment are important sources of nosocomial infections, mainly because surface-associated contaminants express biofilm-specific properties such as increased resistance to biocides, antibiotics, and immunological defences. Although it has long been recognised that the presence of a solid phase can influence many bacterial functions (ZoBell, 1943; Costerton et al., 1987; Van Loosdrecht et al., 1990), we are just beginning to understand the regulatory processes at the molecular level. There is no doubt that the identification of the structures involved in the sensing of the particular microenvironments encountered at interfaces and in developing biofilms and the description of the regulatory networks allowing the appropriate genetic responses will lead to the development of surface coatings and preventive or curative drugs able to deal with these life-threatening infections.

2 BIOFILM FORMATION IS A DEVELOPMENTAL PROCESS

An invidual bacterium present on, or introduced into, the human body can reach the surface of an indwelling medical device by three different mechanisms (Van Loosdrecht et al., 1990): passive transport due to air or liquid flow, diffusive transport resulting from Brownian motion, and active movement involving flagella. Although contact is, therefore, frequently a question of chance, chemotactic processes can direct motile bacteria in response to any concentration gradient that may exist in the interfacial region. Following contact, the next stage may be initial adhesion. This is mainly a physicochemical process based on weak interactions between molecules of the solid phase (or ions and polymers adsorbed on the surface) and appropriate cell surface structures, such as fimbriae and adhesins.

Numerous studies with model bacteria have identified genes and functions required for adhesion, and a picture of the early stages of colonisation has begun to emerge. First, an individual bacterium that approaches a solid surface has to overcome possible repulsive forces and then interact with the solid phase. Depending on the strength of the bonds that the bacterium is able to form with the substrate, motility and gliding properties are often of crucial importance to initiate efficient attachment. Transposon mutations leading to the suppression of the adherent phenotype of *Pseudomonas aeruginosa* (O'Toole and Kolter, 1998a), *P. fluorescens* (O'Toole and Kolter, 1998b), *Vibrio cholerae* (Watnick and Kolter, 1999), *Salmonella enterica* Serovar Typhimurium (Mireles, Togashi, and Harshey, 2001), and the W3110 strain of *Escherichia coli* (Pratt and Kolter, 1998) have been found in genes involved in flagellar motility. Accordingly, a non-adherent phenotype could be detected after transposon inactivation of two types of bacterial gliding movement: twitching motility relying on type IV pilus extension and retraction in *P. aeruginosa* (O'Toole and Kolter, 1998a), and swarming due to overflagellation in *S. enterica* (Mireles et al., 2001). Time-lapse microscopic observations of *P. aeruginosa* adhesion confirmed that the organisms move along the surface before attachment, almost as if they are scanning for an appropriate location for initial contact (O'Toole, Kaplan, and Kolter, 2000).

The next step in early attachment events is an interaction between the surfaces of the bacterium and the material sufficiently strong to prevent disruption by convective forces or Brownian motion. It has been recognised for some time that the introduction of a clean substratum in a natural fluid is immediately followed by fast and efficient adsorption of organic molecules to the surface (ZoBell, 1943), forming a so-called conditioning film. Two types of interactions are then possible: weak chemical bonding between the bacterial envelope and the solid surface or the conditioning film, and bridging mediated by specialised bacterial attachment structures. The first link between the bacterium and the material is presumably a combination of weak chemical bonds, dipole interactions, and hydrophobic interactions (Marshall, 1992).

Different genetic strategies have been used to identify the structural components of the bacterial envelope involved in these interactions. In various species, many natural isolates are not able to adhere to abiotic surfaces under laboratory conditions. But the reservoir of cryptic functions is so large that cyclic flow experiments (which exercise a strong selective pressure in favour of adherent mutant cells) can easily reveal potential adhesion structures (Le Thi et al., 2001). Such an approach was followed to isolate a point mutation in a regulatory gene of *E. coli* K-12, which resulted in the overproduction of curli, a particular type of thin and flexible fimbrium, and allowed the overproducing strains to adhere to any type of material (Vidal et al., 1998). As *E. coli* is the most common bacterium found in biofilms that have developed on catheters introduced into the urinary tract, immunological and genetic studies were undertaken to investigate the role of curli in clinical strains isolated from patients with catheter-related infections. Immunogold labelling with curlin antibodies revealed the constitutive production of these fimbriae at the surface of the bacteria, and transduction of knock-out mutations in the curli-encoding genes demonstrated their essential role in adhesion to biomaterials (Vidal et al., 1998; Prigent-Combaret et al., 2000). Therefore, curli synthesis by other pathogenic strains, such as O157:H7 (Uhlich et al., 2002), has to be regarded as a potential biofilm-forming character.

Identification of bacterial structures of attachment has also been performed by transposon mutagenesis followed by screening for non-adherent clones. In the Gram-positive species *Staphylococcus epidermidis* (Heilmann et al., 1997), *Streptococcus gordonii* (Loo, Corliss, and Ganeshkumar, 2000), and *Staphylococcus aureus* (Cucarella et al., 2001), this approach allowed the characterisation of new envelope proteins involved in surface colonisation. Using a similar type of screening, Vallet and co-workers (2001) detected a new fimbrial adhesin of *P. aeruginosa* and demonstrated the requirement of a periplasmic chaperone involved in pilus synthesis via the so-called chaperone/usher pathway. Similar studies in *E. coli* revealed the importance of accurate lipopolysaccharide synthesis for bacterial attachment (Genevaux et al., 1999a) and identified type I fimbriae as another structure able to promote adhesion (Pratt and Kolter, 1998). Interestingly, the physicochemical processes of *E. coli* adhesion mediated by curli and type I fimbriae are clearly different, since type I pilus-associated attachment requires flagellar motility (Pratt and Kolter, 1998), whereas adhesion mediated by curli is independent of strain motility (Prigent-Combaret et al., 2000).

Recently, Ghigo (2001) demonstrated that the conjugative pili encoded by transferable plasmids (including F) of several incompatibility groups could by themselves act as adhesion factors and promote biofilm development. Furthermore, plasmid transfers by conjugation seemed to be favoured in biofilms (Hausner and Wuertz, 1999; Ghigo, 2001). These observations are of great medical and evolutionary significance because they raise the question of the role of biofilms as a place for the evolution of structures for bacterial interactions and for the horizontal spreading of genes, such as those encoding antibiotic resistance mechanisms.

When an individual bacterium has reached an abiotic surface and is immobilised by relatively firm links, a cascade of physiological changes is initiated. More than 10 years ago, Van Loosdrecht and co-workers (1990) published an exhaustive review of the early literature on the influence of surfaces on microbial behaviour. Changes in global functions, such as growth rate, respiration, and assimilation, could be correlated with substratum-attached growth. More recently, the use of microscopy and reporter gene techniques to quantify gene expression in biofilms clearly established that precise changes in gene expression are triggered during the transition between the free-living and attached states (for a review of these methods, see Prigent-Combaret and Lejeune, 1999).

One of the first features of this transition is of pivotal importance for biofilm development. As fimbriae- and adhesin-mediated interactions are relatively weak, the early stages of the adhesion process are generally reversible. For a bacterium immobilized at the solid–liquid interface, the 'choice' of further surface-associated growth involves multiplication and production of extracellular polymers, resulting in the formation of a first slimy layer on the substratum. In *P. aeruginosa* biofilms, the bacteria are embedded in a polymer matrix that is composed mainly of alginate. Davies and co-workers (1993, 1995) examined the expression of *algC*, a gene required for alginate synthesis, within individual colonising cells. As early as 15 minutes after the initial attachment, they observed an activation of *algC* expression.

By using a library of *lacZ* fusions and a colourimetric assay based on biofilm development in the wells of microtitre plates, Prigent-Combaret and co-workers (1999) showed that the expression of about 40 per cent of the genes of an *E. coli* biofilm-forming strain was modified during the colonisation process. As in *P. aeruginosa*, the synthesis of colanic acid, a major matrix exopolysaccharide, was induced in the biofilm-grown cells. They also observed that the synthesis of the flagella was stopped in the attached cells by downregulation of the *fliC* gene encoding the flagellar structural protein.

Proteome and transcriptome analysis in *P. putida* and *P. aeruginosa* (Table 1.1) recently confirmed the deep physiological changes induced upon bacterial contact with a surface (Sauer and Camper, 2001; Whiteley et al., 2001). These studies also gave further insights into the differences between free-living cells and those in biofilms (discussed later). Analyses of protein and gene expression at different time points suggested that the colonising bacteria undergo a succession of physiological states, which could be

Table 1.1: Selected genes and proteins differentially expressed in biofilms developed on abiotic surfaces

Gene or Locus Number	Protein	Function	Expression[a]	Organism	Reference
Exopolysaccharide and lipopolysaccharide production					
algC		Alginate synthesis	Up	*P. aeruginosa*	Davies, Chakrabarty, and Geesey, 1993
wcaB		Colanic acid synthesis	Up	*E. coli*	Prigent-Combaret et al., 1999
mucC		Negative regulator for alginate synthesis	Up	*P. putida*	Sauer and Camper, 2001
lpxD	UDP-3-0[hydroxylauroyl] glucosamine *N*-acetyltransferase	Lipopolysaccharide synthesis	Up	*P. putida*	Sauer and Camper, 2001
wbpG		Lipopolysaccharide synthesis	Up	*P. putida*	Sauer and Camper, 2001
Motility and attachment					
fliC	Flagellin	Flagellar synthesis	Down	*E. coli*	Prigent-Combaret et al., 1999
csgA	Curlin	Curli synthesis	Up	*E. coli*	Prigent-Combaret, 2000
fleN		Flagellar synthesis regulator	Down	*P. putida*	Sauer and Camper, 2001
flgG	Flagellar basal body rod protein	Flagellar synthesis	Down	*P. putida*	Sauer and Camper, 2001
pilC		Type IV fimbrial synthesis	Up	*P. putida*	Sauer and Camper, 2001
pilK	Methyltransferase CheR homolog	Chemotactism	Up	*P. putida*	Sauer and Camper, 2001
PA2128	Probable fimbrial protein	Fimbrial synthesis	Down	*P. aeruginosa*	Whiteley et al., 2001
pilA	Pilin protein	Fimbrial synthesis	Down	*P. aeruginosa*	Whiteley et al., 2001
flgD		Flagellar basal body modification	Down	*P. aeruginosa*	Whiteley et al., 2001

(*continued*)

Table 1.1 *(continued)*

Gene or Locus Number	Protein	Function	Expression[a]	Organism	Reference
PA2129	Probable pili assembly chaperone	Fimbrial synthesis	Down	*P. aeruginosa*	Whiteley et al., 2001
PA1092	Flagellin type B	Flagellar synthesis	Down	*P. aeruginosa*	Whiteley et al., 2001
fliD	Flagellar capping protein	Flagellar synthesis	Down	*P. aeruginosa*	Whiteley et al., 2001
flgE	Flagellar hook protein	Flagellar synthesis	Down	*P. aeruginosa*	Whiteley et al., 2001
Membrane proteins, secretion, and transport					
ompC		Porin	Up	*E. coli*	Prigent-Combaret et al., 1999
proU	Transport system of glycine betaine	Adaptation to osmotic changes	Up	*E. coli*	Prigent-Combaret et al., 1999
nikA		Transport of nickel	Up	*E. coli*	Prigent-Combaret et al., 1999
nlpD	Outer membrane lipoprotein		Up	*P. putida*	Sauer and Camper, 2001
potB		ABC transporter	Up	*P. putida*	Sauer and Camper, 2001
ybaL		Probable K^+ efflux transporter	Up	*P. putida*	Sauer and Camper, 2001
xcpS	General secretion pathway protein F		Up	*P. putida*	Sauer and Camper, 2001
tatA		Translocation protein	Up	*P. aeruginosa*	Whiteley et al., 2001
tatB		Translocation protein	Up	*P. aeruginosa*	Whiteley et al., 2001
tolA		Involved in lipopolysaccharide (LPS) synthesis	Up	*P. aeruginosa*	Whiteley et al., 2001
omlA	Outer membrane lipoprotein		Up	*P. aeruginosa*	Whiteley et al., 2001

PA3038	Probab e porin	Down	*P. aeruginosa*	Whiteley et al., 2001
PA1710	Type III secretion central regulator	Down	*P. aeruginosa*	Whiteley et al., 2001
PA3234	Probable sodium:solute symporter	Down	*P. aeruginosa*	Whiteley et al., 2001
Carbon and nitrogen catabolism				
pepT	Tripeptidase T	Up	*E. coli*	Prigent-Combaret et al., 1999
PA2015	Probable acyl-coenzyme A dehydrogenase	Down	*P. putida*	Sauer and Camper, 2001
chiC	Chitinase	Up	*P. putida*	Sauer and Camper, 2001
PA4867	Urease β subunit	Up	*P. aeruginosa*	Whiteley et al., 2001
PA3584	Glycerol-3-phosphate dehydrogenase	Down	*P. aeruginosa*	Whiteley et al., 2001
PA0108	Cytochrome *c* oxidase, subunit III	Down	*P. aeruginosa*	Whiteley et al., 2001
PA0105	Cytochrome *c* oxidase, subunit II	Down	*P. aeruginosa*	Whiteley et al., 2001
PA0106	Cytochrome *c* oxidase, subunit I	Down	*P. aeruginosa*	Whiteley et al., 2001
PA3418	Leucine dehydrogenase	Down	*P. aeruginosa*	Whiteley et al., 2001
Transcription and translation				
asnB	Probable asparagine synthetase	Down	*P. putida*	Sauer and Camper, 2001
leuS	Leucyl-tRNA synthase	Down	*P. putida*	Sauer and Camper, 2001
PA5316	50S ribosomal protein L28	Up	*P. aeruginosa*	Whiteley et al., 2001
PA3742	50S ribosomal protein L19	Up	*P. aeruginosa*	Whiteley et al., 2001

(*continued*)

Table 1.1 *(continued)*

Gene or Locus Number	Protein	Function	Expression[a]	Organism	Reference
PA4262	50S ribosomal protein L4		Up	*P. aeruginosa*	Whiteley et al., 2001
PA4247	50S ribosomal protein L18		Up	*P. aeruginosa*	Whiteley et al., 2001
PA4261	50S ribosomal protein L23		Up	*P. aeruginosa*	Whiteley et al., 2001
PA4267	30S ribosomal protein S7		Up	*P. aeruginosa*	Whiteley et al., 2001
PA4744	Translation initiation factor IF-2		Up	*P. aeruginosa*	Whiteley et al., 2001
PA3049		Ribosome modulation factor	Down	*P. aeruginosa*	Whiteley et al., 2001
PA2620	ATP-binding protease component ClpA		Down	*P. aeruginosa*	Whiteley et al., 2001
rpoH		σ-factor	Up	*P. aeruginosa*	Whiteley et al., 2001
rpoS		σ-factor	Down	*P. aeruginosa*	Whiteley et al., 2001
Drug resistance					
ampC	β-lactamase	Ampicillin resistance	Up	*P. putida*	Sauer and Camper, 2001
mexB		Multidrug efflux pump	Up	*P. putida*	Sauer and Camper, 2001
str	Streptomycin 3′-phosphotransferase	Streptomycin resistance	Up	*P. putida*	Sauer and Camper, 2001

[a] Up means activated in biofilm. Down means repressed in biofilm.

compared to a sequential process of development. During the early stages, the individual microorganisms have to sense physicochemical differences at the solid–liquid interface. Later, as the matrix develops and the attached population multiplies, intercellular communications can become progressively operative.

A hallmark of a mature biofilm is its ordered architecture consisting of large mushroom-shaped colonies interspersed among less dense channels in which liquid flow has been measured (for a review, see Costerton et al., 1994). Such organisation suggests that the intercellular signalling takeover is a key to the last episode of biofilm development. In natural conditions, the construction of these ordered multicellular structures involves collective properties, such as positive and negative tropisms, cell aggregation and dispersion, cell-to-cell activation, and repression of subsets of the genome. These processes are obviously very complex and could require an unexpected number of genes and functions.

To date, only one type of signalling process – quorum sensing – has been described in biofilms. In a seminal study, Davies and co-workers (1998) demonstrated that the quorum-sensing molecules, typically acylated homoserine lactones (acyl-HSLs), are involved in biofilm maturation. These signal molecules accumulate in the bacterial environment as a function of cell number and mediate population-density-dependent gene expression. A *P. aeruginosa* mutant defective for the *lasI* system of acyl-HSL production, although still capable of early cell–surface interactions, did not develop the structural organisation of a wild-type biofilm. The *lasI* mutant biofilm was much thinner, more crowded, and sensitive to the biocide sodium dodecyl sulphate (SDS). Addition of the acyl-HSL normally produced by the *lasI* system restored wild-type biofilm architecture and SDS resistance. Furthermore, the loss of SDS resistance by the *lasI* mutant occurred despite the lack of any change in exopolysaccharide production, indicating that the biocide resistance in wild-type biofilms is a result of cellular adaptation rather than a limitation of SDS diffusion in the polymer matrix. Acyl-HSLs have also been detected in natural biofilms developed on urethral catheters removed from patients (Stickler et al., 1998) and on immersed stones from the San Marcos River in Texas (McLean et al., 1997).

3 SURFACE-MODULATED FUNCTIONS

Examples of differential gene and protein expression in biofilms and free-living cells are summarized in Table 1.1. A clear trend is the repression of flagellum synthesis observed in *E. coli*, *P. putida*, and *P. aeruginosa*. In *E. coli*, the loss of flagella in the attached cells was confirmed by electron microscopy (Prigent-Combaret et al., 1999). A second trend is the activation of functions involved in the colonisation process itself, such as exopolysacharide production and oversynthesis of adhesion structures (curli and lipopolysaccharide). These common features strongly suggest that the biofilm-forming bacteria are equipped with specialised recognition structures that enable them to perceive their contact with abiotic materials.

Many genes in Table 1.1 are involved in growth functions (such as assimilation, transport, and ribosome building), indicating strong differences between the planktonic and the biofilm growth modes. However, because of heterogeneous conditions within the biofilms, these differences have probably been underestimated. A complete understanding of the construction of a mature biofilm architecture will require considerably more research. There is no doubt that gradients or local limitations of nutrients and oxygen are responsible for a large number of different gene expression patterns. Interesting processes will be occurring only in particular niches and therefore are impossible to investigate with global methods. The identification of the switching mechanisms allowing bacteria to leave the biofilm and return to the planktonic status as individual cells is an example of an important challenge for future research.

Nevertheless, the results presented in Table1.1 give interesting insights into one of the most detrimental properties of biofilm-grown cells – increased resistance to biocides and antibiotics. As mentioned previously, the SDS resistance of *P. aeruginosa* biofilms has been associated with quorum-sensing mechanisms. In *P. putida,* the expression of *mexB* (encoding a component of the antibiotic efflux system), a streptomycin resistance *str* gene, and the expression of the β-lactamase *ampC* gene were found to be surface induced (Table 1.1). In *P. aeruginosa,* the major aminoglycoside-resistance mechanism is impermeability to antibiotic entrance. This impermeability involves several factors, including the *tolA* gene and terminal electron transport proteins (Whiteley et al., 2001). The *tolA* gene product affects lipopolysaccharide structure and aminoglycoside affinity for the outer membrane. Since mutants that underproduce TolA are hypersensitive to aminoglycoside, *tolA* activation in *P. aeruginosa* biofilms (Table 1.1) could contribute to aminoglycoside resistance. Moreover, repression of cytochrome *c* oxidase in biofilms (Table 1.1) could be regarded as an additional factor contributing to increased resistance to aminoglycosides (Whiteley et al., 2001). Several other surface-modulated functions in Table 1.1, for example, porin synthesis, might also be candidates for antibiotic resistance factors.

4 ABIOTIC SURFACE SENSING

The construction and the physiology of the ordered structure of mature biofilms has to be understood mainly in terms of cell-to-cell signalling. In contrast, the initial stages of surface contamination concern individual cells and have to be described at the level of the intracellular signalling events. As mentioned previously, a pioneer bacterium must sense its contact with the material in order to trigger its transformation from a swimming cell to a surface colonising cell. Two questions have to be answered in this process. What physicochemical parameters are sufficiently different to enable the bacterium to discriminate the liquid phase and the interface? And what cellular structures are able to recognise these differences and transmit the information to the genome?

As it is well established that in *E. coli* there is a quasilinear relationship between the osmolarity of the external medium and the intracellular concentration of potassium (Epstein and Schultz, 1965), Prigent-Combaret and coworkers (1999) compared the intracellular K^+ concentration of planktonic and biofilm cell populations. Ten hours after inoculation into the culture vessel, the attached bacteria displayed a significantly higher internal K^+ concentration than the planktonic cells. This result indicates that the osmolarity of the microenvironment surrounding the two types of bacteria was higher around the attached cells. In my opinion, this observation can have two non-exclusive explanations. First, the electric charges present at the surface and on the organic molecules adsorbed on it decrease water activity (that is, the proportion of water molecules acting as pure solvent) at the liquid–solid interface. This layer of 'different water' is actually very thin with regard to the dimensions of the bacterium and its appendages, but could activate some sensors located on that region of the cell envelope, which faces the abiotic surface. Second, a bacterium that becomes immobilised on a surface is subjected to considerably lower convective forces than a swimming cell. It is therefore conceivable that fimbrial breaking and dispersion could be reduced in these confinement conditions. As a result, fimbrial density would increase around the cell. This situation may have two important consequences: an additional decrease of water activity in the immediate vicinity of the cell, due to the electric charges existing on the fimbriae, and saturation of the processes of fimbrial construction, with subsequent accumulation of monomers in the bacterial periplasm. In all cases, the osmolarity of the periplasm and the microenvironment of the cell would increase.

Recent genetic studies may support these hypotheses and answer the second question about the cellular recognition structures. The curli-mediated adherence of *E. coli* depends on the integrity of two signal transduction systems:

EnvZ/OmpR (Vidal et al., 1998) and CpxA/CpxR (Dorel et al., 1999). Through a complex regulatory network, these systems control the expression of several genes, including those encoding curli (Prigent-Combaret et al., 2001), in response to two types of signals: medium osmolarity and periplasmic accumulation of non-secreted proteins.

Increasing osmolarity of the environment activates the sensor protein EnvZ and leads to improved phosphorylation of OmpR, resulting in modulation of its binding ability to the regulatory sequences of the target genes. In addition to curli synthesis, flagellum and colanic acid production is also controlled by the EnvZ/OmpR two-component system (Prigent-Combaret et al., 1999). Sensing of interfacial osmolarity changes through EnvZ/OmpR could therefore constitute a major part of the transition process that the bacterium undergoes upon contact with a surface.

Although clearly established (Dorel et al., 1999), the role of the CpxA/CpxR sensor-regulator system in colonisation is more difficult to understand. By unknown sensing mechanisms, this system is involved in recognition of periplasm saturation with 'useless' proteins, such as non-secreted monomers or non-addressed outer membrane proteins, and is able to trigger an appropriate scavenging response. For example, the outer membrane lipoprotein NlpE is known to activate the Cpx pathway when overproduced from a multicopy plasmid (Danese et al., 1995). Dorel and co-workers (1999) have observed that transposon insertions in the *cpxA* gene, as well as NlpE overproduction, strongly reduced curli gene expression and adherence. It is therefore conceivable that the CpxA/CpxR system constitutes another part of the *E. coli* surface-sensing machinery and could be activated by periplasmic accumulation of monomers when the external density of fimbriae on the surface of the immobilised cell is increased by electrical and mechanical interactions that remain to be explored.

REFERENCES

Costerton, J. W., Cheng, K. J., Geesey, G. G., Ladd, T. I., Nickel, J. C., Dasgupta, M. and Marrie, T. J. (1987). Bacterial biofilms in nature and disease. *Annual Review of Microbiology*, 41, 435–464.

Costerton, J. W., Lewandowski, Z., De Beer, D., Caldwell, D., Korber, D. and James, C. (1994). Biofilms, the customized microniche. *Journal of Bacteriology*, 176, 2137–2142.

Cucarella, C., Solano, C., Valle, J., Amorena, B., Lasa, I. and Penadés, J. R. (2001). Bap, a *Staphylococcus aureus* surface protein involved in biofilm formation. *Journal of Bacteriology*, 183, 2888–2896.

Danese, P. N., Snyder, W. B., Cosma, C. L., Davis, L. J. and Silhavy, T. J. (1995). The Cpx two-component system signal transduction pathway of *Escherichia coli* regulates transcription of the genes specifying the stress-inducible periplasmic protease, DegP. *Genes and Development*, 9, 387–398.

Davies, D. G., Chakrabarty, A. M. and Geesey, G. G. (1993). Exopolysaccharide production in biofilms: substratum activation of alginate gene expression by *Pseudomonas aeruginosa*. *Applied and Environmental Microbiology*, 59, 1181–1186.

Davies, D. G. and Geesey, G. G. (1995). Regulation of the alginate biosynthesis gene *algC* in *Pseudomonas aeruginosa* during biofilm development in continuous culture. *Applied and Environmental Microbiology*, 61, 860–867.

Davies, D. G., Parsek, M. R., Pearson, J. P., Iglewski, B. H., Costerton, J. W. and Greenberg, E. P. (1998). The involvement of cell-to-cell signals in the development of bacterial biofilm. *Science*, 280, 295–298.

Dorel, C., Vidal, O., Prigent-Combaret, C., Vallet, I. and Lejeune, P. (1999). Involvement of the Cpx signal transduction pathway of *E. coli* in biofilm formation. *FEMS Microbiology Letters*, 178, 169–175.

Epstein, W. and Schultz, S. G. (1965). Cation transport in *Escherichia coli*. V. Regulation of cation content. *Journal of General Physiology*, 49, 221–234.

Genevaux, P., Bauda, P., DuBow, M. S. and Oudega, B. (1999a). Identification of Tn*10* insertions in the *rfaG*, *rfaP*, and *galU* genes involved in lipopolysaccharide core biosynthesis that affect *Escherichia coli* adhesion. *Archives of Microbiology*, 172, 1–8.

Genevaux, P., Bauda, P., DuBow, M. S. and Oudega, B. (1999b). Identification of Tn*10* insertions in the *dsbA* gene affecting *Escherichia coli* biofilm formation. *FEMS Microbiology Letters*, 173, 403–409.

Ghigo, J. M. (2001). Natural conjugative plasmids induce bacterial biofilm development. *Nature*, 412, 442–445.

Hausner, M. and Wuertz, S. (1999). High rates of conjugation in bacterial biofilms as determined by quantitative in situ analysis. *Applied and Environmental Microbiology*, 65, 3710–3713.

Heilmann, C., Hussain, M., Peters, G. and Götz, F. (1997). Evidence for autolysin-mediated primary attachment of *Staphylococcus epidermidis* to a polystyrene surface. *Molecular Microbiology*, 24, 1013–1024.

Le Thi, T. T., Prigent-Combaret, C., Dorel, C. and Lejeune, P. (2001). First stages of biofilm formation: characterization and quantification of bacterial functions involved in colonization process. *Methods in Enzymology*, 336, 152–159.

Loo, C. Y., Corliss, D. A. and Ganeshkumar, N. (2000). *Streptococcus gordonii* biofilm formation: identification of genes that code for biofilm phenotypes. *Journal of Bacteriology*, 182, 1374–1382.

Marshall, K. C. (1992). Biofilms: an overview of bacterial adhesion, activity, and control at surfaces. *ASM News*, 58, 202–207.

McLean, R. J. C., Whitely, M., Stickler, D. J. and Fuqua, W. C. (1997). Evidence of autoinducer activity in naturally occurring biofilms. *FEMS Microbiology Letters*, 154, 259–263.

Mireles, J. R., II, Togushi, A. and Harshey, R. M. (2001). *Salmonella enterica* Serovar

Typhimurium swarming mutants with altered biofilm-forming abilities: surfactin inhibits biofilm formation. *Journal of Bacteriology*, 183, 5848–5854.
O'Toole, G., Kaplan, H. B. and Kolter, R. (2000). Biofilm formation as microbial development. *Annual Review of Microbiology*, 54, 49–79.
O'Toole, G. A. and Kolter, R. (1998a). Flagellar and twitching motility are necessary for *Pseudomonas aeruginosa* biofilm development. *Molecular Microbiology*, 30, 295–304.
O'Toole, G. A. and Kolter, R. (1998b). Initiation of biofilm formation in *Pseudomonas fluorescens* WCS365 proceeds via multiple, convergent signalling pathways: a genetic analysis. *Molecular Microbiology*, 28, 449–461.
Pratt, L. A. and Kolter, R. (1998). Genetic analysis of *Escherichia coli* biofilm formation: roles of flagella, motility, chemotaxis and type I pili. *Molecular Microbiology*, 30, 285–293.
Prigent-Combaret, C. (2000). Processus de régulations métaboliques au cours de la colonisation des surfaces inertes par *Escherichia coli* K-12. Ph.D. Thesis, University of Paris, 7.
Prigent-Combaret, C., Brombacher, E., Vidal, O., Ambert, A., Lejeune, P., Landini, P. and Dorel, C. (2001). Complex regulatory network controls initial adhesion and biofilm formation in *Escherichia coli* via regulation of the *csgD* gene. *Journal of Bacteriology*, 183, 7213–7223.
Prigent-Combaret, C. and Lejeune, P. (1999). Monitoring gene expression in biofilms. *Methods in Enzymology*, 310, 56–79.
Prigent-Combaret, C., Prensier, G., Le Thi, T. T., Vidal, O., Lejeune, P. and Dorel, C. (2000). Developmental pathway for biofilm formation in curli-producing *Escherichia coli* strains: role of flagella, curli, and colanic acid. *Environmental Microbiology*, 2, 450–464.
Prigent-Combaret, C., Vidal, O., Dorel, C. and Lejeune, P. (1999). Abiotic surface sensing and biofilm-dependent gene expression in *Escherichia coli*. *Journal of Bacteriology*, 181, 5993–6002.
Sauer, K. and Camper, A. K. (2001). Characterization of phenotypic changes in *Pseudomonas putida* in response to surface-associated growth. *Journal of Bacteriology*, 183, 6579–6589.
Stickler, D. J., Morris, N. A., McLean, R. J. C. and Fuqua, C. (1998). Biofilms on indwelling urethral catheters produce quorum-sensing molecules in-situ. *Applied and Environmental Microbiology*, 64, 3486–3490.
Uhlich, G. A., Keen, J. E. and Elder, R. O. (2002). Variations in the *csgD* promoter of *Escherichia coli* O157:H7 associated with increased virulence in mice and increased invasion of HEp-2 cells. *Infection and Immunity*, 70, 395–399.
Vallet, I., Olson, J. W., Lory, S., Ladzunski, A. and Filloux, A. (2001). The chaperone/usher pathways of *Pseudomonas aeruginosa*: identification of fimbrial gene clusters (*cup*) and their involvement in biofilm formation. *Proceedings of the National Academy of Sciences of the USA*, 98, 6911–6916.
Van Loosdrecht, M. C. M., Lyklema, J., Norde, W. and Zehnder, A. J. B. (1990). Influence of interfaces on microbial activity. *Microbiological Reviews*, 54, 75–87.
Vidal, O., Longin, R., Prigent-Combaret, C., Dorel, C., Hooreman, M. and Lejeune, P. (1998). Isolation of an *Escherichia coli* mutant strain able to form biofilms on inert

surfaces: involvement of a new *ompR* allele that increases curli expression. *Journal of Bacteriology*, 180, 2442–2449.

Watnick, P. I. and Kolter, R. (1999). Steps in the development of a *Vibrio cholerae* El Tor biofilm. *Molecular Microbiology*, 34, 586–595.

Whiteley, M., Bangera, M. G., Bumgarner, R. E., Parsek, M. R., Teltzel, G. M., Lory, S. and Greenberg, E. P. (2001). Gene expression in *Pseudomonas aeruginosa* biofilms. *Nature*, 413, 860–864.

ZoBell, C. E. (1943). The effect of solid surfaces upon bacterial activity. *Journal of Bacteriology*, 46, 39–56.

CHAPTER TWO

Quorum Sensing and Microbial Biofilms

Teresa R. de Kievit and Barbara H. Iglewski

1 INTRODUCTION

For a long time bacteria were believed to exist as unicellular organisms; however, it is now realized that in nature bacteria are more often found clustered in communities. Within these communities, bacteria are capable of coordinated activity through the use of a sophisticated intercellular communication mechanism called quorum sensing (QS). The capacity to behave collectively as a group has obvious advantages, for example, migration to a more suitable environment/better nutrient supply or adopting a more favourable mode of growth such as sporulation. Recently, QS was discovered to play a role in the formation of biofilms. This latter phenomenon will be the focus of this chapter as we review our current understanding of how QS affects the complex processes of biofilm development. With respect to intercellular communication and biofilms, *Pseudomonas aeruginosa* is one of the most intensely studied organisms, and therefore, much of this chapter will concentrate on this bacterium.

2 QUORUM SENSING

QS exists in both Gram-positive and Gram-negative bacteria with obvious differences between the two systems (for reviews, see Dunny and Leonard, 1997; Fuqua, Winans, and Greenberg, 1996). Here, we will focus on Gram-negative bacteria, where the two primary components of QS systems are the autoinducer (AI) signal molecule and the transcriptional activator, or R-protein. In general, the 'language' used for intercellular communication is based on small, diffusible, self-generated signal molecules called AIs. The premise of

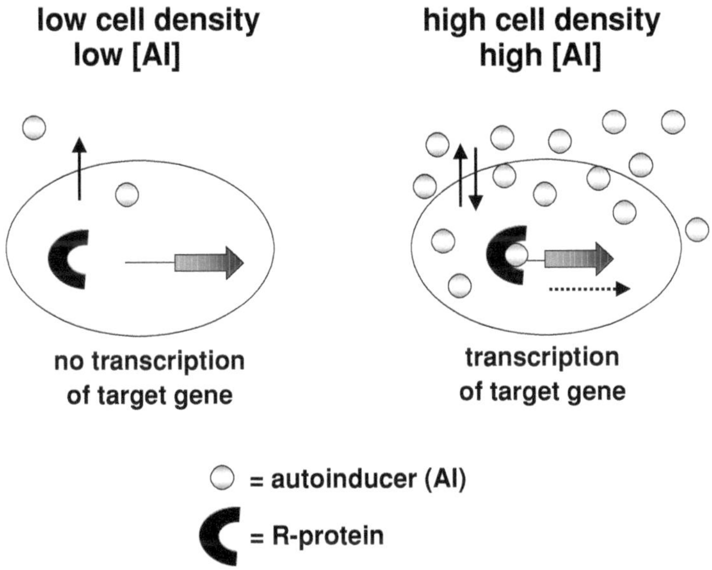

low cell density
low [AI]

high cell density
high [AI]

no transcription
of target gene

transcription
of target gene

= autoinducer (AI)

= R-protein

Figure 2.1: Quorum sensing in Gram-negative bacteria involves two regulatory components: the transcriptional activator or R-protein and the autoinducer (AI) signal molecule, the product of the AI synthase enzyme. Accumulation of AI occurs in a population-dependent fashion until a threshold level is reached. At this time, the AI can bind to and activate its cognate R-protein, which in turn induces expression of target genes. Although Gram-negative AI signals are typically *N*-acyl homoserine lactones, other types of signal molecules exist.

cell-to-cell communication or QS is based on the fact that when a single bacterium releases AIs into the environment, their concentration is too low to be detected. However, when sufficient bacteria are present, the AI reaches a threshold concentration that allows it to bind to and activate its cognate R-protein, which can then induce transcription of target genes (Figure 2.1). Although the majority of Gram-negative AIs are *N*-acyl homoserine lactones (AHL), alternative signal molecules exist (for a review, see de Kievit and Iglewski, 2000).

QS has been well studied in *P. aeruginosa*, where it is used to regulate expression of numerous virulence factors. *P. aeruginosa* contains two complete QS systems, *las* and *rhl* (Figure 2.2). The *las* QS system consists of a transcriptional activator, LasR, and an AI synthase enzyme, LasI, which directs the synthesis of *N*-(3-oxododecanoyl) homoserine lactone (3O-C_{12}-HSL) (Passador et al., 1993; Pearson et al., 1994). Similarly, the *rhl* system is comprised of the transcriptional activator RhlR together with RhlI, which synthesizes *N*-butyryl homoserine lactone (C_4-HSL) (Ochsner et al., 1994; Ochsner and Reiser, 1995;

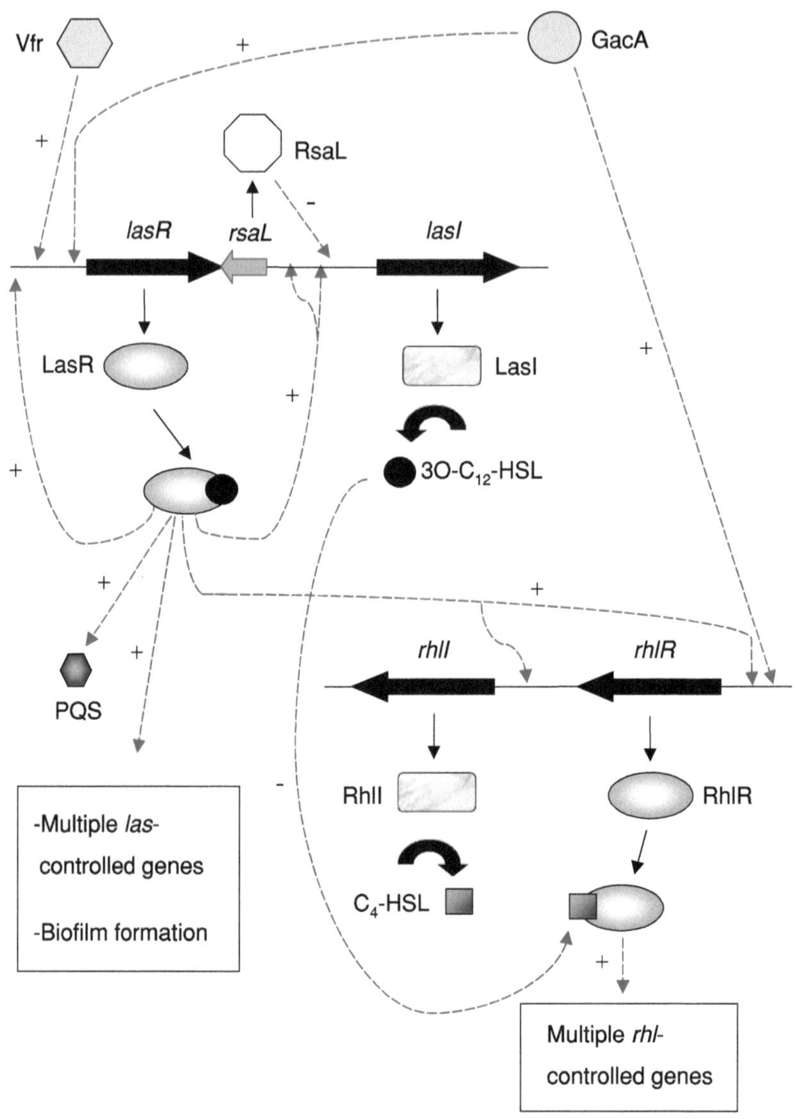

Figure 2.2: *P. aeruginosa* contains two complete QS systems: *las* and *rhl*. Expression of the *lasR* gene is subject to at least two levels of control: the global regulators Vfr and GacA (Albus et al., 1997; Reimmann et al., 1997) and the *las* QS system, which regulates expression of both *lasR* and *lasI*. The *lasI* gene is also subject to negative regulation by the repressor protein RsaL (de Kievit et al., 1999). Regulation of the *rhl* system is similar to *las* in that GacA positively regulates *rhlR* (Reimman et al., 1997) and the *rhlR* and *rhlI* genes are controlled to some degree by the *las* QS system. The *las* QS system exhibits an additional level of control over the *rhl* system because the *las* signal molecule, $3O-C_{12}$-HSL, can post-translationally block RhlR activation by C_4-HSL. Together, the two QS systems regulate expression of numerous virulence factors that contribute to *P. aeruginosa* pathogenicity. In addition, the *las* QS system is important for the formation of mature, differentiated biofilms. PQS = *Pseudomonas* quinolone signal.

Pearson et al., 1995). A number of genes and gene products are regulated by the *las* QS system, including *lasI* itself (Seed, Passador, and Iglewski, 1995), *lasB* (elastase) (Passador et al., 1993), *lasA* (LasA protease) (Gambello, Kaye, and Iglewski, 1993), *toxA* (exotoxin A) (Gambello et al., 1993), the *xcpPQ* and *xcpR-Z* operons (type II secretion apparatus) (Chapon-Herve et al., 1997), as well as *rhlR* and *rhlI* (Latifi et al., 1996; Pesci et al., 1997). Similarly, the *rhl* system controls expression of *rhlAB* (required for rhamnolipid production) (Ochsner et al., 1994; Pearson, Pesci, and Iglewski, 1997) and *rhlI* (Latifi et al., 1996), and it enhances expression of LasB elastase, LasA protease, pyocyanin, alkaline protease, and cyanide (Brint and Ohman, 1995; Latifi et al., 1995; Pearson et al., 1995, 1997; Reimmann et al., 1997). Recently, a third *P. aeruginosa* AI molecule was identified (Pesci et al., 1999). This molecule is structurally very different from the other two *P. aeruginosa* AIs in that it is a 2-heptyl-3-hydroxy-4-quinolone, designated PQS (for *Pseudomonas* quinolone signal). At present, many aspects of PQS have yet to be uncovered, including the R-protein with which it interacts.

3 QUORUM SENSING (QS) AND BIOFILM FORMATION

In nature, *P. aeruginosa* is frequently found growing in biofilms. Microscopic analysis of *P. aeruginosa* biofilm communities reveals that they are not just sugar-encased masses of cells. Rather, distinct mushroom- and stalk-like structures are present that contain intervening water channels to allow nutrients to flow in and waste products to flow out (Figure 2.3). Until recently, the means by which bacteria could coordinate their activity to form these elaborate structures remained a mystery. Work done by Davies et al. (1998) revealed that, in a flowing system, the *las* QS signal molecule, 3O-C_{12}-HSL, is a key component for creating the three-dimensional architecture of the biofilm. Although 3O-C_{12}-HSL signal-deficient mutants were able to attach to the substrate and form microcolonies, they developed biofilms that were much thinner and lacked the structure observed in that of the parent (Davies et al., 1998). Even more important was the fact that the functional integrity of the mutant biofilm was severely impaired. In these studies, the parental biofilm remained intact after 24 hours exposure to the detergent sodium dodecyl sulphate (SDS). In contrast, the mutant biofilm rapidly dispersed from the underlying surface after 5 minutes of SDS exposure. Addition of the missing 3O-C_{12}-HSL signal molecule restored both the structure and the detergent resistance of the mutant biofilm. Intriguingly, Davies et al. (1998) found that although the *las* system played a pivotal role in *P. aeruginosa* biofilm development, the *rhl* system was not involved.

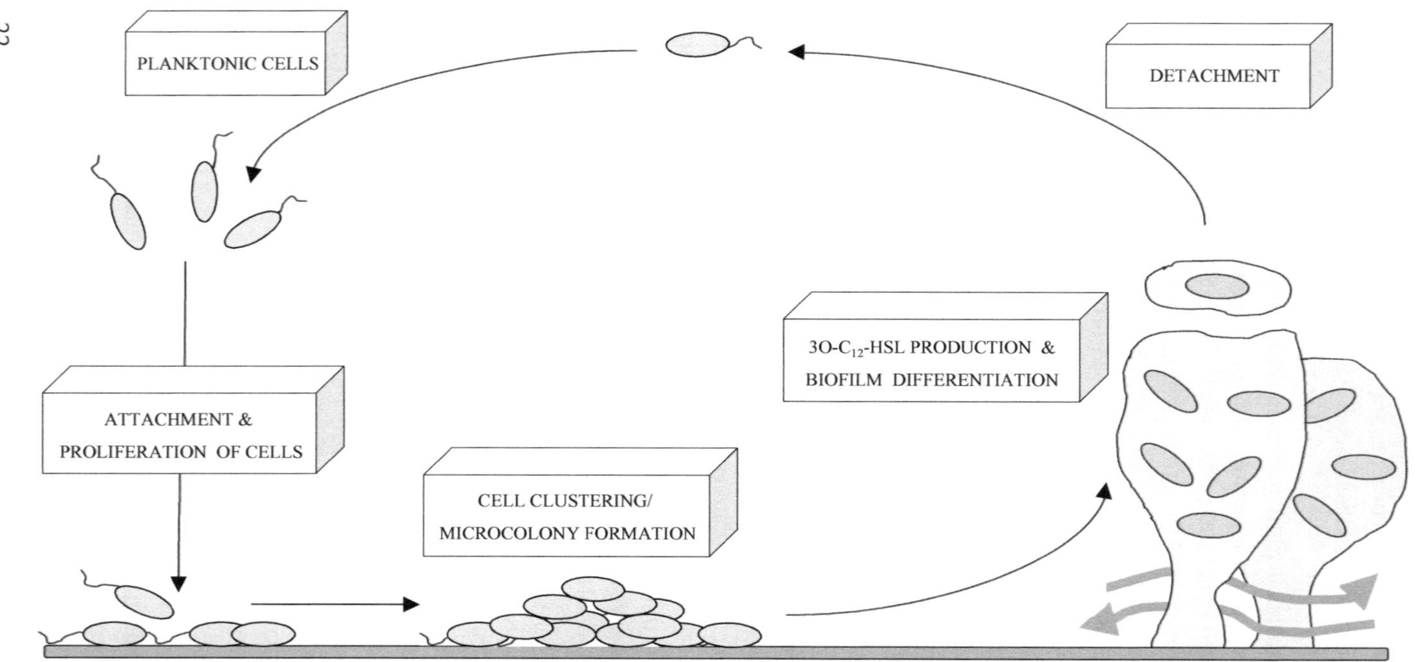

Figure 2.3: In *P. aeruginosa,* biofilm formation begins with the attachment of planktonic bacteria to a surface by means of bacterial adhesins. Cells form a monolayer across the surface, followed by cell clustering or microcolony formation. Production of the *las* QS signal molecule $3O\text{-}C_{12}$-HSL is important during the next stage, when cells are differentiating into mushroom- and stalk-like structures. Cells periodically leave the biofilm matrix and resume the planktonic mode of growth. Detachment is thought to occur by two processes. The first involves separation of cells due to shear forces in the surrounding liquid. It is hypothesized that there is also a programmed release of cells from the biofilm population; however, the molecular basis for this second mechanism remains an enigma. The released planktonic bacteria are able to disperse and can attach to a surface to reinitiate biofilm formation.

The discovery that cell-to-cell communication is linked to the development of mature *P. aeruginosa* biofilms is extremely provocative and leads to a number of questions. For example, which *las*-controlled genes are required for biofilm development? Do other bacteria use QS during biofilm formation, or is this finding unique to *P. aeruginosa*? Does QS take place in biofilms formed *in vivo*? Research has begun to address some of these questions. For example, studies looking at naturally occurring biofilms indicate that QS does occur *in vivo*. Analysis of aquatic biofilms formed on stones isolated from the San Marcos River in Texas revealed the first proof that AHL molecules are present in natural biofilms. Using a cross-feeding assay and an AHL-sensitive reporter strain (*Agrobacterium tumefaciens* containing a *traI::lacZ* fusion), McLean and co-workers (1997) detected AHL molecules in living biofilms formed on stones, as well as extracts of those biofilms. In a second study examining clinical biofilms, patient-derived catheter tubing was assayed for the presence of AHL molecules (Stickler et al., 1998). AHL signals were detected in four of the nine catheters, providing evidence that QS occurs in biofilms found in a medical setting. Clearly, this is not a mechanism restricted to the laboratory. Rather, QS takes place in a wide spectrum of biofilms found in nature. As such, it becomes even more important that we gain a better understanding of its role in biofilm development. These studies will not only provide fundamental knowledge regarding the complex physiology of the biofilm, but will have practical applications, for instance, the development of novel means of preventing/eradicating biofilm formation.

4 *PSEUDOMONAS AERUGINOSA* QS AND BIOFILM INITIATION

During the initial stages of biofilm development, cells attach to a surface and begin to form microcolonies (Figure 2.3). Research has shown that, at least for static (non-flowing) biofilms, the *P. aeruginosa las* and *rhl* QS systems play a role during these early phases (de Kievit et al., 2001). Biofilms formed by *P. aeruginosa* wild-type strain PAO1 and a panel of QS mutants including PAO-JP1 (Δ*lasI*), PDO-100 (Δ*rhlI*), and PAO-JP2 (Δ*lasIrhlI*) were examined after 18 hours of growth in a static environment. Moreover, two different media commonly used in biofilm studies, namely, M9 supplemented with glucose and FAB containing citrate, were employed to see what effect, if any, medium composition had on biofilm development. When glucose was used as a C-source, differences were observed in the ability of the PAO1 wild type and the QS mutants to initiate biofilm formation. After 18 hours, the wild-type strain had formed thick, multilayered biofilms, whereas the QS mutants were significantly impaired in their ability to attach to a surface. Conversely, in

FAB medium containing citrate, both the parent and the mutants exhibited poor attachment. Lipopolysaccharide (LPS), type IV pili, and flagella have all been shown to influence *P. aeruginosa* surface attachment (Flemming et al., 1998; O'Toole and Kolter, 1998; Rocchetta, Burrows, and Lam, 1999). To uncover how the C-source influenced biofilm initiation, each of these features was examined in cells grown in the two media. The results of these analyses demonstrated that the greatest difference occurred with type IV twitching motility. In FAB-citrate, strain PAO1 was greatly decreased in twitching motility compared to the M9 medium. Moreover, the QS mutants were completely twitching deficient in the FAB medium. These results indicated that type IV twitching plays a major role in static biofilm development, which is consistent with an earlier report (O'Toole and Kolter, 1998). Furthermore, these findings suggest that, at least with static biofilms, medium composition can markedly influence attachment and biofilm initiation.

5 EXPRESSION OF QS GENES DURING BIOFILM DEVELOPMENT

There is mounting evidence to suggest QS plays an integral role during the development of mature, differentiated *P. aeruginosa* biofilms. However, it is unclear how signalling facilitates construction of these elaborate mushroom- and pillar-like structures. To attempt to address this question, studies aimed at tracking the spatial and temporal expression patterns of *lasI* and *rhlI* throughout the course of *P. aeruginosa* biofilm development were undertaken (de Kievit et al., 2001). The *lasI* and *rhlI* genes were chosen because they encode enzymes that ultimately generate the AI signal molecules. Therefore, understanding their gene expression patterns should reveal important clues about the role of intercellular communication during biofilm formation. For this analysis, *lasI* and *rhlI* promoters were independently fused to a green fluorescent protein (GFP) reporter. Because the wild-type GFP and original mutant GFPs are very stable, with half-lives in excess of 24 hours, they are less than ideal for monitoring real-time gene expression (Tombolini et al., 1997). To circumvent this problem, an unstable GFP construct (LVAgfp) with a half-life of approximately 40 minutes was utilized (Anderson et al., 1998). The *gfp* gene on these plasmids was manipulated to contain a C-terminal peptide tail that is recognized and rapidly degraded by indigenous cellular proteases, imparting a very short half-life (Anderson et al., 1998). *P. aeruginosa* strain PAO1 containing either the *lasI*-LVAgfp or *rhlI*-LVAgfp fusion was grown in flow cells and analyzed for gene expression on days 4, 6, and 8 using scanning laser confocal microscopy (SCLM). Expression of *lasI* was found to be maximal on day 4 and decreased progressively throughout the course of biofilm development. Intriguingly, there was a notable change in biofilm morphology that took place between days 4 and 6. On day 4, the biofilm consisted of predominantly microcolonies, whereas by day 6 mushroom-like structures began to emerge. It may be that elevated expression of *lasI* prior to day 6 results in induction of genes required for maturation and differentiation into these three-dimensional structures. In contrast, the *rhlI* gene, which was previously found to be of little significance in biofilm differentiation (Davies et al., 1998), was expressed in a fewer number of cells and oscillated very little throughout the course of biofilm development. Examination of the spatial expression patterns revealed that *lasI* and *rhlI* gene activity was maximal at the substratum and decreased with increasing biofilm height. Both *lasI* and *rhlI* are autoregulated to some degree; the *lasI* gene requires the presence of 3O-C_{12}-HSL for gene activation and *rhlI* requires both 3O-C_{12}-HSL and C_4-HSL (Latifi et al., 1996; Pesci et al., 1997). In light of this, it is not surprising that maximal gene expression occurred at the substratum where cells are in close proximity with one another and the surface to which they are attached. Presumably, decreased diffusion at this location in the biofilm leads to a higher concentration of AI, resulting in increased expression of *lasI* and *rhlI*.

6 QUANTIFICATION OF AHLs IN BIOFILMS

An assay was recently developed that enables 3-oxoacyl homoserine lactones (3O-AHLs), one of the predominant classes of AHLs produced by Gram-negative bacteria, to be quantified in both biofilm and planktonic cultures (Charlton et al., 2000). The assay is based on the conversion of 3O-AHLs to their pentafluorobenzyloxime derivatives followed by gas chromatography-mass spectrometry (GC-MS). This technique was used to analyse both the biofilm and the effluent from *P. aeruginosa* grown in flow cells. The *P. aeruginosa* strain used, 6294, was expressing GFP, which enabled the biofilm volume to be ascertained using SCLM. After measuring the volume, the biofilm sample was extracted, derivatised, and then examined using GC-MS to determine the concentration of 3O-AHLs present. The most predominant 3O-AHL produced by *P. aeruginosa*, 3O-C_{12}-HSL, was found in the biofilm at 632 ± 381 μM and in the effluent at a concentration of 14 ± 3 nM. The next most abundant 3O-AHL, *N*-3-oxo-tetradecanoyl homoserine lactone (3O-C_{14}-HSL), was found at a concentration of 40 ± 15 μM and 1.5 ± 0.7 nM in the biofilm and effluent, respectively. Intriguingly, 3O-C_{14}-HSL has not been reported previously in *P. aeruginosa* culture supernatants (Pearson et al., 1994, Shaw et al., 1997). At present, it is not known whether this AHL is specific for strain 6294 or whether it is produced by other strains of *P. aeruginosa*.

The concentration of 3O-C_{12}-HSL in the biofilm is about sixty-fold greater than that reported for *P. aeruginosa* culture supernatants (Pearson et al., 1994) and the highest ever reported for a bacterial system. This concentration of AHL is well in excess of that which is required for activation of *P. aeruginosa* QS-controlled genes. For example, the concentration of 3O-C_{12}-HSL required for half-maximal activation of *lasB* is 1 μ*M*, whereas *lasI* requires only 100 n*M* (Pearson et al., 1995; Seed et al., 1995). The authors speculated that the level of 3O-C_{12}-HSL in the biofilm is indicative of other biological functions outside of QS (Charlton et al., 2000). Previously, AHL signal molecules have been shown to act directly with eukaryotic cells to modulate host immune response. Telford and co-workers (1998) demonstrated that 3O-C_{12}-HSL suppressed release of interleukin-12 and tumor necrosis factor alpha from macrophages stimulated with LPS. Furthermore, at high concentrations (>70 μ*M*), 3O-C_{12}-HSL inhibited antibody production in spleen cells that had been stimulated with keyhole limpet haemocyanin, whereas at lower concentrations (<70 m*M*), antibody production was increased. Taken together, these findings indicate that 3O-C_{12}-HSL may affect the Th1-Th2 response during infections. Furthermore, 3O-C_{12}-HSL may not only contribute to pathogenicity through regulating virulence factor expression, but may be a virulence factor itself. In a second study, DiMango et al. (1995) showed that 3O-C_{12}-HSL was able to stimulate airway epithelial cells to produce the proinflammatory cytokine interleukin-8 in a dose-dependent manner. This cytokine is a chemoattractant for neutrophils, and it seems unlikely that this response is advantageous for *P. aeruginosa*. It is more conceivable that 3O-C_{12}-HSL is having the undesirable effect of warning the host of the presence of these bacteria. In the two aforementioned studies, concentrations of 30 to 100 μ*M* 3O-C_{12}-HSL gave the greatest effect on eukaryotic cells in levels that were previously considered too high to be biologically relevant (DiMango et al., 1995; Telford et al., 1998). The finding that AHL concentrations exceed 600 μ*M* in *P. aeruginosa* biofilms, however, implies that these levels are readily achieved in certain environments.

7 AHL SIGNALLING DURING CHRONIC LUNG INFECTIONS

In individuals afflicted with the recessive genetic disorder cystic fibrosis (CF), the leading cause of mortality is respiratory failure due to chronic *P. aeruginosa* pulmonary infection. During CF lung infections, *P. aeruginosa* can be found encased in a polysaccharide matrix, much like that of a biofilm. In light of the fact that QS plays an important role in both biofilm formation and virulence (Pearson et al., 2000; Rahme et al., 1995, 1997; Rumbaugh et al., 1999a,b; Tan et al., 1999; Tang et al., 1996), it would not be surprising to discover that QS occurs during chronic lung infections. Studies designed to address this question established chronic lung infections in mice by intratracheally instilling bacteria trapped in alginate beads (Wu et al., 2000). Mice were challenged with *P. aeruginosa* strain PA0579 together with an *Escherichia coli* AHL monitor. This strain of *E. coli* harbours a reporter plasmid containing *luxR* and the LuxR-controlled *luxI* promoter fused to GFP, which is activated in the presence of *P. aeruginosa* signal molecules. On days 1 to 3, green fluorescent bacteria could be detected by epifluoresence microscopy in the lumen of the bronchi and infected lung tissues, indicating that QS signalling was occurring. Even more noteworthy was the fact that green fluorescent bacteria were predominantly found in lung tissue exhibiting severe pathological changes, but only rarely found in tissues with little change. Therefore, it appears that QS takes place during chronic *P. aeruginosa* lung infections in mice, and it may be associated with increased lung damage.

These findings set the stage for an even more important question: Does QS occur during human infections? There are three lines of evidence to suggest that it does. First, a study examining sputa from the lungs of CF patients infected with *P. aeruginosa* revealed a correlation between *lasA, lasB,* and *toxA* transcript accumulation, signifying that the three genes are coordinately regulated. Furthermore, *lasR* transcript accumulation correlated with that of the other genes. Together, these findings indicated that the *las* QS system actively regulates gene expression during chronic pulmonary infections in humans. In the second study, sequential *P. aeruginosa* isolates from chronically infected CF patients were examined for AHL production (Geisenberger et al., 2000). Using thin-layer chromatography, it was discovered that neither the amounts nor the types of AHL molecules changed significantly during colonisation periods of up to 11 years. It has been demonstrated that during chronic pulmonary infections, *P. aeruginosa* undergoes phenotypic changes that may include synthesis of the exopolysaccharide alginate, decreased extracellular virulence factor production (Woods et al., 1991), and decreased expression of LPS O-antigen (Hancock et al., 1983; Lam et al., 1989). These phenotypic modifications are believed to represent an adaptive response to the lung environment. The results of this study suggest that a changing AHL profile does not constitute part of this adaptation process, rather AHL signalling remains relatively constant throughout the course of lung infection. Finally, in a study carried out by Singh et al. (2000), analysis of sputum from CF patients revealed the presence of the two predominant *P. aeruginosa* signal molecules: 3O-C_{12}-HSL and C_4-HSL. Intriguingly, the relative ratio of these signals was opposite to that of planktonic cultures of *P. aeruginosa* strain PAO1, where the level of 3O-C_{12}-HSL exceeds that of C_4-HSL. To determine whether differences in growth conditions, that is, biofilm versus broth culture, accounted for the differing AHL profiles, nine clinical isolates from the sputum samples were grown in broth culture and were analysed for AHL production. While six of the samples remained unchanged in terms of the predominant AHL, three produced more 3O-C_{12}-HSL than C_4-HSL, similar to planktonic PAO1 cultures. Growth of these three isolates, as well as strain PAO1, in a laboratory biofilm resulted in a switch in the relative abundance of AHL molecules produced; C_4-HSL levels now exceeded 3O-C_{12}-HSL, resembling the CF sputum. Together, these findings indicate that when *P. aeruginosa* is growing as a biofilm, C_4-HSL is the predominant AHL molecule expressed. Moreover, *P. aeruginosa* adopts the biofilm mode of growth during chronic pulmonary infections.

8 QS IN *BURKHOLDERIA CEPACIA* BIOFILM DEVELOPMENT

For individuals with CF, *Burkholderia cepacia* represents another important pathogen that can lead to fatal infections. In most cases, infection with *B. cepacia* occurs in patients already colonised with *P. aeruginosa*, and it is believed that *P. aeruginosa* may precondition the lung environment to facilitate *B. cepacia* colonisation. Recently, *B. cepacia* was discovered to have components of a QS system, called CepR and CepI, which regulate protease and siderophore production (Lewenza et al., 1999). The predominant AHL molecule used by the *cep* QS system is *N*-octanoylhomoserine lactone (C_8-HSL). Using random transposon mutagenesis, it was discovered that the *B. cepacia* transcriptional regulator CepR is important for biofilm formation (Huber et al., 2001). Further analysis using CepR and CepI null mutants revealed that, in a flowing system, both the mutants and the wild-type strains were able to attach to the surface and form microcolonies. However, the mutant strains colonised the surface less efficiently than the wild type and were unable to differentiate from microcolonies into a mature biofilm. Adding of 200 n*M* of C_8-HSL and providing *cepR in trans* to the Δ*cepI* and Δ*cepR* mutants, respectively, restored the biofilms to that of the wild-type strain. Therefore, it appears that, like *P. aeruginosa*, QS is important for biofilm differentiation in *B. cepacia*. Another interesting finding from this study was that, compared to the wild-type strain, *B. cepacia cepR* and *cepI* mutants exhibited markedly reduced swarming. Addition of the biosurfactant serrawettin W2 or surfactin to the media restored swarming in both mutants, indicating that a biosurfactant is the QS-controlled factor that is required for swarming motility. Addition of serrawettin or surfactin to the QS mutants, however, did not markedly increase biofilm formation, suggesting that swarming motility is not involved in the formation of *B. cepacia* biofilms.

9 BIOFILM PREVENTION STRATEGIES

Biofilms produced by *P. aeruginosa* and other bacteria are hugely problematic in both medical and industrial environments. In the clinical setting, biofilms formed on medical devices and in bacterial infections can wreak havoc, largely because bacteria growing as a biofilm are refractile to host defenses, including antibodies, phagocytes, and complement (Costerton, Stewart, and Greenberg, 1999). Furthermore, the recalcitrance of biofilm bacteria to antibiotics renders standard chemotherapy strategies relatively ineffectual. Currently, much attention is focused on new ways of preventing biofilm formation on both industrial and medical surfaces. One strategy that has received much attention focusses on coating or embedding biofilm-retardant compounds onto the surface of the material. In nature, the seaweed *Delisea pulchra* produces furanones that are structurally very similar to AHL molecules (Givskov et al., 1996) and have antifouling characteristics (de Nys et al., 1995; Reichelt and Borowitzka, 1984). Studies using these furanone compounds, either coated onto surfaces or preincubated with cells, have demonstrated that they are able to inhibit AHL-regulated phenotypes, including bacterial swarming in *Serratia liquifaciens* and bioluminescence production by *Vibriot fischeri* and *V. harveyi* (Givskov et al., 1996). In *P. aeruginosa* biofilms, expression of the AI synthase genes is highest at the substratum where AI concentrations are presumably high. If surface-associated AI analogs or other QS-inhibitory compounds could shut down *lasI* and *rhlI* expression, then all of the other genes in the QS cascade would be shut down as well, including those involved in biofilm formation. Our observation that *lasI* and *rhlI* expression is maximal at the substratum supports the idea that treatment of surfaces with compounds capable of interfering with QS could be an effective means of preventing *P. aeruginosa* biofilm formation. The GFP-based AHL sensor strains of bacteria that have been generated (Anderson et al., 2001; Wu et al., 2000) represent one potential means of determining both the half-life and the ability of these inhibitory compounds to penetrate the biofilm structure. Using these AHL sensors in conjunction with confocal microscopy to monitor green fluorescence, it should be possible to ascertain the functionality of inhibitors in model biofilm systems.

During biofilm development, cells periodically leave the biofilm matrix and resume the planktonic mode of growth. Currently, research directed towards uncovering the molecular cues that stimulate this behaviour is underway. Understanding what triggers cells to transition back to planktonic growth may enable us to stimulate the entire biofilm population to detach. As such, this may represent another potential strategy for eradicating refractory biofilms.

10 CONCLUSIONS

Over the last few years, a combination of microscopy, molecular biology, and chemical analysis has allowed us to learn a great deal about the complex physiology of biofilms. The elaborate design of biofilms erected such that nutrients and waste products are easily exchanged has been compared to a primitive circulatory system (Kolter and Losick, 1998). Clearly, the activity of the bacterial population in the biofilm must be coordinated to enable development of these highly ordered structures. In view of this, the discovery that QS signalling is involved in biofilm differentiation is not surprising. Although substantial progress has been made towards understanding how QS directs biofilm formation, fundamental questions remain. For example, which QS-regulated genes are responsible for biofilm differentiation? Furthermore, the mechanisms underlying the increased antimicrobial resistance of biofilms remain an enigma. Several theories have been put forth to explain this resistance, including reduced antimicrobial diffusion, the presence of slow-growing or sessile bacteria, and an altered 'biofilm phenotype'. However, these hypotheses are either largely unproven or are insufficient to explain resistance to such a broad range of compounds with diverse modes of action. A recent study by Spoering and Lewis (2001) suggests that biofilms are no more resistant to antimicrobial agents than stationary phase cultures of planktonic cells. These researchers claim that the antimicrobial recalcitrance of biofilms is not related to the mode of growth, but rather to the presence of 'persisters' in the cell population, and these bacteria are present in both biofilms and stationary planktonic cultures. This presumption is based on the observation that stationary-phase cultures of *P. aeruginosa* are either as resistant as, or more resistant than, biofilm populations to three different classes of antibiotics as well as the oxidant peracetic acid. The authors speculate that if biofilm resistance can be attributed to persister cells, drugs that are able to kill persisters should be effective in eradicating biofilms (Spoering and Lewis, 2001). As such, stationary-phase planktonic cultures, which are easier to manipulate than biofilms, can be used to search for drugs that target persisters. The biofilms examined in this study were formed over 18 hours in a closed system with an exhaustible nutrient supply. Future investigations should include biofilms grown in a chemostat, which are more representative of natural biofilms. In spite of the fact that these findings are somewhat preliminary, they present a provocative hypothesis regarding biofilm resistance and warrant further investigation.

In the future, uncovering the molecular mechanisms governing biofilm formation will be a huge undertaking. However, studying the expression of genes critical for these processes will give us a better understanding of the genotypic and phenotypic changes associated with the biofilm mode of growth, and these findings may ultimately lead to novel strategies for controlling problematic and recalcitrant biofilms.

ACKNOWLEDGEMENTS

This work was supported by National Institutes of Health research grant AI133713 (to B.H.I.).

REFERENCES

Albus, A. M., Pesci, E. C., Runyen-Janecky, L. J., West, S. E. and Iglewski, B. H. (1997). Vfr controls quorum sensing in *Pseudomonas aeruginosa*. *Journal of Bacteriology*, 179, 3928–3935.

Anderson, J. B., Heydorn, A., Hentzer, M., Eberl, L., Geisenberger, O., Christensen, B. B., Molin, S. and Givskov, M. (2001). *gfp*-Based *N*-acyl homoserine-lactone sensor systems for detection of bacterial communication. *Applied and Environmental Microbiology*, 67, 575–585.

Andersen, J. B., Sternberg, C., Poulsen, L. K., Bjorn, S. P., Givskov, M. and Molin, S. (1998). New unstable variants of green fluorescent protein for studies of transient gene expression in bacteria. *Applied and Environmental Microbiology*, 64, 2240–2246.

Brint, J. M. and Ohman, D. E. (1995). Synthesis of multiple exoproducts in *Pseudomonas aeruginosa* is under the control of RhlR-RhlI, another set of regulators in strain PAO1 with homology to the autoinducer-responsive LuxR-LuxI family. *Journal of Bacteriology*, 177, 7155–7163.

Chapon-Herve, V., Akrim, M., Latifi, A., Williams, P., Lazdunski, A. and Bally, M. (1997). Regulation of the *xcp* secretion pathway by multiple quorum-sensing modulons in *Pseudomonas aeruginosa*. *Molecular Microbiology*, 24, 1169–1178.

Charlton, T. S., de Nys, R., Netting, A., Kumar, N., Hentzer, M., Givskov, M. and Kjelleberg, S. (2000). A novel and sensitive method for the quantification of *N*-3-oxoacyl homoserine lactones using gas chromatograph-mass spectrometry: application to a model bacterial biofilm. *Environmental Microbiology*, 2, 530–541.

Costerton, J. W., Stewart, P. S. and Greenberg, E. P. (1999). Bacterial biofilms: a common cause of persistant infections. *Science*, 284, 1318–1322.

Davies, D. G., Parsek, M. R., Pearson, J. P., Iglewski, B. H., Costerton, J. W. and Greenberg, E. P. (1998). The involvement of cell-to-cell signals in the development of a bacterial biofilm. *Science*, 280, 295–298.

de Kievit, T. R., Gillis, R., Marx, S., Brown C. and Iglewski, B. H. (2001). Quorum-sensing genes in *Pseudomonas aeruginosa* biofilms: their role and expression patterns. *Applied and Environmental Microbiology*, 67, 1865–1873.

de Kievit, T. R. and Iglewski, B. H. (2000). Bacterial quorum sensing in pathogenic relationships. *Infection and Immunity*, 68, 4839–4849.

de Kievit, T., Seed, P. C., Nezezon, J., Passador, L. and Iglewski, B. H. (1999). RsaL, a novel repressor of virulence gene expression in *Pseudomonas aeruginosa*. *Journal of Bacteriology*, 181, 2175–84.

de Nys, R., Steinberg, P. D., Willemsen, P., Dworjanyn, S. A., Gabelish, C. L. and King, R. J. (1995). Broad spectrum effects of secondary metabolites from the red alga *Delisea pulchra* in antifouling assays. *Biofouling*, 8, 259–271.

DiMango, E., Zar, H. J., Bryan, R. and Prince A. (1995). Diverse *Pseudomonas aeruginosa* gene products stimulate respiratory epithelial cells to produce interleukin-8. *Journal of Clinical Investigation*, 96, 2204–2210.

Dunny, G. M. and Leonard, B. A. B. (1997). Cell–cell communication in Gram-positive bacteria. *Annual Review of Microbiology*, 51, 527–564.

Flemming, C. A., Palmer, R. J., Jr., Arrage, A. A., Van de Mei, H. C. and White, D. C. (1998). Cell surface physiochemistry alters biofilm development of *Pseudomonas aeruginosa* lipopolysaccharide mutants. *Biofouling*, 13, 213–231.

Fuqua, W. C., Winans, S. C. and Greenberg, E. P. (1996). Census and consensus in bacterial ecosystems: the LuxR-LuxI family of quorum-sensing transcriptional regulators. *Annual Review of Microbiology*, 50, 727–751.

Gambello, M. J., Kaye, S. and Iglewski, B. H. (1993). LasR of *Pseudomonas aeruginosa* is a transcriptional activator of the alkaline protease gene (*apr*) and an enhancer of exotoxin A expression. *Infection and Immunity*, 61, 1180–1184.

Geisenberger, O., Givskov, M., Riedel, K., Høiby, N., Tümmler, B. and Eberl, L. (2000). Production of *N*-acyl-*L*-homoserine lactones by *P. aeruginosa* isolates from chronic lung infections associated with cystic fibrosis. *FEMS Microbiology Letters*, 184, 273–278.

Givskov, M., de Nys, R., Manefield, M., Gram, L., Maximilien, R., Eberl, L., Molin, S., Steinberg, P. D. and Kjelleberg, S. (1996). Eukaryotic interference with homoserine lactone-mediated prokaryotic signalling. *Journal of Bacteriology*, 178, 6618–6622.

Hancock, R. W., Mutharia, L. M., Chan, M., Darveau, R. P., Speert, D. P. and Pier, G. B. (1983). *Pseudomonas aeruginosa* isolates from patients with cystic fibrosis: a class of serum-sensitive, nontypable strains deficient in lipopolysaccharide O side chains. *Infection and Immunity*, 42, 170–177.

Huber, B., Riedel, K., Hentzer, M., Heydorn, A., Gotschlich, A., Givskov, M., Molin, S. and Eberl, L. (2001). The *cep* quorum-sensing system of *Burkholderia cepacia* H111 controls biofilm formation and swarming motility. *Microbiology*, 147, 2517–2528.

Kolter, R. and Losick, R. (1998). One for all and all for one. *Science*, 280, 226–227.

Lam, M. Y. C., McGroarty, E. J., Kropinski, A. M., MacDonald, L. A., Peterson, S. S., Høiby, N. and Lam, J. S. (1989). Occurrence of a common lipopolysaccharide antigen in standard and clinical strains of *Pseudomonas aeruginosa*. *Journal of Clinical Microbiology*, 27, 962–967.

Latifi, A., Foglino, M., Tanaka, K., Williams, P. and Lazdunski, A. (1996). A hierarchical quorum-sensing cascade in *Pseudomonas aeruginosa* links the transcriptional activators LasR and RhlR to expression of the stationary-phase sigma factor RpoS. *Molecular Microbiology*, 21, 1137–1146.

Latifi, A., Winson, M. K., Foglino, M., Bycroft, B. W., Stewart, G. S. A. B., Lazdunski, L. and Williams, P. (1995). Multiple homologues of LuxR and LuxI control expression of virulence determinants and secondary metabolites through quorum sensing in *Pseudomonas aeruginosa* PAO1. *Molecular Microbiology*, 17, 333–343.

Lewenza, S., Conway, B., Greenberg, E. P. and Sokol, P. A. (1999). Quorum sensing in *Burkholderia cepacia*: identification of the LuxRI homologs CepRI. *Journal of Bacteriology*, 181, 748–756.

McLean, R. J., Whiteley, M., Stickler, D. J. and Fuqua, W. C. (1997). Evidence of autoinducer activity in naturally occurring biofilms. *FEMS Microbiology Letters*, 154, 259–263.

Ochsner, U. A., Koch, A. K., Fiechter, A. and Reiser, J. (1994). Isolation and characterization of a regulatory gene affecting rhamnolipid biosurfactant synthesis in *Pseudomonas aeruginosa*. *Journal of Bacteriology*, 176, 2044–2054.

Ochsner, U. A. and Reiser, J. (1995). Autoinducer-mediated regulation of rhamnolipid biosurfactant synthesis in *Pseudomonas aeruginosa*. *Proceedings of the National Academy of Sciences of the USA*, 92, 6424–6428.

O'Toole, G. A. and Kolter, R. (1998). Flagellar and twitching motility are necessary for *Pseudomonas aeruginosa* biofilm development. *Molecular Microbiology*, 30, 295–304.

Passador, L., Cook, J. M., Gambello, M. J., Rust, L. and Iglewski, B. H. (1993). Expression of *Pseudomonas aeruginosa* virulence genes requires cell-to-cell communication. *Science*, 260, 1127–1130.

Pearson, J. P., Feldman, M., Iglewski, B. H. and Prince, A. (2000). *Pseudomonas aeruginosa* cell-to-cell signaling is required for virulence in a model of acute pulmonary infection. *Infection and Immunity*, 68, 4331–4334.

Pearson, J. P., Gray, K. M., Passador, L., Tucker, K. D., Eberhard, A., Iglewski, B. H. and Greenberg, E. P. (1994). Structure of the autoinducer required for expression of *Pseudomonas aeruginosa* virulence genes. *Proceedings of the National Academy of Sciences of the USA*, 91, 197–201.

Pearson, J. P., Passador, L., Iglewski, B. H. and Greenberg, E. P. (1995). A second *N*-acyl homoserine lactone signal produced by *Pseudomonas aeruginosa*. *Proceedings of the National Academy of Sciences of the USA*, 92, 1490–1494.

Pearson, J. P., Pesci, E. C. and Iglewski, B. H. (1997). Roles of *Pseudomonas aeruginosa las* and *rhl* quorum-sensing systems in control of elastase and rhamnolipid biosynthesis genes. *Journal of Bacteriology*, 179, 5756–5767.

Pesci, E. C., Milbank, J. B., Pearson, J. P., McKnight, S., Kende, A. S., Greenberg, E. P., and Iglewski B. H. (1999). Quinolone signaling in the cell-to-cell communication system of *Pseudomonas aeruginosa*. *Proceedings of the National Academy of Sciences of the USA*, 96, 11229–11234.

Pesci, E. C., Pearson, J. P., Seed, P. C. and Iglewski, B. H. (1997). Regulation of *las* and *rhl* quorum sensing in *Pseudomonas aeruginosa*. *Journal of Bacteriology*, 179, 3127–3132.

Rahme, L. G., Stevens, E. J., Wolfort, S. F., Shao, J., Tompkins, R. G. and Ausubel, F. M. (1995). Common virulence factors for bacterial pathogenicity in plants and animals. *Science*, 268, 1899–1902.

Rahme, L. G., Tan, M.-W., Le, L., Wong, S. M., Tompkins, R. G., Calderwood, S. B. and Ausubel, F. M. (1997). Use of model plant hosts to identify *Pseudomonas aeruginosa* virulence factors. *Proceedings of the National Academy of Sciences of the USA*, 94, 13245–13250.

Reichelt, J. L. and Borowitzka, M. A. (1984). Antimicrobial activity from marine algae: results of a large-scale screening programme. *Hydrobiologia*, 116/117, 158–168.

Reimmann, C., Beyeler, M., Latifi, A., Winteler, H., Foglino, M., Lazdunski, A. and Haas D. (1997). The global activator GacA of *Pseudomonas aeruginosa* PAO positively controls the production of the autoinducer *N*-butyryl-homoserine lactone and the formation of the virulence factors pyocyanin, cyanide, and lipase. *Molecular Microbiology*, 24, 309–319.

Rocchetta, H. L., Burrows, L. L. and Lam, J. S. (1999). Genetics of O-antigen biosynthesis in *Pseudomonas aeruginosa. Microbiology and Molecular Biology Reviews*, 63, 523–553.

Rumbaugh, K. P., Griswold, J. A. and Hamood A. N. (1999a). Contribution of the regulatory gene *lasR* to the pathogenesis of *Pseudomonas aeruginosa* infection of burned mice. *Journal of Burn Care and Rehabilitation*, 20, 42–49.

Rumbaugh, K. P., Griswold, J. A., Iglewski, B. H. and Hamood, A. N. (1999b). Contribution of quorum sensing to the virulence of *Pseudomonas aeruginosa* in burn wound infections. *Infection and Immunity*, 67, 5854–5862.

Seed, P. C., Passador, L. and Iglewski, B. H. (1995). Activation of the *Pseudomonas aeruginosa lasI* gene by LasR and the *Pseudomonas* autoinducer PAI: an autoinduction regulatory hierarchy. *Journal of Bacteriology*, 177, 654–659.

Shaw, P. D., Ping, G., Daly, S. L., Cha, C., Cronan J. E., Jr., Renehart, K. L., et al., (1997). Detecting and characterizing *N*-acyl-homoserine lactone signal molecules by thin-layer chromatography. *Proceedings of the National Academy of Sciences of the USA*, 94, 6036–6041.

Singh, P. K., Schaefer, A. L., Parsek, M. R., Moninger, T. O., Welsh, M. J. and Greenberg, E. P. (2000). Quorum-sensing signals indicate that cystic fibrosis lungs are infected with bacterial biofilms. *Nature*, 407, 762–764.

Spoering, A. L. and Lewis, K. (2001). Biofilms and planktonic cells of *Pseudomonas aeruginosa* have similar resistance to killing by antimicrobials. *Journal of Bacteriology*, 183, 6746–6751.

Stickler, D. J., Morris, N. S., McLean, R. J. and Fuqua, C. (1998). Biofilms on indwelling urethral catheters produce quorum-sensing signal molecules in situ and in vitro. *Applied and Environmental Microbiology*, 64, 3486–3490.

Storey, D. G., Ujack, E. E., Rabin, H. R. and Mitchell, I. (1998). *Pseudomonas aeruginosa lasR* transcription correlates with the transcription of *lasA, lasB,* and *toxA* in chronic lung infections associated with cystic fibrosis. *Infection and Immunity*, 66, 2521–2528.

Tan, M.-W., Rahme, L. G., Sternberg, J. A., Tompkins, R. G. and Ausubel, F. M. (1999). *Pseudomonas aeruginosa* killing of *Caenorhabditis elegans* used to identify *P. aeruginosa* virulence factors. *Proceedings of the National Academy of Sciences of the USA*, 96, 2408–2413.

Tang, H. B., DiMango, E., Bryan, R., Gambello, M., Iglewski, B. H., Goldberg, J. B. and Prince, A. (1996). Contribution of specific *Pseudomonas aeruginosa* virulence factors to pathogenesis of pneumonia in a neonatal mouse model of infection. *Infection and Immunity*, 64, 37–43.

Telford, G., Wheeler, D., Williams, P., Tomkins, P. T., Appleby, P., Sewell, H., Stewart, G. S. A. B., Bycroft, B. W. and Pritchard, D. I. (1998). The *Pseudomonas aeruginosa* quorum-sensing signal molecule *N*-(3–oxododecanoyl)-L-homoserine lactone has immunomodulatory activity. *Infection and Immunity*, 66, 36–42.

Tombolini, R., Unge, A., Davey, M. E., de Bruijn, F. J. and Jansson, J. K. (1997). Flow cytometric and microscopic analysis of GFP-tagged *Pseudomonas fluorescens* bacteria. *FEMS Microbiology Ecology*, 22, 17–28.

Woods, D. E., Sokol, P. A., Bryan, L. E., Storey, D. G., Mattingly, S. J., Vogel, H. J. and Ceri, H. (1991). In vivo regulation of virulence in *Pseudomonas aeruginosa* associated with genetic rearrangement. *Journal of Infectious Diseases*, 163, 143–149.

Wu, H., Song, Z., Hentzer, M. andersen, J. B., Heydorn, A., Mathee, K., Moser, C., Eberl, L., Molin, S., Høiby, N. and Givskov, M. (2000). Detection of *N*-acylhomoserine lactones in lung tissue of mice infected with *Pseudomonas aeruginosa*. *Microbiology*, 146, 2481–2493.

CHAPTER THREE

Antimicrobial Agents and Biofilms

Michael R. W. Brown and Anthony W. Smith

1 INTRODUCTION: THE PROBLEMS

There is increasing concern over the role played by microbial biofilms in infection. These include well-known examples of medical device-related infections such as those associated with artificial joints, prosthetic heart valves, and catheters. Indeed, recent surveys indicate that catheter-associated bacteraemia, consequent from catheter-related infection, is by far the leading cause of nosocomial bloodstream infection in intensive care units (Brub-Buisson, 2001). Many chronic infections, not related to medical devices, are now recognised to be due to bacteria either not growing and relatively dormant or growing slowly as biomasses or adherent biofilms on mucosal surfaces. Thus, the question of how to treat biofilm infections extends to many aspects of medicine. Indeed, the issue of biofilm eradication extends way beyond the infected patient, since bacteria in the environment typically exist as biofilms. These are commonly complex multispecies ecosystems associated with protozoa (Brown and Barker, 1999). The biofilm mode of growth greatly enhances the survival of the constituent microbes.

Growth as a biofilm almost always leads to a large increase in resistance to antimicrobial agents, including antibiotics, biocides, and preservatives, compared with cultures grown in suspension (planktonic) in conventional liquid media (Gilbert, Collier, and Brown, 1990; Stewart and Costerton, 2001). However, a recent paper with high density planktonic cultures indicated similar resistance to antimicrobials as did biofilm cultures (Spoering and Lewis, 2001). Currently, there is no generally agreed mechanism to account for the broad resistance to chemical agents. We suggest that dormancy, related to the general stress response and associated survival responses, offers an explanation of the overall general resistance of biofilm microbes (Brown and Smith, 2001; Foley et al., 1999).

2 ANTIMICROBIALS AND BIOFILM DEVELOPMENT

The role of specific adhesin/receptor binding to host cell surfaces, with consequent complex signal transduction cascades in the host cell, are reviewed elsewhere (Boland, Latour, and Stutzenber, 2000; Hopelman and Tuomanen, 1992; Wilson, 2002). Detailed accounts of the physics of attachment can be studied elsewhere (An, Dickinson, and Doyle, 2000; Gilbert, Hodgson, and Brown, 1995).

An unadulterated surface does not exist. Any surface, synthetic or otherwise, is coated with constituents of the local environment: first water and salt ions, then organic material. This conditioning film exists before the arrival of the first microbe. Next, there is a weak and reversible contact between microbe and conditioning film resulting from Brownian motion, gravitation, diffusion, or microbial motility and involving electrostatic interactions. The surface interaction is a function of the cell surface (determined by the cell physiology) and the nature of the film. For example, recent work with a *Staphylococcus aureus* mutant bearing a stronger negative charge, due to the lack of D-alanine esters in its teichoic acids, could no longer colonise polystyrene or glass surfaces, highlighting the contribution of electrostatic forces to biofilm formation, particularly on medical devices (Gross et al., 2001). Charge attraction or repulsion could also contribute to interaction between bacteria and the substratum (Gottenbos et al., 2001). Moreover, growth in the presence of sub-inhibitory antibiotic concentrations can influence cell surface hydrophobicity (Gottenbos et al., 2001), as can the physiological state of the cell (Allison et al., 1990a,b; Domingue, Lambert, and Brown, 1989; Williams et al., 1986). Specific interactions with bacterial surface structures can also be important in establishing a biofilm. For example, flagella and pilus-mediated twitching motility are required/important for biofilms of *Escherichia coli* and *Pseudomonas aeruginosa* (O'Toole and Kolter, 1998; Pratt and Kolter, 1998). The penultimate step is when the adsorption becomes irreversible. This is partly due to surface appendages overcoming the repulsive forces between the two surfaces and also because of sticky exopolymers secreted by the cells. Commonly, the entire biofilm may be coated with a hydrophilic exopolymer (the glycocalyx), which is itself a complex and dynamic structure (Sutherland, 2001). When the host is unable to opsonise this hydrophilic glycocalyx, the entire biofilm is resistant to phagocytosis (Pier et al., 1987). The final stage is when the biofilm population increases, typically as a result of adherent cells replicating, but including a contribution from fresh cells adhering to the biofilm (Al-Bakri, Gilbert, and Allisen, 1999; Gilbert et al., 1997; Stoodley et al., 2001). In staphylococcal species, commonly associated with device-related infections, biofilm formation requires cell–cell adhesion mediated by the *ica* locus following adhesion to the substratum (Cramton et al., 1999). Subinhibitory concentrations of tetracycline and the semisynthetic streptogramin antibiotic quinupristin-dalfopristin enhanced *ica* expression and biofilm formation (Rachid et al., 2000b).

The biofilm structure/phenotype depends on numerous factors, notably, the organism and its physiology, the substratum, the surrounding nutrient environment, and the rate of flow of any liquid over its surface (Costerton et al., 1995; Karthikeyan et al., 2000). The resulting biofilms may vary from sparse amorphous masses to highly structured consortia with mushroom-like cell stacks surrounded by channels with rapid aqueous movement (Costerton, Stewart, and Greenberg, 1999). Signal molecules (discussed later) which influence cell physiology, including virulence and the general stress response (and thus dormancy), have also been shown to influence biofilm structure and susceptibility to antimicrobials.

3 ASSESSMENT OF BIOFILM RESISTANCE: COMPARISON OF BIOFILM AND PLANKTONIC CULTURES

There are numerous biofilm culture models, and these have been reviewed by others (Allison, Maira-Litrán, and Gilbert, 1999; Dibdin and Wimpenny, 1999; Kharazmi, Giwercman, and Høiby, 1999; Sissons, Wong, and An, 2000; Yasuda, Koga, and Fukoka, 1999). The majority of the literature examining the resistance of bacterial biofilms is driven by comparisons with cells grown planktonically. Although there is general acceptance that there are numerous planktonic phenotypes, many papers refer to 'the biofilm phenotype', implicitly assuming (wrongly) that there is only one. Those same parameters known to influence planktonic physiology, including growth rate and/or specific nutrient limitation (Brown, 1977), also apply to biofilm physiology. Valid comparisons between biofilm and planktonic cultures are therefore difficult to make, especially if such key parameters are not comparable between the two states. A model which controls biofilm growth rate in a defined, nutrient-limited medium consists of surface growth on the underside of a bacteria-proof cellulose membrane (Gilbert et al., 1989). The membrane is perfused with fresh, defined medium from the sterile side, and cells eluted from the biofilm are collected. The growth rate of the nutrient-limited, adherent population is controlled by the rate of perfusion of fresh medium. A method of controlling density and nutrient limitation of biofilm growth consists of membrane culture on the surface of a defined agar medium, with growth being limited by any specified major nutrient (Bühler et al., 1998, Desai et al., 1998). These methods enable comparison of planktonic and biofilm cultures at similar stages of growth and under similar eventual nutrient restriction (as well as temperature and pH). However, it is not possible to obtain *comparable* planktonic cells at a density equivalent even to that of a sparse biofilm. Such planktonic cultures would require growth in extremely high concentrations of nutrients, including massive sparging with oxygen. Thus, the price of comparable density would be a lack of comparison in terms of osmolarity and the damaging consequences of oxygen.

4 RESISTANCE TO ANTIMICROBIAL AGENTS

Eradication of infection by antibiotic treatment requires elimination of all the bacteria, typically assisted by the host defences. Otherwise, infection recurs and chronicity is established. In other words, biofilm resistance can be determined by the susceptibility of the most resistant cell. It is not the case that all cells within a biofilm are always highly resistant (Brooun, Liu, and Lewis, 2000; Lewis, 2001). But the most resistant members of a biofilm population are typically orders of magnitude more resistant than similar members of a planktonic population; yet subculture rarely shows the existence of resistant mutants. Hence, biofilm resistance is characteristically phenotypic. But what are the phenotypic mechanism(s)? We will examine if any of the conventionally recognised mechanisms of antibiotic resistance play an enhanced role in biofilms, thus contributing to their extraordinary resistance.

4.1 Lack of Antibiotic Penetration

Many papers investigate the possible lack of antibiotic/biocide penetration as an explanation of biofilm resistance (Gilbert et al., 1995; Lewis, 2001; Mah and O'Toole, 2001; Stewart, 1996; Xu, McFeters, and Stewart, 2000). Given, in some cases, biofilms consisting largely of stacks of cells with flowing aqueous channels (even though coated with glycocalyx), impenetrability seems highly unlikely (Nichols, 1991), a finding confirmed with biofilms of *Klebsiella pneumoniae* (Anderl, Franklin, and Stewart, 2000). Where the antimicrobial agent either reacts chemically with components of the exopolymer or is significantly adsorbed by these typically anionic polymers, then the net effect is as if there is a penetration barrier. There will be a similar effect if such interactions occur with cells, perhaps dead ones, in the outer parts of the biofilm. Heterogeneity has also been given as a reason for biofilm resistance. But, in terms of eradication/sterilisation, resistance is caused by the most resistant members of the biofilm. Hence, the question still remains as to the mechanism. Heterogeneity per se is not a mechanism. Given that biofilms are indeed typically heterogeneous, these generalisations do not preclude the possibility of areas in a biofilm where diffusion is restricted (Lewis, 2001).

4.2 Antimicrobial Modification by Enzymes

Antimicrobial agents can frequently induce the production of inactivating enzymes in microbes. The relatively large amounts of antibiotic-inactivating enzymes such as β-lactams which accumulate within the glycocalyx produce concentration gradients of antimicrobial across it, and the underlying cells have been shown to be thus protected (Bagge et al., 2000; Giwercman et al., 1991), although in other systems enzyme inactivation appears not to contribute to the resistance of biofilm bacteria (Anderl et al., 2000).

4.3 Efflux

The contribution of efflux systems to the resistance of bacteria to antimicrobial agents has been studied extensively in recent years (Lewis, 2001; Zgurskaya and Nikaido, 2000). Several classes of antibiotics are substrates for the pumps and include the tetracyclines, macrolides, β-lactams, and fluoroquinolones (Van Bambeke, Balzi, and Tulkens, 2000). Their role within biofilms have been studied. In a comprehensive study of multidrug efflux pumps in *P. aeruginosa*, none of the four systems present in the genome contributed to resistance in a biofilm. Temporal and spatial analyses using fusions to *gfp* indicated that expression of *mexAB-oprM* and *mexCD-oprJ* decreased over time in the developing biofilm, with maximal expression occurring at the biofilm substratum (De Kievit et al., 2001). Also, Gilbert and co-workers (Maira-Litrán, Allison, and Gilbert, 2000) have addressed the contribution of the **m**ultiple **a**ntibiotic **r**esistance (*mar*) operon and *acrAB* in biofilms of *E. coli*. The *mar* operon is present in a number of Gram-negative bacteria, and the antibiotic resistance phenotype is mediated by upregulation of *acrAB* (Moken, McMurray, and Levy, 1997). Mutants deleted for *mar* showed similar sensitivity to the fluoroquinolone antibiotic ciprofloxacin as wild-type cells grown in a biofilm, whereas a constitutive *mar* mutant showed decreased susceptibility. Isolates in which the *acrAB* efflux pump was deleted also showed similar sensitivity, whereas constitutive expression of *acrAB* protected biofilms at low but not high antibiotic concentrations, leading the authors to conclude that ciprofloxacin resistance in *E. coli* biofilms is not mediated by upregulation of the *mar* and *acrAB* operons (Maira-Litrán, Allison, and Gilbert, 2000).

The contribution of efflux pumps to quorum sensing and biofilm formation is now becoming apparent, and thus, there is potential for efflux pumps to contribute to resistance of biofilm cells through mechanisms relating to cell density, stress responses, and dormancy rather than drug efflux per se (discussed later).

4.4 Repair

Enhanced repair systems operating within a biofilm could contribute to the decreased susceptibility of cells within a biofilm to antimicrobial agents. One example is the inducible SOS system which increases the survival of bacteria exposed to damaging agents by increasing the capacity of error-free and error-prone repair mechanisms (Sutton et al., 2000). Although there are no reports of this system in biofilms per se, induction has been reported in ageing colonies on agar plates (Taddei, Matic, and Radman, 1995). Also, the SOS system can be induced by quinolone antibiotics and trimethoprim (Kimmitt, Harwood, and Barer, 2000). Enzymes involved with detoxification of reactive oxygen species, notably, superoxide dismutase and catalase, have been extensively studied in biofilms. Hassett and co-workers have shown that levels of the manganese- and iron-cofactored superoxide dismutases and the major catalase KatA are decreased in mutants of *P. aeruginosa* devoid of one or both quorum sensing molecules grown planktonically, with a concomitant increase in sensitivity to hydrogen peroxide and phenazine methosulphate (Hassett et al., 1999). Perhaps surprisingly, biofilm-grown cells had less catalase activity and yet were more resistant to hydrogen peroxide than their planktonic counterparts. Catalase levels were even lower in quorum-sensing deficient mutants and yet they were also resistant to hydrogen peroxide. One resistance mechanism appears to be the prevention of hydrogen peroxide penetration fully into the biofilm (Stewart et al., 2000).

The susceptibility to reactive oxygen species may also be related to repair of sub-lethal injury following antimicrobial treatment. It has been proposed that sublethal injury by antimicrobial agents leads to an imbalance in anabolism and catabolism and a burst of damaging oxygen free-radical production (Dodd et al., 1997) on attempting to recover treated cultures. Viable competitive microflora at high density can protect exponential phase cells of another organism at lower density from the lethal effects of heat (chemical antimicrobials have not been tested). The authors propose that this addition creates an immediate reduction in the oxygen tension of the culture and oxidative metabolism is reduced (Dodd et al., 1997). Although not tested, variation in oxygen tension gradients within a biofilm, together with the varying metabolic activities within a mixed microbial biofilm, could be conceived to contribute to resistance.

5 SLOW GROWTH RATE AND DORMANCY: ASSOCIATED RESISTANCE MECHANISMS?

None of the conventional mechanisms outlined previously satisfactorily explains the general resistance of biofilms. We suggest that dormancy and associated survival responses do offer an explanation.

5.1 Slow/No Growth

Decreased culture growth rate is associated with decreased susceptibility to almost all chemical antimicrobial agents, some of which have a requirement for replication (Brown, Allison, and Gilbert, 1988; Brown, Collier, and Gilbert, 1990; Mah and O'Toole, 2001). Also, in any stress response by a growing culture, a reduction in growth rate and even growth cessation are associated. This makes it difficult to separate out the individual contributions to the start of a resistance cascade of growth rate per se and/or an enforced *change* in rate, cell density (and/or quorum sensing), and the nature of nutrient starvation or other stress. In the case of a biofilm, density is high at an early stage relative to the same number of cells growing in a conventional planktonic culture. It is also difficult to make a valid comparison between biofilm cells and planktonic cells when they have been cultured often in different media and harvested in different physiological states. Using growth-rate-controlled cultures of *P. aeruginosa, E. coli,* and *S. epidermidis,* in both planktonic and biofilm modes of growth, there was a definite growth rate effect. In both cases, sensitivity increased with increased growth rate (Duguid et al., 1992a,b; Evans et al., 1991). However, increases in growth rate caused bigger changes in sensitivity with planktonic cells, indicating factors operating in addition to growth rate. Using chemically and nutritionally defined cultures (Brown and Gilbert, 1995; Brown et al., 1995; Bühler et al., 1998), which ultimately entered a stationary phase because of iron limitation, susceptibility to ciprofloxacin and to ceftazidime was measured along the exponential phase of batch cultures, planktonic and biofilm (Desai et al., 1998). In both growth modes there were dramatic changes in resistance throughout the exponential phase and before measurable growth reduction for the stationary phase. Increases in resistance to both agents occurred in planktonic culture about three to four generations before onset of the stationary phase and with biofilms about ten generations before the stationary phase. Cell density may have played a part, although effects were noted well below the commonly recorded quorum sensing density. Events underpinning changes in growth rate are complex. The alarmone ppGpp, part of the stringent response, has been shown to play a role in regulating growth rate (Cashel et al., 1996), as has the accumulation of inorganic polyphosphate (Kornberg, Rao, and Ault-Richie, 1999; Rao, Liu, and Kornberg, 1998), and serves to highlight the complexity of overlapping regulatory networks that could all operate within a biofilm (discussed later).

5.2 Dormancy and Stress Responses

Considering the available literature, stress responses are always accompanied by a reduction in growth rate. Even static suspensions used for susceptibility assays are from cultures that have ceased growth due to starvation or the presence of inhibitory substances. Also, handling techniques such as centrifugation and resuspension can contribute to stress (Gilbert, Coplan, and Brown, 1991). Consequently, a reduction in growth rate is an indication of a stress, and a specific slow growth rate may well maintain the cells in the initial stages of a stress response. Thus, a slow growth rate is a likely contributor to, but not the main reason for, reduced susceptibility.

There is a large literature on an aspect of the behaviour of the stationary phase of planktonic bacteria known as the general stress response (GSR) (Hengge-Aronis, 1999). This stress response has been implicated directly (Foley et al., 1999) in chronic infection involving biofilms and could clearly occur in circumstances where high density and quorum sensing have been reported (Singh et al., 2000; Stickler et al., 1998; Wu et al., 2001). Whereas most of the work is with Gram-negative bacteria, there are well-characterised stress response systems in Gram-positive bacteria. Systems under the control of alternative sigma factors are found in *Bacillus subtilis* (Scott, Mitchell, and Haldenwang, 2000); *S. aureus* (Chan et al., 1998; Clements and Foster, 1999), where biofilm formation is also affected (Rachid et al., 2000a); and *Mycobacterium tuberculosis* (DeMaio et al., 1996).

5.3 The GSR

The GSR involves a late log and stationary phase cascade during which structures are protected and the cells become quiescent. It resembles sporulation in its physiological consequences. The result is an ability to survive prolonged periods of nutrient starvation and multiple environmental stresses, such as heat, oxidising agents, and hyperosmolarity. Unlike sporulation, the GSR does not involve an all-or-nothing switch or an irreversible commitment. Also, some genes involved exhibit expression, which is inversely related to growth rate, and are already partially induced under conditions of slow growth

(Hengge-Aronis, 1996). The final, slow/non-replicating stages have been described as quiescent, resting, or dormant. The expression of many proteins is regulated on entry into the stationary phase, of which a core set is induced regardless of the cause of cessation of growth, for example, the nature of the depleted nutrient (Matin, 1991). In *E. coli*, the *rpoS*-encoded sigma factor σ^s is a master regulator of the GSR (Hengge-Aronis, 1996). There are numerous papers that show a general tendency for nutrient depletion and slow or no growth to be associated with antibiotic and biocide resistance (Brown, 1997; Brown et al., 1990; Brown and Williams, 1985). In retrospect, it seems probable that, in addition to the consequences of adaptation to the specific nutrient depletion and reduced growth rate, a major role in resistance is played by the GSR. There is also evidence that specific nutrient depletion has major effects on the sensitivity of microorganisms to host defences (Anwar, Brown, and Lambert, 1983; Finch and Brown, 1978). Nevertheless, there is as yet little work on the effects of the GSR per se on susceptibility to antibiotics (McLeod and Spector, 1996). Its role in biofilm formation is emerging. For example, *E. coli* biofilm density was reduced in an *rpoS* mutant grown in a modified Robbins device (Adams and McLean, 1999). Although infections caused by Gram-negative bacteria are not treated with glycopeptide antibiotics, the gene encoding D-alanine-D-alanine dipeptidase, part of the vancomycin resistance cluster, is transcribed in the stationary phase by RpoS (Lessard and Walsh, 1999). In *E. coli*, it is thought that the D-alanine could be used as an energy source for cell survival under starvation conditions. Recent work with microarrays and *P. aeruginosa* biofilms has shown reduced transcription of *rpoS*, whereas biofilms established with *rpoS* mutants were thicker and had larger structured groups of bacteria than those formed by the isogenic wild type (Whiteley et al., 2001). Interestingly, biofilm cells of a *P. aeruginosa rpoS*-deficient mutant were more resistant to tobramycin than wild-type cells. The authors also noted that tobramycin induced the differential expression of twenty genes.

5.4 Quorum Sensing, Biofilms, and Antimicrobial Susceptibility

There is now clear evidence that quorum sensing systems contribute significantly to biofilm development. Quorum sensing systems comprise a transcriptional activator protein that acts in concert with a low molecular weight autoinducer (AI) signalling molecule to alter the expression of target genes. As the cell population density increases, so does AI density, providing a means to monitor cell density. By definition, cell density will be high in a compact, adherent biofilm population, and, consequently, relatively small biofilm populations probably demonstrate signal-driven, stationary-phase survival responses, which equivalent numbers of free-growing planktonic counterparts would not (Gilbert et al., 1995). This could partially explain the general high resistance of biofilm organisms to exogenous stress. Using the example of *P. aeruginosa*, biofilm structure has been shown to be dependent on quorum sensing, which may itself be dependent on accumulation of inorganic polyphosphate (discussed later). *LasI* mutants deficient in production of the *N*-(3-oxododecanoyl) homoserine lactone (3O-C_{12}-HSL) AI molecule formed flat, undifferentiated biofilms that were sensitive to treatment with sodium dodecyl sulphate (Davies et al., 1998). Evidence suggests that this quorum-sensing molecule is a substrate for the MexAB-OprM multidrug efflux system, since mutants which hyperexpress this system showed reduced levels of extracellular virulence factors known to be regulated by quorum sensing (Evans et al., 1998). Also, a defined mutant lacking the pump accumulated more 3O-C_{12}-HSL, comparable with wild-type cells treated with cytoplasmic membrane proton gradient inhibitors (Pearson, Van Delden, and Iglewski, 1999). *P. aeruginosa* produces two AI molecules, 3O-C_{12}-HSL and *N*-butyryl-L-homoserine lactone (C_4-HSL). In conventional planktonic culture, *P. aeruginosa* produces 3O-C_{12}-HSL at a rate between three and ten times that of C_4-HSL, whereas more C_4-HSL was produced by biofilm-grown cells (Singh et al., 2000). Greater C_4-HSL levels were noted in extracts of sputum from cystic fibrosis patients, indicating that bacteria are perhaps growing as a biofilm in the lungs of these patients (Singh et al., 2000).

Signal diffusion will be influenced by the nature of the biofilm matrix and the substratum. Thus, an impermeable substratum for biofilm growth would concentrate any signal, whereas the degree of hydrophobicity of a boundary could influence entrapment or diffusion, depending on the chemistry of the signal. A hydrophobic signal could be trapped as aggregates/micelles within the cell exopolymer and maintain an equilibrium concentration of hydrophobic monomer close to the cell, while hydrophilic molecules could diffuse away. Consistent with this hypothesis, expression of *P. aeruginosa* AI synthase genes was greatest at the interface with the impermeable substratum (De Kievit et al., 2001), permitting rapid amplification of cell-density-dependent responses since the AI synthase genes themselves are subject to autoregulation. The *N*-acyl substituted homoserine lactones vary with respect to the state of oxidation at the C-3 position and the fatty acid chain length. Relative hydrophobicity/lipophilicity for a bioactive compound can be predicted by the parameter log P (the logarithm of the octanol/water partition coefficient). Optimum permeation through a Gram-negative envelope is commonly at about log P of 4. Longer chain length AI molecules, such as 3O-C_{12}-HSL from

P. aeruginosa, having a log P of approximately 3, would be predicted to enter or exit cells by simple diffusion (Heys et al., 1997), but their amphipathic nature may well result in a tendency to form large aggregates under some conditions. Indeed, in at least the case of *P. aeruginosa* 3O-C_{12}-HSL, active efflux appears to be required (Pearson et al., 1999).

Acylated homoserine lactones do not appear to operate in members of the enterobacteriaceae such as *E. coli* and *Salmonella enteritidis*, although AI activity has been reported (Surette, Miller, and Bassler, 1999) and quorum sensing regulates activity of the locus of enterocyte attachment and effacement and intestinal colonisation (Sperandio et al., 1999). However, there are no studies on the influence of antimicrobial agents.

A number of quorum sensing or cell-density-dependent systems also operate in Gram-positive species, although here the AI molecules are typically small peptides. Examples include regulation of streptococcal competence for genetic transformation (Havarstein and Morrison, 1999); cell-density control of gene expression and sporulation in *Bacillus* species (Lazazzera et al., 1999); the sex pheromone systems regulating conjugative plasmid transfer in *Enterococcus* species, some of which have associated antibiotic resistance (Clewell, 1999); and regulation of *Staphylococcus aureus* pathogenicity (Novick, 1999). Agents such as chloramphenicol and tetracycline at subinhibitory concentrations can perturb signalling cascades through inhibition of AI peptide synthesis. Studies with the *agr* quorum sensing system in *S. aureus* have shown that quorum sensing-deficient mutants were more able to form a biofilm on polystyrene than wild-type strains, leading the authors to question the utility of antiquorum-sensing molecules to eradicate biofilm infections (Vuong et al., 2000).

Evidence for the direct contribution of AI or quorum-sensing molecules to the antimicrobical susceptibility of bacteria is lacking. Nevertheless, given their central role in virulence factor production and biofilm formation, they are themselves attractive targets for antimicrobial drug design. However, it is already clear that there are fundamental differences in the contribution of these systems to biofilm formation between species. Such differences are not surprising. It is necessary to bear in mind the plethora both of bacterial phenotypes and of potential surfaces for colonisation.

6 OVERLAPPING REGULATORY NETWORKS IN BIOFILMS

Although as yet little studied in developing and mature biofilms, it is clear that there are many overlapping regulatory networks with high degrees of interaction operating in slow- or non-growing, relatively dormant bacteria that likely contribute to their susceptibility to antimicrobials. The GSR regulator RpoS is itself subject to complex transcriptional, translational, and posttranslational control by factors including (p)ppGpp and the stringent response (Cashel et al., 1996), inorganic polyphosphate (Shiba et al., 1997), OxyR (Pomposiello and Demple, 2001), SOS (Sutton et al., 2000), and cAMP-CRP (Spector, 1998), which are themselves subject to complex regulatory control. Moreover, RpoS is not solely responsible for stress responses in the many Gram-negative species in which it is found (Jørgensen et al., 1999). Examples of complex overlapping networks include the *rhl* quorum sensing system in *P. aeruginosa*, one of two quorum sensing systems which themselves interact, which is regulated by RpoS (Whiteley, Parsek, and Greenberg, 2000). Again, in studies with *P. aeruginosa*, recent evidence indicates an important role for polyphosphate kinase and the accumulation of inorganic polyphosphate in biofilm development (Rashid et al., 2000). Inorganic polyphosphate (poly P) is a linear polymer of many orthophosphate residues. In bacteria, the highly conserved poly P kinase (PPK) polymerises the terminal phosphate of ATP to the poly P chain. PPK-deficient mutants of *Escherichia coli* have been shown to be unable to adapt to nutritional stringencies and environmental stress, attributed in part to the failure to express *rpoS* (Shiba et al., 1997). Now a *P. aeruginosa ppk* mutant has been shown to be unable to form thick, differentiated biofilms. Synthesis of both quorum-sensing AI molecules was reduced by 50 per cent in the *ppk* mutant, and the expression of two quorum-sensing target genes was reduced by more than 90 per cent. These data led the authors to suggest that PPK and/or poly P may affect AI complex formation or perhaps interaction between RNA polymerase and promoter sequences (Rashid et al., 2000). Further complexity lies in the regulation of SOS genes by PPK and poly P (Tsutsumi, Munekata, and Shiba, 2000). Again, no studies have been reported on the contribution of PPK to antimicrobial susceptibility; however *P. aeruginosa ppk* mutants showed reduced virulence in a mouse model of infection (Rashid et al., 2000). These data clearly imply a role in susceptibility to host defences, which, as noted earlier in this chapter, will likely always make a significant contribution to 'eradication' of biofilm infections even when treated with antibiotics. Given that PPK is conserved in bacteria and not present in eukaryotes, it has been proposed as an attractive target for antimicrobial drugs (Rashid et al., 2000).

7 CONCLUDING REMARKS – WHICH WAY FORWARD FOR TREATMENT OF BIOFILM INFECTION?

Increasing knowledge of the physiology of biofilm bacteria, some of it now underpinned by detail at the genomic and proteomic levels, affords an answer to an often-posed question. Why does treatment fail in the patient in spite of antibiotic sensitivity in laboratory tests? Although biofilm growth offers an explanation, it does not offer a solution so readily. Much effort is being focussed on the surface properties of materials, antibiotic impregnated materials (van de Belt et al., 2001), addition of probiotic microorganisms (van der Mei et al., 2000), and delivery and targeting issues (Sihorkar and Vyas, 2001), but agents that disrupt biofilm bacteria must be sought. They might be used in isolation or in combination with conventional antibiotics.

The general resistance of biofilms is clearly phenotypic. The well-characterised resistance mechanisms – lack of antibiotic penetration, inactivation, efflux, and repair – make contributions in some circumstances. However, compelling evidence that they are uniquely responsible for biofilm resistance is lacking. The suspicion that reduced growth rate has an involvement is true in that it is associated with responses to stress. Key structures are protected and cellular processes close down to a state of dormancy. Such stress responses, linked with reduced growth rate, will be driven, at least in part, by the high density and quorum-sensing events occurring within the biofilm. It is not surprising that exceptional vegetative cell dormancy is the basic explanation of biofilm resistance. In biology, dormancy is a widespread survival response to stress. It would be well to remember that antibiotics are not man's invention. Microorganisms have spent billions of years in attack and counterattack against each other. Dismantling the results of all those years of evolution will not be an easy task.

REFERENCES

Adams, J. L. and McLean, R. J. C. (1999). Impact of *rpoS* deletion on *Escherichia coli* biofilms. *Applied and Environmental Microbiology*, 65, 4285–4287.

Al-Bakri, A. G., Gilbert, P. and Allison, D. G. (1999). Mixed species biofilms of *Burkholderia cepacia* and *Pseudomonas aeruginosa*. In: *Biofilms: The Good, the Bad and the Ugly*, 327–337. Edited by Wimpenny, J., Gilbert, P., Walker, J., Brading, M. and Bayston, R. Cardiff: BioLine.

Allison, D., Maira-Litrán, T. and Gilbert, P. (1999). Perfused biofilm fermenters. *Methods in Enzymology*, 310, 232–248.

Allison, D. G., Brown, M. R. W., Evans, D. J. and Gilbert, P. (1990a). Surface hydrophobicity and dispersal of *Pseudomonas aeruginosa* from biofilms. *FEMS Microbiology Letters*, 71, 101–104.

Allison, D. G., Evans, D. J., Brown, M. R. W. and Gilbert, P. (1990b). Possible involvement of the division cycle in dispersal of *Escherichia coli* from biofilms. *Journal of Bacteriology*, 172, 1667–1669.

An, Y. H., Dickinson, R. B. and Doyle, R. J. (2000). Mechanisms of bacterial adhesion and pathogenesis of implant and tissue infections. In *Handbook of Bacterial*

Adhesion: Principles, Methods, and Applications, 1–27. Edited by An, Y. H. and Friedman, R. J. Totowa, NJ: Humana Press.
Anderl, J. N., Franklin, M. J. and Stewart, P. S. (2000). Role of antibiotic penetration limitation in *Klebsiella pneumoniae* biofilm resistance to ampicillin and ciprofloxacin. *Antimicrobial Agents and Chemotherapy*, 44, 1818–1824.
Anwar, H., Brown, M. R. W. and Lambert, P. A. (1983). Effect of nutrient depletion on sensitivity of *Pseudomonas cepacia* to phagocytosis and serum bactericidal activity at different temperatures. *Journal of General Microbiology*, 129, 2021–2027.
Bagge, N., Ciofu, O., Skovgaard, L. T. and Høiby, N. (2000). Rapid development in vitro and in vivo of resistance to ceftazidime in biofilm-growing *Pseudomonas aeruginosa* due to chromosomal beta-lactamase. *APMIS*, 108, 589–600.
Boland, T., Latour, R. A. and Stutzenberger, F. J. (2000). Molecular basis of bacterial adhesion. In: *Handbook of Bacterial Adhesion: Principles, Methods, and Applications*, 29–41. Edited by An, Y. H. and Friedman, R. J. Totowa, NJ: Humana Press.
Brooun, A., Liu, S. and Lewis, K. (2000). A dose-response study of antibiotic resistance in *Pseudomonas aeruginosa* biofilms. *Antimicrobial Agents and Chemotherapy*, 44, 640–646.
Brown, M. R. W. (1977). Nutrient depletion and antibiotic susceptibility. *Journal of Antimicrobial Chemotherapy*, 3, 198–201.
Brown, M. R. W., Allison, D. and Gilbert, P. (1988). Resistance of bacterial biofilms to antibiotics: a growth-rate related effect? *Journal of Antimicrobial Chemotherapy*, 22, 777–780.
Brown, M. R. W. and Barker, J. (1999). Unexplored reservoirs of pathogenic bacteria: protozoa and biofilms. *Trends in Microbiology*, 7, 46–50.
Brown, M. R. W., Collier, P. J., Courcol, R. J. and Gilbert, P. (1995). Definition of phenotype in batch culture. In: *Microbiological Quality Assurance. A Guide towards Relevance and Reproducibility of Inocula*, 13–20. Edited by Brown, M. R. W. and Gilbert, P. Boca Raton: CRC Press.
Brown, M. R. W., Collier, P. J. and Gilbert, P. (1990). Influence of growth rate on susceptibility to antimicrobial agents: modification of the cell envelope and batch and continuous culture studies. *Antimicrobial Agents and Chemotherapy*, 34, 1623–1628.
Brown, M. R. W. and Gilbert, P. (1995). *Microbiological Quality Assurance: A Guide towards Relevance and Reproducibility of Inocula*, Boca Raton: CRC Press.
Brown, M. R. W. and Smith, A. W. (2001). Dormancy and persistence in chronic infection: role of the general stress response in resistance to chemotherapy. *Journal of Antimicrobial Chemotherapy*, 48, 141–142.
Brown, M. R. W. and Williams, P. (1985). The influence of environment on envelope properties affecting survival of bacteria in infections. *Annual Review of Microbiology*, 39, 527–556.
Brub-Buisson, C. (2001). New technologies and infection control practices to prevent intravascular catheter-related infections. *American Journal of Respiratory and Critical Care Medicine*, 164, 1557–1558.
Bühler, T., Ballestero, S., Desai, M. and Brown, M. R. W. (1998). Generation of a reproducible nutrient-depleted biofilm of *Escherichia coli* and *Burkholderia cepacia*. *Journal of Applied Bacteriology*, 85, 457–462.

Cashel, M., Gentry, D. R., Hernandez, V. J. and Vinella, D. (1996). The stringent response. In *Escherichia coli and Salmonella*, 1458–1496. Edited by Neidhardt, F. C., Curtiss, R., III, Ingraham, J. L., Lin, E. C. C., Low, K. B., Magasanik, B., Reznikoff, W. S., Riley, M., Schaechter, M. and Umbarger, H. E. Washington DC: ASM Press.

Chan, P. F., Foster, S. J., Ingham, E. and Clements, M. O. (1998). The *Staphylococcus aureus* alternative sigma factor sigma B controls the environmental stress response but not starvation survival or pathogenicity in a mouse abscess model. *Journal of Bacteriology*, 180, 6082–6089.

Clements, M. O. and Foster, S. J. (1999). Stress resistance in *Staphylococcus aureus*. *Trends in Microbiology*, 7, 458–462.

Clewell, D. B. (1999). Sex pheromone systems in Enterococci. In *Cell-Cell Signaling in Bacteria*, 47–66. Edited by Dunny, G. M. and Winans, S. C. Washington DC: ASM Press.

Costerton, J. W., Lewandowski, Z., Caldwell, D. E., Korber, D. R. and Lappin-Scott, H. M. (1995). Microbial biofilms. *Annual Review of Microbiology*, 49, 711–745.

Costerton, J. W., Stewart, P. S. and Greenberg, E. P. (1999). Bacterial biofilms: a common cause of persistent infections. *Science*, 284, 1318–1322.

Cramton, S. E., Gerke, C., Nichols, W. W. and Gotz, F. (1999). The intercellular adhesion (*ica*) locus is present in *Staphylococcus aureus* and is required for biofilm. *Infection and Immunity*, 67, 5427–5433.

Davies, D. G., Parsek, M. R., Pearson, J. P., Iglewski, B. H., Costerton, J. W. and Greenberg, E. P. (1998). The involvement of cell-to-cell signals in the development of a bacterial biofilm. *Science*, 280, 295–298.

De Kievit, T. R., Gillis, R., Marx, S., Brown, C. and Iglewski, B. H. (2001). Quorum-sensing genes in *Pseudomonas aeruginosa* biofilms: their role and expression patterns. *Applied and Environmental Microbiology*, 67, 1865–1873.

De Kievit, T. R., Parkins, M. D., Gillis, R. J., Srikumar, R., Ceri, H., Poole, K., Iglewski, B. H. and Storey, D. G. (2001). Multidrug efflux pumps: expression patterns and contribution to antibiotic resistance in *Pseudomonas aeruginosa* biofilms. *Antimicrobial Agents and Chemotherapy*, 45, 1761–1770.

DeMaio, J., Zhang, Y., Ko, C., Young, D. B. and Bishai, W. R. (1996). A stationary-phase stress-response sigma factor from *Mycobacterium tuberculosis*. *Proceedings of the National Academy of Sciences of the USA*, 93, 2790–2794.

Desai, M., Bühler, T., Weller, P. H. and Brown, M. R. W. (1998). Increasing resistance of planktonic and biofilm cultures of *Burkholderia cepacia* to ciprofloxacin and ceftazidime during exponential growth. *Journal of Antimicrobial Chemotherapy*, 42, 153–160.

Dibdin, G. and Wimpenny, J. (1999). Steady-state biofilm: practical and theoretical models. *Methods in Enzymology*, 310, 296–322.

Dodd, C. E. R., Sharman, R. L., Bloomfield, S. F., Booth, I. R. and Stewart, G. S. A. B. (1997). Inimical processes: bacterial self destruction and sub-lethal injury. *Trends in Food Science & Technology*, 8, 238–241.

Domingue, P. A. G., Lambert, P. A. and Brown, M. R. W. (1989). Iron depletion alters surface-associated properties of *Staphylococcus aureus* and its association to human neutrophils in chemiluminescence. *FEMS Microbiology Letters*, 59, 265–268.

Duguid, I. G., Evans, E., Brown, M. R. W. and Gilbert, P. (1992a). Growth-rate-independent killing by ciprofloxacin of biofilm-derived *Staphylococcus epidermidis*: evidence for cell-cycle dependency. *Journal of Antimicrobial Chemotherapy*, 30, 791–802.

Duguid, I. G., Evans, E., Brown, M. R. W. and Gilbert, P. (1992b). Effect of biofilm culture upon the susceptibility of *Staphylococcus epidermidis* to tobramycin. *Journal of Antimicrobial Chemotherapy*, 30, 803–810.

Evans, D. J., Allison, D. G., Brown, M. R. W. and Gilbert, P. (1991). Susceptibility of *Pseudomonas aeruginosa* and *Escherichia coli* biofilms towards ciprofloxacin: effect of specific growth rate. *Journal of Antimicrobial Chemotherapy*, 27, 177–184.

Evans, K., Passador, L., Srikumar, R., Tsang, E., Nezezon, J. and Poole, K. (1998). Influence of the MexAB-OprM multidrug efflux system on quorum sensing in *Pseudomonas aeruginosa*. *Journal of Bacteriology*, 180, 5443–5447.

Finch, J. E. and Brown, M. R. W. (1978). Effect of growth environment on *Pseudomonas aeruginosa* killing by rabbit polymorphonuclear leukocytes and cationic proteins. *Infection and Immunity*, 20, 340–346.

Foley, I., Marsh, P., Wellington, E. M. H., Smith, A. W. and Brown, M. R. W. (1999). General stress response master regulator *rpoS* is expressed in human infection: a possible role in chronicity. *Journal of Antimicrobial Chemotherapy*, 43, 164–165.

Gilbert, P., Allison, D. G., Evans, D. J., Handley, P. S. and Brown, M. R. W. (1989). Growth rate control of adherent bacterial populations. *Applied and Environmental Microbiology*, 55, 1308–1311.

Gilbert, P., Allison, D. G., Jacob, A., Korner, D., Wolfaa, G. and Foley, I. (1997). Immigration of planktonic *Enterococcus faecalis* cells into mature *E. faecalis* biofilms. In *Biofilms: Community Interactions and Control*, 133–142. Edited by Wimpenny, J. T., Handley, P., Gilbert, P. and Lappin-Scott, H. M. Cardiff: Bioline.

Gilbert, P., Collier, P. J. and Brown, M. R. W. (1990). Influence of growth rate on susceptibility to antimicrobial agents: biofilms, cell cycle, dormancy, and stringent response. *Antimicrobial Agents and Chemotherapy*, 34, 1865–1868.

Gilbert, P., Coplan, F. and Brown, M. R. W. (1991). Centrifugation injury of Gram-negative bacteria. *Journal of Antimicrobial Chemotherapy*, 27, 550–551.

Gilbert, P., Hodgson, A. E. and Brown, M. R. W. (1995). Influence of the environment on the properties of microorganisms grown in association with surfaces. In *Microbiological Quality Assurance: A Guide towards Relevance and Reproducibility of Inocula*, 61–82. Edited by Brown, M. R. W. and Gilbert, P. Boca Raton: CRC Press.

Giwercman, B., Jensen, E. T., Høiby, N., Kharazmi, A. and Costerton, J. W. (1991). Induction of beta-lactamase production in *Pseudomonas aeruginosa* biofilm. *Antimicrobial Agents and Chemotherapy*, 35, 1008–1010.

Gottenbos, B., Grijpma, D. W., van der Mei, H. C., Feijen, J. and Busscher, H. J. (2001). Antimicrobial effects of positively charged surfaces on adhering Gram-positive and Gram-negative bacteria. *Journal of Antimicrobial Chemotherapy*, 48, 7–13.

Gross, M., Cramton, S. E., Gotz, F. and Peschel, A. (2001). Key role of teichoic acid net charge in *Staphylococcus aureus* colonization of artificial surfaces. *Infection and Immunity*, 69, 3423–3426.

Hassett, D. J., Ma, J. F., Elkins, J. G., McDermott, T. R., Ochsner, U. A., West, S. E. H., Huang, C. T., Fredericks, J., Burnett, S., Stewart, P. S., McFeters, G., Passador, L.

and Iglewski, B. H. (1999). Quorum sensing in *Pseudomonas aeruginosa* controls expression of catalase and superoxide dismutase genes and mediates biofilm susceptibility to hydrogen peroxide. *Molecular Microbiology*, 34, 1082–1093.

Havarstein, L. S. and Morrison, D. A. (1999). Quorum sensing and peptide pheromones in Streptococcal competence for genetic transformation. In *Cell-Cell Signaling in Bacteria*, 9–26. Edited by Dunny, G. M. and Winans, S. C. Washington DC: ASM Press.

Hengge-Aronis, R. (1996). Regulation of gene expression during entry into stationary phase. In *Escherichia coli and Salmonella. Cellular and Molecular Biology*, 1497–1512. Edited by Neidhardt, F. C., Curtiss, R., III, Ingraham, J. K., Lin, E. C. C., Low, K. B., Magasanik, B., Reznikoff, W. S., Riley, M., Schaechter, M. and Umbarger, H. E. Washington DC: ASM Press.

Hengge-Aronis, R. (1999). Interplay of global regulators and cell physiology in the general stress response of *Escherichia coli. Current Opinion in Microbiology*, 2, 148–152.

Heys, S. J. D., Gilbert, P., Eberhard, A. and Allison, D. G. (1997). Homoserine lactones and bacterial biofilms. In *Biofilms: Community Interactions and Control*, 103–112. Edited by Wimpenny, J., Handley, P., Gilbert, P., Lappin-Scott, H. M. and Jones, M. Cardiff: Bioline.

Hopelman, A. I. M. and Tuomanen, E. (1992). Consequences of microbial attachment: directing host cell functions with adhesins. *Infection and Immunity*, 60, 1729–1733.

Jørgensen, F., Bally, M., Chapon-Herve, V., Stewart, G. S. A. B., Michel, G., Lazdunski A. and Williams, P. (1999). RpoS-dependent stress tolerance in *Pseudomonas aeruginosa. Microbiology*, 145, 835–844.

Karthikeyan, S., Korber, D. R., Wolfaardt, G. M. and Caldwell, D. E. (2000). Monitoring the organization of microbial biofilm communities. In *Handbook of Bacterial Adhesion: Principles, Methods, and Applications*, 171–188. Edited by An, Y. H. and Friedman, R. J. Totowa, NJ: Humana Press.

Kharazmi, A., Giwercman, B. and Høiby, N. (1999). Robbins device in biofilm research. *Methods in Enzymology*, 310, 207–215.

Kimmitt, P. T., Harwood, C. R. and Barer, M. R. (2000). Toxin gene expression by shiga toxin-producing *Escherichia coli*: the role of antibiotics and the bacterial SOS response. *Emerging Infectious Diseases*, 6, 458–465.

Kornberg, A., Rao, N. N. and Ault-Richie, D. (1999). Inorganic polyphosphate: a molecule of many functions. *Annual Review of Biochemistry*, 68, 89–125.

Lazazzera, B. A., Palmer, T., Quisle, J. and Grossman, A. D. (1999). Cell-density control of gene expression and development in *Bacillus subtilis*. In *Cell-Cell Signaling in Bacteria*, 27–46. Edited by Dunny, G. M. and Winans, S. C. Washington DC: ASM Press.

Lessard, I. A. and Walsh, C. T. (1999). VanX, a bacterial D-alanyl-D-alanine dipeptidase: resistance, immunity, or survival function? *Proceedings of the National Academy of Sciences of the USA*, 96, 11028–11032.

Lewis, K. (2001). Riddle of biofilm resistance. *Antimicrobial Agents and Chemotherapy*, 45, 999–1007.

Mah, T.-F. C. and O'Toole, G. A. (2001). Mechanisms of biofilm resistance to antimicrobial agents. *Trends in Microbiology*, 9, 34–39.

Maira-Litrán, T., Allison, D.G. and Gilbert, P. (2000). Expression of the multiple antibiotic resistance operon (*mar*) during growth of *Escherichia coli* as a biofilm. *Journal of Applied Microbiology*, 88, 243–247.

Matin, A. (1991). The molecular basis of carbon-starvation-induced general resistance in *Escherichia coli*. *Molecular Microbiology*, 5, 3–10.

McLeod, G. I. and Spector, M. P. (1996). Starvation- and stationary-phase-induced resistance to the antimicrobial peptide polymyxin B in *Salmonella typhimurium* is RpoS (sigma(S)) independent and occurs through both phoP-dependent and -independent pathways. *Journal of Bacteriology*, 178, 3683–3688.

Moken, M. C., McMurray, L. M. and Levy, S. B. (1997). Selection of multiple-antibiotic-resistant (Mar) mutants of *Escherichia coli* by using the disinfectant pine oil: roles of the *mar* and *acrAB* loci. *Antimicrobial Agents and Chemotherapy*, 41, 2770–2772.

Nichols, W. W. (1991). Biofilms, antibiotics and penetration. *Reviews of Medical Microbiology*, 2, 177–181.

Novick, R. P. (1999). Regulation of pathogenicity in *Staphylococcus aureus* by a peptide-based density-sensing system. In *Cell-Cell Signaling in Bacteria*, 129–146. Edited by Dunny, G. M. and Winans, S. C. Washington DC: ASM Press.

O'Toole, G. A. and Kolter, R. (1998). Flagella and twitching motility are necessary for *Pseudomonas aeruginosa* biofilm development. *Molecular Microbiology*, 30, 295–304.

Pearson, J. P., Van Delden, C. and Iglewski, B. H. (1999). Active efflux and diffusion are involved in transport of *Pseudomonas aeruginosa* cell-to-cell signals. *Journal of Bacteriology*, 181, 1203–1210.

Pier, G. B., Saunders, J. M., Ames, P., Edwards, M. S., Auerbach, H., Speert, D. P. and Hurwitch, S. (1987). Opsonophagocytic killing antibody to *Pseudomonas aeruginosa* mucoid exopolysaccharide in older noncolonized patients with cystic fibrosis. *New England Journal of Medicine*, 317, 793–798.

Pomposiello, P. J. and Demple, B. (2001). Redox-operated genetic switches: the SoxR and OxyR transcription factors. *Trends in Biotechnology*, 19, 109–114.

Pratt, L. A. and Kolter, R. (1998). Genetic analysis of *Escherichia coli* biofilm formation: roles of flagella, motility, chemotaxis and type I pili. *Molecular Microbiology*, 30, 285–293.

Rachid, S., Ohlsen, K., Wallner, U., Hacker, J., Hecker, M. and Ziebuhr, W. (2000a). Alternative transcription factor sigma B is involved in regulation of biofilm expression in a *Staphylococcus aureus* mucosal isolate. *Journal of Bacteriology*, 182, 6824–6826.

Rachid, S., Ohlsen, K., Witte, W., Hacker, J. and Ziebuhr, W. (2000b). Effect of subinhibitory antibiotic concentrations on polysaccharide intercellular adhesin expression in biofilm-forming *Staphylococcus epidermidis*. *Antimicrobial Agents and Chemotherapy*, 44, 3357–3363.

Rao, N. N., Liu, S. J. and Kornberg, A. (1998). Inorganic polyphosphate in *Escherichia coli*: the phosphate regulon and the stringent response. *Journal of Bacteriology*, 180, 2186–2193.

Rashid, M. H., Rumbaugh, K., Passador, L., Davies, D. G., Hamood, A. N., Iglewski, B. H. and Kornberg, A. (2000a). Polyphosphate kinase is essential for biofilm development, quorum sensing, and virulence of *Pseudomonas aeruginosa*. *Proceedings of the National Academy of Sciences of the USA*, 97, 9636–9641.

Scott, J. M., Mitchell, T. and Haldenwang, W. G. (2000). Stress triggers a process that limits activation of the *Bacillus subtilis* stress transcription factor sigma B. *Journal of Bacteriology*, 182, 1452–1456.

Shiba, T., Tsutsumi, K., Yano, H., Iahara, Y., Yameda, A., Tanaka, T., Takahashia, M., Munekata, M., Rao, N. N. and Kornberg, A. (1997). Inorganic polyphosphate and the induction of *rpoS* expression. *Proceedings of the National Academy of Sciences of the USA*, 94, 11210–11215.

Sihorkar, V. and Vyas, S. P. (2001). Biofilm consortia on biomedical and biological surfaces: delivery and targeting strategies. *Pharmaceutical Research*, 18, 1247–1254.

Singh, P. K., Schaefer, A. L., Parsek, M. R., Moninger, T. O., Welsh, M. J. and Greenberg, E. P. (2000). Quorum-sensing signals indicate that cystic fibrosis lungs are infected with bacterial biofilms. *Nature*, 407, 762–764.

Sissons, C. H., Wong, L. and An, Y. H. (2000). Laboratory culture and analysis of biofilms. In *Handbook of Microbial Adhesion: Principles, Methods, and Applications*, 133–169. Edited by An, Y. H. and Friedman, R. J. Totowa, NJ: Humana Press.

Spector, M. P. (1998). The starvation-stress response (SSR) of Salmonella. *Advances in Microbial Physiology*, 40, 233–279.

Sperandio, V., Mellies, J. L., Nguyen, W., Shin, S. and Kaper, J. B. (1999). Quorum sensing controls expression of the type III secretion gene transcription and protein secretion in enterohemorrhagic and enteropathogenic *Escherichia coli*. *Proceedings of the National Academy of Sciences of the USA*, 96, 15196–15201.

Spoering, A. L. and Lewis, K. (2001). Biofilms and planktonic cells of *Pseudomonas aeruginosa* have similar resistance to killing by antimicrobials. *Journal of Bacteriology*, 182, 6746–6751.

Stewart, P. S. (1996). Theoretical aspects of antibiotic diffusion into microbial biofilms. *Antimicrobial Agents and Chemotherapy*, 38, 2125–2133.

Stewart, P. S. and Costerton, J. W. (2001). Antibiotic resistance of bacteria in biofilms. *Lancet*, 358, 135–138.

Stewart, P. S., Roe, F., Rayner, J., Elkins, J. G., Lewandowski, Z., Ochsner, U. A. and Hassett, D. J. (2000). Effect of catalase on hydrogen peroxide penetration into *Pseudomonas aeruginosa* biofilms. *Applied and Environmental Microbiology*, 66, 836–838.

Stickler, D. J., Morris, N. S., McLean, R. J. and Fuqua, C. (1998). Biofilms on indwelling urethral catheters produce quorum-sensing signal molecules in situ and in vitro. *Applied and Environmental Microbiology*, 64, 3486–3490.

Stoodley, P., Wilson, S., Hall-Stoodley, L., Lappin-Scott, H. M. and Costerton, J. W. (2001). Growth and detachment of cell clusters from mature mixed-species biofilms. *Applied and Environmental Microbiology*, 67, 5608–5613.

Surette, M. G., Miller, M. B. and Bassler, B. L. (1999). Quorum sensing in *Escherichia coli*, *Salmonella typhimurium*, and *Vibrio harveyi*: a new family of genes responsible for autoinducer production. *Proceedings of the National Academy of Sciences of the USA*, 96, 1639–1644.

Sutherland, I. W. (2001). The biofilm matrix – an immobilised but dynamic environment. *Trends in Microbiology*, 9, 222–227.

Sutton, M. D., Smith, B. T., Godoy, V. G. and Walker, G. C. (2000). The SOS response: recent insights into umuDC-dependent mutagenesis and DNA damage tolerance. *Annual Review of Genetics*, 34, 479–497.
Taddei, F., Matic, I. and Radman, M. (1995). cAMP-dependent SOS induction and mutagenesis in resting bacterial populations. *Proceedings of the National Academy of Sciences of the USA*, 92, 11736–11740.
Tsutsumi, K., Munekata, M. and Shiba, T. (2000). Involvement of inorganic polyphosphate in expression of SOS genes. *Biochimica Biophysica Acta*, 1493, 73–81.
Van Bambeke, F., Balzi, E. and Tulkens, P. M. (2000). Antibiotic efflux pumps. *Biochemical Pharmacology*, 60, 457–470.
van de Belt, H., Neut, D., Schenk, W., van der Horn, J. W., van der Mei, H. C. and Busscher, H. J. (2001). *Staphylococcus aureus* biofilm formation on different gentamicin-loaded polymethylmethacrylate bone cements. *Biomaterials*, 22, 1607–1611.
van der Mei, H. C., Free, R. H., Elving, G. J., Van Weissenbruch, R. and Busscher, H. J. (2000). Effect of probiotic bacteria on prevalence of yeasts in oropharyngeal biofilms on silicone rubber voice prostheses *in vitro*. *Journal of Medical Microbiology*, 49, 713–718.
Vuong, C., Saenz, H. L., Gotz, F. and Otto, M. (2000). Impact of the *agr* quorum-sensing system on adherence to polystyrene in *Staphylococcus aureus*. *Journal of Infectious Diseases*, 182, 1688–1693.
Whiteley, M., Bangera, M. G., Bumgarner, R. E., Parsek, M. R., Teitzel, G. M., Lory, S. and Greenberg, E. P. (2001). Gene expression in *Pseudomonas aeruginosa* biofilms. *Nature*, 413, 860–864.
Whiteley, M., Parsek, M. R. and Greenberg, E. P. (2000). Regulation of quorum sensing by RpoS in *Pseudomonas aeruginosa*. *Journal of Bacteriology*, 182, 4356–4360.
Williams, P., Lambert, P. A., Haigh, C. G. and Brown, M. R. W. (1986). The influence of the O and K antigens of *Klebsiella aerogenes* on surface hydrophobicity and susceptibility to phagocytosis and antimicrobial agents. *Journal of Medical Microbiology*, 21, 125–132.
Wilson, M. (2002). *Bacterial Adhesion to Host Tissues: Mechanisms and Consequences*. Cambridge: Cambridge University Press.
Wu, H., Song, Z., Givskov, M., Doring, G., Worlitzsch, D., Rygaard, J. and Høiby, N. (2001). *Pseudomonas aeruginosa* mutations in *lasI* and *rhlI* quorum sensing systems result in milder chronic lung infection. *Microbiology*, 147, 1105–1113.
Xu, K. D., McFeters, G. and Stewart, P. S. (2000). Biofilm resistance to antimicrobial agents. *Microbiology*, 146, 547–549.
Yasuda, H., Koga, T. and Fukoka, T. (1999). *In vitro* and *in vivo* models of bacterial biofilms. *Methods in Enzymology*, 310, 577–595.
Zgurskaya, H. I. and Nikaido, H. (2000). Multidrug resistance mechanisms: drug efflux across two membranes. *Molecular Microbiology*, 37, 219–225.

PART TWO

BIOFILMS ON PROSTHETIC DEVICES

CHAPTER FOUR

Animal Models of Orthopaedic Implant Infection

Yuehuei H. An, Christopher M. Hill, and
Richard J. Friedman

1 INTRODUCTION

More than 200,000 primary hip and 200,000 primary knee arthroplasties are performed each year in the United States alone. Between 0.5 and 3.9 per cent of them will become infected within 10 years (Table 4.1) (An and Friedman, 1996; Stocks and Janssen, 2000). For revision total joint replacements, the infection rates can be much higher, with numbers as high as 3.2 per cent reported by Sperling et al. (2001), 12.5 per cent reported by Itasaka et al. (2001), and 17 per cent reported by Spangehl et al. (1999). Infection often causes complete failure of a total joint arthroplasty. Sepsis following total joint replacement can have catastrophic results, both physically and psychologically, for the patient, leading to failure of the arthroplasty, prolonged hospitalisation, possible amputation, and even death (Cheatle, 1991). In addition, the management of infected cases, especially those of joint replacements, is very costly (Hebert et al., 1996). Although the use of prophylactic antibiotics and greatly improved surgical techniques have decreased the infection rate of joint replacement from an average of 5.9 per cent in 1975 to 1.2 per cent in 1993 (An and Friedman, 1996), challenges still remain for better preventive and therapeutic measures. In addition to joint replacement, implant infections also occur in other orthopaedic subspecialties, such as trauma (Eijer et al., 2001) and spine (Wimmer and Gluch, 1996), with a significant impact on the patient and society.

Adhesion of bacteria to human tissue surfaces and implanted biomaterial surfaces is an important step in the pathogenesis of implant infection. The exact mechanism by which these foreign body infections occur still remains

Table 4.1: Rates of joint prosthetic infection

Reference	Joint	Total Cases	Infected Cases	% Infected[a]
Sperling et al., 2001	Shoulder	2,279	19	0.08
Josefsson and Kolmert, 1993	Hip	835	13	1.6
Josefsson and Kolmert, 1993	Hip	853	9	1.1
Lidwell et al., 1982	Hip/knee	5,831	34	0.6
Lidwell et al., 1982	Hip/knee	2,221	52	2.3
Andrews et al., 1981	Hip	1,746	68	3.9
Nelson et al., 1980	Hip	711	16	2.3
Fitzgerald et al., 1977	Hip	3,215	42	1.3
Eftekhar, Kiernan, and Stinchfield, 1976	Hip	800	4	0.5

[a] Number of infection cases/total cases $\times 100 =$ % infected.

unclear (An and Friedman, 1998c; An, Dickinson, and Doyle, 2000). It is thought that certain strains of bacteria, such as *Staphylococcus epidermidis* and *Pseudomonas aeruginosa*, secrete a layer of slime. Once adhesion occurs on the implant surface, they form a biofilm, making themselves less accessible to human defence systems (Gray et al., 1984) and significantly decreasing antibiotic susceptibility (Gristina et al., 1987; Donlan, 2000). They can remain dormant on the material surface for a long period of time until conditions occur which allow them to overgrow, such as a decreased systemic immune function or reduced local defence (such as a poor tissue ingrowth around the implant, severe inflammation, or osteolysis).

Animal models of implant infection have played an important role in the study of the mechanisms of orthopaedic implant infection as well as in the search for new preventive and treatment methodologies (Cremieux and Carbon, 1997; An and Friedman, 1998a, 1998b). This chapter will present an overview of the mechanisms and pathogenesis of orthopaedic implant infection, major animal models of orthopaedic implant infection and their applications, and how to design an animal model of implant infection.

2 MECHANISMS AND PATHOGENESIS OF ORTHOPAEDIC IMPLANT INFECTION

2.1 Bacterial Colonisation and Common Pathogens

Bacteria initially attach to material surfaces by physical forces (phase 1). If the local environment is hospitable to bacteria, such as abnormal tissue integration and a weak host defence, bacteria will remain viable on the material surface and complete the second phase (phase 2) of adhesion by secreting exopolysaccharides and eventually forming a biofilm (An and Friedman, 1998c; An et al., 2000).

Bacterial colonisation of biomaterial surfaces is the initial step in the pathogenesis of prosthetic infections (Gristina and Costerton, 1985; An and Friedman, 1998c; An et al., 2000). Bacteria come from two routes. One is direct contamination of the wound and implant at surgery, when bacteria have a chance to reach these surfaces from the patient's skin and the air. The other type of contamination is haematogenous or lymphatic seeding from other infection foci in the body (Fitzgerald et al., 1977; Maderazo, Judson, and Pasternak, 1988; Gillespie, 1990).

Theoretically, bacteria can reach the implant surface as early as the time of operation and as late as several years after implantation when an infection, such as a periodontal abscess or a urinary tract infection, occurs elsewhere in the body. In one study, four cases of prosthetic infection caused by strains of *Streptococcus sanguis* were reported (Bartzokas et al., 1994). In each patient, the strain of *S. sanguis* isolated from the mouth (periodontal disease and caries) was indistinguishable from that isolated from the infected implant. In another report, two cases of hip arthroplasties were infected by *Mycobacterium tuberculosis* 1.5 and 14 years after the surgery, with the primary source being pulmonary tuberculosis (Ueng, Shih, and Hseuh, 1995). These reports make it clear that the implant infection can be caused by haematogenous seeding.

Staphylococci are members of the Micrococcaceae family, characterised as Gram-positive, non-motile, catalase positive, coagulase negative, aerobic, or anaerobic cocci. Strains are distinguished by the coagulase and mannitol fermentation tests. Staphylococci are the most important pathogens of implant infection. *Staphylococcus epidermidis* is the cause of a large percentage of the late or chronic implant infections, whereas *S. aureus* remains a common pathogen of those occurring relatively early on (Charnley, 1972; Andrews et al., 1981; Kamme and Lindberg, 1981). It has been shown that there is a roughly equal incidence of *S. epidermidis* and *S. aureus* causing prosthetic hip joint infections, accounting for 50–60 per cent of all infections since 1980 (Sanderson, 1991; An and Friedman, 1996). A trend of increasing coagulase-negative staphylococci (CNS) infections is noted, rising from 13 per cent in the 1970s to 25 per cent in the 1980s and to 33 per cent in the 1990s. *S. aureus* is coagulase positive and *S. epidermidis* is coagulase negative. CNS are a normal component of the skin flora, and *S. epidermidis* is the most common species and the most predominant (Lowy and Hammer, 1983). CNS are widely recognised as significant pathogens in patients with infections associated with orthopaedic prostheses or implants, prosthetic heart valves, vascular prostheses, cerebrospinal fluid shunts, urinary tract catheters, peritoneal dialysis catheters, and others. *S. aureus* causes more severe and more rapid infection than *S. epidermidis*, and therefore, its effects may be more clinically obvious at an earlier stage after surgery. *S. epidermidis* is less virulent, and the clinical features are less severe than those of *S. aureus*.

Other bacteria isolated from orthopaedic implant infections include *Escherichia coli*, *Klebsiella* spp., Micrococcaceae, *Pseudomonas* spp., *Proteus* spp., peptococci, streptococci, and anaerobes (Benson and Hughes, 1975; Fitzgerald et al., 1977; Grogan et al., 1986; Hughes, 1988; Sanderson, 1988).

2.2 Tissue Integration and Bacterial Invasion

Plasma proteins, albumin, fibronectin, vitronectin, and other proteins rapidly coat any biomaterial introduced into the body – this is often termed a 'conditioning film' – and modify the extent of bacterial adhesion (Simpson, Courtney, and Ofek, 1987). The famous phrase 'race for the surface' refers to a contest between tissue cell integration and bacterial adhesion to an available implant surface (Gristina, Naylor, and Myrvik, 1990, 1991). Once established on the surface, this conditioning film is not easily traumatised or altered. If tissue integration occurs first, the implant becomes relatively resistant to bacterial colonisation. If bacterial adhesion occurs first, then host cells can seldom displace the primary colonisers on that portion of the implant surface, establishing a condition for eventual infection (Costerton, Marrie, and Cheng, 1985).

The initial conditioning film provides a favourable surface for colonisation by either the bacteria or eukaryotic cells, whichever makes contact first (Tollefson et al., 1987; Blenkinsopp and Costerton, 1991). On biomaterials such as metallic alloys, binding sites are further modified by ionic and glycoproteinaceous constituents from the host environment. In addition to these specific interactions, exposed surfaces may also act as catalytic surfaces for close-range molecular and cellular activities. These interactions can be dramatically changed by variations in local pH and inflammation and tissue damage caused by such factors as surgery, trauma, and infection.

The susceptibility of biomaterials to infection is a function not only of the number and type of bacteria, but also the time needed for tissue integration on the implanted surface versus the time needed for adhesion of bacteria to the same surface. When prosthetic loosening occurs due to technical or mechanical reasons, the local tissue integration is disrupted by the process of loosening, and the local environment may become susceptible to bacterial duplication or haematogenous bacterial seeding and subsequent infection.

The presence of bacterial growth can have a significant effect on the corrosion of stainless steel (Merritt et al., 1991). Significant destruction of a hydroxyapatite (HA) coating after exposure to bacteria has also been reported, and this may lead to a better understanding of prosthetic loosening (Arizono, Oga, and Sugioka, 1992; Kieswetter, Merritt, and Myers, 1993). These events can change the local physiological environment, such as the chemical composition and pH of the tissue fluid, and, consequently, stimulate bacterial duplication or aggravate existing infections. Due to their unique characteristics, especially when embedded in a biofilm, the adherent bacteria can remain on the material surface for a long period of time and display increased resistance to antibiotics (Gristina et al., 1987, 1989; Arizono et al., 1992; Vergères and Blaser, 1992; Pascual et al., 1993) and the human immune system (Vaudaux et al., 1985; Dobbins, Seligson, and Ratt, 1988; Sugarman and Young, 1989).

2.3 Effects of Implants

Dougherty reviewed the effects of an implant on the incidence of bacterial colonisation and subsequent infection (Dougherty, 1988; Dougherty and Simmons, 1989), which include foreign body reactivity leading to local tissue damage and inflammation, harmful effects of the implant on local host defences, and the effects of trapping and sequestering of bacteria. Foreign body reaction has long been recognised as a very important infection-promoting factor (James and MacLeod, 1961), which has been verified both experimentally and clinically. The determinants of implant reactivity include chemical composition of the implant material (Chu and Williams, 1984), surface characteristics such as surface configuration or particle size (Klock and Bainton, 1976), implantation site, and mechanical interactions with host tissues (Merritt, Shafer, and Brown, 1979).

Implant wear debris stimulates inflammation and facilitates prosthetic loosening. If sufficient stimulus exists, inflammation will occur (DiCarlo and Bullough, 1992). The products of inflammation will trigger the local defence system to release tissue-toxic enzymes and oxygen-free radicals. The latter will further damage the local tissues. If precolonised or haematogenous seeded bacteria are present, infection may occur.

Leukocyte bactericidal capacity can be impaired by contact with the implant surface (Merritt et al., 1979; Zimmerli et al., 1982). Using a tissue cage model, Zimmerli et al. (1982) demonstrated that when compared to neutrophils from peripheral blood, polymorphonuclear neutrophils from sterile tissue cages showed decreased phagocytic and microbicidal activity. The mechanism of this effect is not clear, but some evidence exists suggesting that the local release of lysosomal enzymes and oxygen-free radicals from leukocytes triggered by contact with the implant surface may damage the leukocytes themselves or that certain metal ions like nickel or cobalt can interfere with bacterial phagocytosis by neutrophils (Rae, 1983).

Implant surface configurations (pores or grooves) provide bacteria with harbours which protect them from the impact of host defence systems and antibiotics. Porous and multifilament surfaces are examples of this effect, and these surfaces have a much higher implant site infection rate (Jerry, Rand, and Ilstrup, 1988; Lidwell, 1988).

2.4 Host Factors

Tissue inflammation caused by a surgical procedure provides a favourable local environment for bacterial colonisation. If the immune system function is normal and the local tissues are in a healthy state, there will be no bacterial colonisation, bacterial aggregation, or implant infection. Certain individuals are more predisposed to prosthetic infection, such as those with rheumatoid arthritis (Charnley, 1972; Fitzgerald et al., 1977; Lidwell, 1988; Maderazo et al., 1988), previous joint surgery (Canner *et al.,* 1984), previous joint sepsis (Jerry et al., 1988; Kim et al., 1988), remote infection at the time of surgery (Canner et al., 1984), or diabetes mellitus (Canner et al., 1984), or those with an immune deficiency. These patients are especially susceptible to haematogenous infections (Gillespie, 1990).

3 ANIMAL MODELS OF IMPLANT INFECTION AND THEIR APPLICATIONS

3.1 An Overview

Animal models of osteomyelitis (Norden, Myerowitz, and Keleti, 1980; Norden, 1988; Rissing, 1990; Curtis et al., 1995; Clasper et al., 1999) and foreign body infection (Varma et al., 1974; Zimmerli et al., 1982; Christensen et al., 1983; Mayberry-Carson et al., 1984; Gallimore et al., 1991; Mayberry-Carson et al., 1992; Espersen et al., 1993) have been established using dogs, sheep, goats, chickens, rabbits, guinea pigs, rats, and mice. The use of these models, especially the models of foreign body infection, has been very helpful in designing an *in vivo* prosthetic infection model. In fact, an orthopaedic prosthetic infection model is an extension of the models used for foreign body infection, with more attention to the effect of the implant material, the use of bone tissue (instead of soft tissue), and the imitation of a human internal bone fixation or total joint replacement.

Several animal models have been reported for the study of orthopaedic prosthetic infection, usually as a joint replacement, skeletal implant, or soft tissue model (Table 4.2). The latter is actually a foreign body infection model. Animal models have been used to study the virulence and pathogenesis of different strains of bacteria, the effect of antibiotics on infection, and the *in vivo* behaviour of bacterial biofilms (Mayberry-Carson et al., 1984; Isiklar et al., 1996; Smeltzer et al., 1997; Shirtliff, Mader, and Calhoun, 1999). Other animal models have focussed on the effect of implant material and design on the infection rate as well as the effect of infection on biomaterial surfaces (Buret et al., 1991; Curtis et al., 1995; Melcher et al., 1995).

3.2 Virulence of Bacteria

When an animal model is employed, consideration must be given to the number and type of bacteria used to cause prosthetic infection, the duration of infection, and the method of inoculation. Zimmerli et al. (1982) tested the effectiveness of different numbers of colony-forming units (cfu) of *S. aureus* on the infection rate of a foreign body (tissue cage: a perforated tube). Smeltzer et al. (1997) found that an inoculum as small as 2×10^3 cfu of *S. aureus* strain UAMS-1 can be used to create infection in devascularised bone. However, in order to induce significant radiographic and histologic changes within 4 weeks, an inoculum of at least 2×10^4 cfu was needed. They also found that the heavily encapsulated *S. aureus* strain Smith diffuse was much less likely to produce disease despite the use of a larger inoculum. This indicates these two strains of *S. aureus* express different virulence factors that significantly affect their ability to induce osteomyelitis. To find the effect of bacterial slime on the infection rate, Christensen et al. (1983) reported a mouse foreign body infection model. Animals challenged with the slime-producing *S. epidermidis* developed foreign body infections three times as often as animals challenged with the strain that did not produce slime. Animal models can also be used to produce a bacterial biofilm for pathobiological study of a biofilm on biomaterial surfaces (Buret et al., 1991).

The duration of infection also plays a role in determining the severity of osteomyelitis. Saleh Mghir et al. (1998) noted that at 7 days postinoculation with *S. aureus* in a rabbit joint prostheses model there was a periprosthetic infection that involved primarily the joint and adjacent bone marrow. This is similar to human prosthetic joint infections, where extensive cortical bone

Table 4.2: Animal models of implant or prosthetic infection

Reference	Animal	Model Description	Inoculation Route	Bacteria	Number of Bacteria Used[a]	Incubation Time
Joint replacement models						
Belmatoug et al., 1996	Rabbit	Partial knee arthroplasty (tibial plateau) with Silastic implant	Intraarticular injection	*S. aureus*	0.5 ml 10^{5-8} cfu/ml[b]	—
Southwood et al., 1985	Rabbit	Prosthetic femoral head replacement	Local inoculation or haematogenous	*S. aureus*	10^{6-7} cfu[c] local 10^{8-9} cfu IV	—
Blomgren, 1981	Rabbit	Femur defect filled by cement or total knee replacement	Local inoculation or haematogenous	*S. aureus*	10^{8-9} cfu	—
Skeletal models						
Poelstra et al., 2000	Rabbit	Spinal implant model using threaded wire	Local inoculation	*S. aureus*	10^{2-5} cfu	—
Johansson et al., 1999	Rabbit	Tibial diaphyseal plating	Local inoculation	*S. aureus*	$10^8 - 10^9$ cfu	—
Johansson et al., 1999	Rabbit	Tibial diaphyseal plating	Haematogenous	*S. aureus*	$10^8 - 2 \times 10^9$ cfu	—
Clasper et al., 1999	Sheep	Tibial external fixator pin track infection	Local inoculation	*S. aureus*	0.5 ml 2.5×10^5 cfu/ml	—
An et al., 1997	Rabbit	Cylindrical implants inserted into lateral femoral condyle	*In vitro* colonisation before implantation	*S. epidermidis*	In suspension of 10^6 cfu/ml	1 hour
Arens et al., 1996	Rabbit	Tibial diaphyseal plating	Local bacteria injection	*S. aureus*	$4 \times 10^{3-6}$ cfu	—
Curtis et al., 1995	Goat	Tibial external fixator or tibial IM nail	Local inoculation with gelatin sponge	*S. aureus*	10^3 cfu	—

Melcher et al., 1995	Rabbit	Tibial intramedullary nailing	Local inoculation	*S. aureus*	$2 \times 10^3 - 4 \times 10^7$ cfu	—
Collinge et al., 1994	Sheep	Stainless steel pins inserted into iliac crest	Local inoculation	*S. aureus*	10^5 cfu	—
Isiklar et al., 1996	Rabbit	Femoral intercondylar notch, drill hole, cancellous SS screw	Local inoculation	*S. epidermidis*	At least 10^7 cfu per implant site	—
Petty et al., 1985; Penny, Spanier, and Shuster, 1988	Dog	Cylindrical implants inserted into proximal femoral canal	Local inoculation	Three different bacteria	10^8 cfu	—
Soft tissue models						
Rediske et al., 1999	Rabbit	Polyethylene disk placed subcutaneously	Precolonisation of disk	*E. coli*	Biofilm formation	16 hours
Nakamoto et al., 1995	Hamster	Coated stainless steel wires placed subcutaneously	Incubation with bacteria before implantation	*S. epidermidis*	In suspension of 10^7 cfu/ml	15 minutes
Chang and Merritt, 1994	Hamster	Subcutaneous cylindrical implants on the back	*In vitro* colonisation Local inoculation	*S. epidermidis*	10^7 cfu in TSB[d] 10^8 cfu/implant	Overnight
Buret et al., 1991	Rabbit	Silastic placed in subdermal tissue for biofilm study	*In vitro* colonisation before implantation	*P. aeruginosa*	In suspension of 10^7 cfu/ml	3 hours

[a] Number of bacteria injected or the concentration of bacteria in the incubation suspension.
[b] cfu/ml = the concentration of the bacterial suspension.
[c] cfu = the number of bacteria in total or the number of bacteria used at each implant site.
[d] TSB = trypticase soy broth.

destruction is usually lacking (Saleh Mghir et al., 1998). In contrast, the rabbit tibial osteomyelitis model described by Norden et al. (1980) showed that by 14 days postinoculation with *S. aureus* there were advanced cortical bone lesions. Using a rabbit total joint model, Blomgren (1981) verified the possibility of haematogenous bacterial dissemination of a total joint prosthesis and the subsequent infection. He also found that *S. aureus* and *Propionibacterium acnes* have the same ability to cause haematogenous infection of a total joint replacement. Subsequent models of haematogenous orthopaedic infection have been developed utilising rabbit tibial bone plating and mouse tibial fractures (Chadha et al., 1999; Johansson et al., 1999).

3.3 *In Vivo* Behaviour of Biofilms

Buret et al. (1991) studied the morphology, ultrastructure, and microbiology of intact biofilms developing on an implant surface harvested from an implant colonised with *Pseudomonas aeruginosa* inserted into the peritoneal cavity of rabbits. Also in a rabbit model, Isiklar et al. (1996) examined the penetration of antibiotics into biofilm formed by *S. epidermidis* following local and parenteral administration of vancomycin. This method is closer to the human situation because of the use of tibial bone, stainless steel as the implant, and the introduction of *S. epidermidis*, which is a common pathogen for prosthetic infection. A clinical report by Smith, Vasseur, and Saunders (1989) suggests that biofilms may be responsible for cryptic infections where bacteria are colonised at an implant site for extended periods of time without any clinical or radiographic signs of infection.

3.4 Effect of Prophylactic and Therapeutic Antibiotics

Animal models are excellent for studying the effects of prophylactic and therapeutic antibiotics on prosthetic infection because of the homogeneity of the animals, allowing good comparison, reproducibility, and easily controlled time periods for observation (Petty, Spanier, and Shuster, 1988; Mayberry-Carson et al., 1990; Espersen et al., 1993; Isiklar et al., 1996). For example, Petty et al. (1988) studied the preventive effectiveness of wound irrigation with normal saline, irrigation with saline containing neomycin, antibiotic-impregnated polymethylmethacrylate, and systemic antibiotic administration on prosthetic infection. The results showed that saline irrigation had no effect on the infection rate, systemic use of cefazolin and neomycin irrigation slightly reduced the infection rate, and the use of bone cement containing gentamicin caused a statistically significant reduction of the infection rate.

Since the microenvironment of the biofilm protects the sequestered bacteria from normal levels of antibiotic therapy, conventional *in vitro* antibiotic sensitivity testing is often inaccurate for treating *in vivo* infections. As new *in vitro* antibiotic sensitivity tests are developed, they must be validated using *in vivo* models of orthopaedic implant infection. These models must ensure that the biofilm is formed on the implant prior to the initiation of the experimental antibiotic therapy (Gracia et al., 1998; Rediske et al., 1999; Monzon et al., 2001). In an effort to increase the bactericidal efficacy of antibiotics, Rediske et al. (1999) found that gentamicin combined with low-frequency ultrasound was superior to gentamicin alone in reducing the number of viable bacteria in an implanted *E. coli* biofilm.

Several models are also available for testing the antiseptic or therapeutic effects of locally implanted devices coated or impregnated with antiseptic agents or antibiotics, such as the rabbit tibial and femoral models (Darouiche et al., 1998; Nijhof et al., 2000, 2001), the canine tibial model (Garvin et al., 1994), the goat tibial model (DeJong et al., 2001), and the sheep iliac crest model (Collinge et al., 1994; Voos et al., 1999).

3.5 Effect of Biomaterials On the Prosthetic Infection Rate

Several *in vivo* investigations on the effects of orthopaedic implants on the incidence of infection have been reported. Merritt et al. (1979) designed a soft tissue model in the mouse and tested the implant site infection rates with porous and dense materials. They found that the infection rate with the porous implant was greater in the acute model and the infection rate with the dense materials was greater in the chronic model. Recently, a rabbit tibial model of osteomyelitis was used to investigate the relationship between bioactivity of the implant material and susceptibility to infection (Vogely et al., 2000). HA-coated titanium implants were shown to produce more severe infections than non-coated implants.

Petty et al. (1985) established a dog model to evaluate the influence of skeletal implants on the incidence of infection when challenged with *S. aureus*, *S. epidermidis*, and *E. coli* and showed that all of the implants (including stainless steel alloy, cobalt-chromium alloy, high-density polyethylene, prepolymerised polymethylmethacrylate, and polymethylmethacrylate polymerised *in vivo*) were significantly more likely than the controls (medullary reaming only, without implantation and bacterial challenge) to be associated with infection. Polymethylmethacrylate polymerised *in vivo* was found to be significantly more likely than all other implants to be associated with an

S. epidermidis infection. In another study, rabbit tibiae were plated with 6-hole, 2.0-mm stainless steel or titanium plates and then challenged with local injection of *S. aureus*. Under otherwise identical experimental conditions, the rate of infection for stainless steel plates (75 per cent) was significantly higher than for titanium plates (35 per cent) (Arens et al., 1996).

Using a rabbit model of tibial intramedullary nailing, it was found that the physical configuration of the nail had a significant effect on the implant infection rate. A much higher rate of infection was found following implantation of slotted nails (59 per cent) compared to using solid nails (27 per cent) (Melcher et al., 1995). A similar study examining bone plates found that both material and design properties of the implant have an effect on infection rate (Johansson et al., 1999).

The use of heparin and fibrinolytic agents to coat steel wires to decrease infection has been reported (Nakamoto et al., 1995). In this study both the heparin-coated wires and the urokinase-coated wires exhibited significantly decreased infection rates compared with uncoated wires. The mechanism of this effect may be the inhibition of bacterial adherence by heparin and fibrinolytic agents. A recently developed sheep model using a tibial external fixator has been proposed for studying the antimicrobial effects of different treatments of external-fixator pins (Clasper et al., 1999). Silver coating and antimicrobial agent coating (chlorhexidine and chloroxylenol) have been shown to decrease the infection rate of orthopaedic implants (Collinge et al., 1994; Darouiche et al., 1998).

In a study from the authors' laboratory, coated (albumin) and uncoated titanium implants were exposed to a suspension of *S. epidermidis* prior to implantation into the lateral femoral condyle. We found that rabbits with albumin-coated implants had a much lower infection rate (27 per cent) than those with uncoated implants (62 per cent). This finding may represent a new method for preventing orthopaedic implant infection (An et al., 1996).

3.6 Effect of Infection On Biomaterial Surfaces

It has been noted that an implant-centred infection *in vitro* has certain effects on the behaviour of a prosthetic material. Merritt et al. (1991) found that the presence of bacterial growth can have a significant corrosion effect on stainless steel surfaces. Verheyen et al. (1993) and Kieswetter et al. (1993) studied the destructive effect of *S. aureus, S. epidermidis,* and *Proteus* on the integrity of HA coating *in vitro* or using an animal model (subcutaneous implantation of prosthetic materials in hamsters). Their results showed that significant destruction of HA coating can occur due to the growth of *S. aureus, S. epidermidis,* or *Proteus.* Damage to HA-coated implants appeared to be more severe *in vitro* than *in vivo.*

4 DESIGNING AN ANIMAL MODEL OF ORTHOPAEDIC IMPLANT INFECTION

4.1 Selection of Research Animal

Theoretically, to imitate the human situation it is better to use large animals such as sheep, goats, or dogs, especially when attempting to use a joint prosthesis as the implant. A dog femoral model has been used to test the influence of skeletal implants on the incidence of infection and the preventative effect of prophylactic antibiotics (Petty et al., 1985, 1988). The shortcomings of using large animals include the need for a large housing space and high costs.

Rabbit joint replacement models have been reported, including femoral head replacement using specially designed prostheses (Southwood et al., 1985), total knee replacement using human interphalangeal joint prostheses (Blomgren, 1981), or partial knee replacement using human first metatarsophalangeal joint prostheses (Belmatoug et al., 1996). A rabbit implant infection model using bone screws was used to evaluate the interaction between bacterial biofilm and antibiotics (Isiklar et al., 1996). Implant site infection models such as rabbit subcutaneous implantation were also reported (Buret et al., 1991; Chang and Merritt, 1994; Nakamoto et al., 1995). More recently, rabbit implant models such as a femoral condyle cylindrical implant (An et al., 1996), tibial intramedullary nailing (Melcher et al., 1995), and tibial plating have been reported. A new rabbit model of spinal implant infection has demonstrated that multiple sites can be used in the same animal without fear of cross-contamination. This is significant because it allows collection of data from multiple locations in one animal, which should decrease statistical variance and, consequently, the number of animals (and funding) needed for a given study (Poelstra et al., 2000). This concept has been validated by another study using a standardised foreign body infection model (Rediske et al., 1999). The number of viable bacteria retrieved varied little from site to site in the same rabbit, but it varied significantly from rabbit to rabbit, despite having the same inoculum. This suggests that, whenever possible, allowing the animal to serve as its own control provides a better comparison than using a separate control group of animals.

According to the literature and our own experience, rabbits are excellent for implant infection models because (1) they are large enough to easily perform common surgical procedures, including joint replacement, with the use of conventional orthopaedic implants; (2) they are easier to infect compared to dogs or rats; and (3) they are relatively economical. Small animals (rodents) are good for studying bacterial pathogenesis, implant site infection rate (normally need large numbers of animals), or antibiotic effects. They have been widely used for foreign body infection models (Zimmerli, 1993). Small animals, which have been used include guinea pigs (Zimmerli et al., 1982), hamsters (Merritt et al., 1991; Verheyen et al., 1993), and mice (Merritt et al., 1979; Christensen et al., 1983; Mayberry-Carson et al., 1990; Gallimore et al., 1991; Espersen et al., 1993). Nevertheless, animal selection is a very important step for *in vivo* prosthetic infection studies. One should be very careful to choose an animal that will fit the purpose of the project and to use the minimum number possible that will allow collection of meaningful data.

4.2 Implant Fabrication

Based on specific purposes, implants made from different kinds of materials and with different sizes and shapes have been used. Implants can be fabricated into, or obtained as, a cylinder (Merritt et al., 1979; Petty et al., 1985; An et al., 1996), a metal wire (Nakamoto et al., 1995), a screw (Isiklar et al., 1996), a tissue cage (Kieswetter et al., 1993), a bone plate (Arens et al., 1996), an intramedullary nail (Melcher et al., 1995), a small joint prosthesis designed for humans (Blomgren, 1981; Belmatoug et al., 1996), or even a prosthetic joint component (Southwood et al., 1985). Implants can be made with smooth or porous surfaces. Most of the implants made for bone ingrowth studies are cylindrical and may be used as implants for prosthetic infection.

4.3 Bacterial Inoculation and Cultivation of Biofilms

Common pathogens isolated from human prosthetic infections should be used for most of the *in vivo* infection models, such as *S. aureus, S. epidermidis,* or less frequently *E. coli, Proteus,* or some common anaerobes. Bacteria can be delivered onto the implant surface or the area surrounding it by (1) pre-colonisation of bacteria on the implant *in vitro* before implantation, (2) direct inoculation of a bacterial suspension at the implant site, (3) a combination of pre-colonisation and local inoculation, or (4) injection of the bacteria into the bloodstream (Saleh Mghir et al., 1998; Johansson et al., 1999).

The number of bacteria needed to produce an experimental infection varies from one bacterium to another. For example, fewer slime-producing *S. epidermidis* will be needed to produce an infection than non-slime producers. The relationship between the dose of inoculum and the development of infection after prosthetic replacement has been studied in a rabbit model

(Southwood et al., 1985). Contamination of the implanted wound site with only a few bacteria (less than fifty *S. epidermidis*) will be likely to result in infection. It is more difficult to induce infection when the operation is performed without insertion of a prosthesis (10^4 bacteria), which may suggest that the implant inhibits the defence system for dealing with the insult. It is also possible to purposely inhibit the local defence system by injecting a sclerosing agent, usually sodium morrhuate, in order to reduce the number of bacteria needed to produce infection (Gratz et al., 2001; Schulz, Steinhart, and Mutters, 2001).

The route of inoculation also makes a significant difference. It is more difficult to produce an infection by inoculating the bacteria intravenously than locally, and this is more noticeable if the inoculation is given 3 weeks after the operation (Blomgren, 1981; Southwood et al., 1985). Johansson et al. (1999) found that a postoperative delay of approximately 5 days was best for inducing haematogenous implant infections in a rabbit tibial bone plate model. If the inoculation was given at the time of surgery there was a high mortality rate, and if given 10 days after surgery it was ineffective at producing clinical infection. In a parallel study, it was suggested that although both material and design considerations are important in determining the risk of haematogenous infection, these considerations can be negated when a large number of locally inoculated bacteria overwhelm the host's defence systems (Johansson et al., 1999). Therefore, care should be taken to appropriately select the route and number of bacteria in order to sufficiently challenge, yet not overwhelm, the local or systemic defence systems of the animal. Based on the data shown in Table 4.2, at least 10^3 cfu, or an average of 10^{5-8} cfu, are needed to produce an implant site infection.

If the study needs a bacterial biofilm preformed on the implant surface, a static culture or a flow chamber method can be used. For both methods, the implant can be left in the culture well or in the flow chamber for the desired period of time (hours to days) before implantation (An and Friedman, 1997; Gracia et al., 1998; Rediske et al., 1999; An et al., 2001; Monzon et al., 2001).

4.4 Surgery and Necropsy

Surgery should be performed under strict aseptic conditions. Normally, the skin area of the incision should be clipped carefully and cleaned with 70 per cent isopropyl alcohol before transferring the animal to the operating table. The area of the incision should be prepared with solutions of chlorhexidine gluconate/alcohol-chlorhexidine or povidone-iodine before draping. Standard aseptic surgical techniques should be employed throughout the

Table 4.3: Appearance of the wound and grading

Grade	Appearance	Score	Clinical Meaning
A	No abnormal signs	0	No infection
B	Erythema and moderate soft swelling	0.5	Early infection or due to surgery
C	Large soft swelling or pus exudation	1.0	Definite infection
D	Pus exudation and systemic illness	1.5	Definite infection

Note: This table was reprinted from page 144 from An, Y. H. and Friedman, R. J. (1998b). *Journal of Investigative Surgery*, 11, 139–46. With permission.

operation. After completion of the implantation, the wound should be washed with normal saline and the skin and underlying tissues should be closed securely in layers. Harvesting of specimens should also be done under the same sterile conditions as the surgery in order to obtain a valid microbiological evaluation.

4.5 Methods of Evaluation

4.5.1 Clinical Evaluation

In contrast to human subjects, it is difficult to get subjective findings from an animal. Certain abnormal behaviours, such as lethargy, less appetite, lack of normal weight gain, or weight loss, may indicate a severe local or systemic infection. Physical signs are very important for judging an infection. After surgery, the animal's temperature should be taken and the wound observed daily for the first week and twice a week thereafter. A change in body temperature is a sensitive parameter of a local or systemic infection. The appearance of the wound can be recorded and graded using a grading system modified from the one introduced by Petty et al. (1985) (Table 4.3). Normally, infection can be confirmed by soft tissue swelling with pus exudate.

4.5.2 Radiography

For models with implants attached to or penetrating the bone, radiographs should be taken immediately after the operation to check the implant position and at 1–2 week intervals thereafter, until the animal is sacrificed. The development and progression of infection can be assessed using five criteria (Table 4.4) modified from previous radiological descriptions of osteomyelitis in rabbits (Norden et al., 1980). These are (1) diaphyseal periosteal reaction, (2) osteolysis, (3) sequestrum formation, (4) joint effusion, and (5) soft tissue

Table 4.4: Radiographic criteria

Variable	Definition	Score
1. Diaphyseal periosteal reaction	+/present	1
	−/absent	0
2. Osteolysis	+/present	1
	−/absent	0
3. Sequestrum formation	+/present	1
	−/absent	0
4. Joint effusion	+/present, widening of joint space	1
	−/absent	0
5. Soft tissue swelling	+/present	1
	−/absent	0

Note: This table was reprinted from page 144 from An, Y. H. and Friedman, R. J. (1998b). *Journal of Investigative Surgery*, 11, 139–46. With permission.

swelling. Using these criteria, a numerical score can be assigned, and the six scores are added together to give an overall ranking for radiographic severity. Radiography is a very useful method for diagnosing an infection in a diaphyseal area, but is less favourable when a prosthetic infection is present at an epiphyseal/metaphyseal location or in the very acute stages of disease. Recently, more sensitive techniques utilising scintigraphy and radiopharmaceuticals targeting neutrophils have provided an easy and relatively quick method for evaluating acute osteomyelitis in a rabbit model (Gratz et al., 2001). In antibiotic efficacy investigations, it is also helpful to study the diffusion pattern of a specific antibiotic using ^{14}C radiolabelling and quantitative autoradiographic techniques (Saleh Mghir et al., 1998).

4.5.3 Laboratory Tests and Bacteriological Analyses

For abnormal wound exudate or sinus discharge, a swab is taken for bacterial culture. One should realise that there is no direct relationship between the results of swab culture and clinical signs of infection. In a previous study by the authors using a rabbit femoral condyle implant infection model, there were only two positive swab cultures in eleven histologically diagnosed infections (An et al., 1997). By contrast, Clasper et al. (1999, 2001) reported that all swabs in one of their studies were positive for *S. aureus* (which was the organism introduced into the implant sites) and in another study many positive cultures were reported. Using broth culture may yield more positive results.

A common method for identifying the existence of bacteria on explant surfaces is agar plate culture after rolling or touching the agar surface with the harvested implant. Implants with attached bacteria can be cultured in broth directly (DeJong et al., 2001). Attached bacteria can also be dislodged from the implant and seeded onto an agar plate for culture. The harvested implants are placed in phosphate-buffered saline (PBS) and agitated on a vortex mixer for 5 minutes or sonicated for 30 minutes to dislodge the adherent bacteria for culture. Serial dilutions are made, and the diluted suspensions are then spread on a tryptic soy agar plate and/or added to tryptic soy broth for bacterial culture. Subcultures will be prepared at 24 hours if needed. The cultures will be considered positive for infection as described by Petty et al. (1985) if (1) the primary or subculture yields any bacteria that had been inoculated or (2) the primary or subculture yields any bacteria. These criteria are subject to change according to the individual situation. If the implant is cylindrical, it can also be rolled over an agar surface for overnight culture.

When measuring antibiotic levels in tissue homogenates, it is important to check for the possibility of a heterogeneous diffusion pattern that could give inaccurate results. For example, Norden, Niederreiter, and Shinners (1986) found homogenates of bone to contain levels of teicoplanin that were eight times higher than the minimum inhibitory concentration (MIC) for *S. aureus*, yet the antibiotic was not efficacious in a model of chronic osteomyelitis with extensive cortical bone involvement. This can be explained by the fact that high concentrations of teicoplanin can be found in the periosteum and bone marrow, but it is almost absent from cortical bone (Saleh Mghir et al., 1998). A later study also showed that the ^{14}C-labelled sparfloxacin content was very low in compact bone (Cremieux et al., 1996).

4.5.4 Histological Study

For histological study, the specimens are embedded in paraffin (if the specimens are bone, they should be decalcified first) and 4–6 μm thick sections are cut. Implants should be removed before embedding. According to the authors' experience, this will not destroy the specimen if the implant surface is smooth. We found that abscesses were still intact after the implant was removed (An et al., 1997). Adjacent soft tissues should be evaluated histologically because soft tissue abscesses or drainage tracts may exist. Draining lymph nodes should also be evaluated.

Histological sections are stained with haematoxylin and eosin and examined under a light microscope. Selected sections should be stained with the Gram stain for detecting the existence of Gram-positive bacteria in the pus, the capsule of the pus, or in any inflamed tissues. The histological parameters in Table 4.5 can be used to evaluate samples with the implant inserted into

Table 4.5: Histological criteria of animal models of implant infection

Histologic Findings	Diagnostic Meaning
Inflammation with abscess formation	Infected
Presence of sequestrum within abscess or near an abscess	Infected
Inflammation with fibrosis	Reparative response (no bacteria found) or infection (bacteria found in inflamed tissues)
Intracellular bacteria (Gram stain)	Infected when found in an abscess, abscess capsule, or in inflamed tissues (often neutrophil- or macrophage-infiltrated area) or not infected when found without evident inflammation or abscess

Note: This table was reprinted from page 145 from An, Y. H. and Friedman, R. J. (1998b). *Journal of Investigative Surgery*, 11, 139–46. With permission.

cancellous bone. These are useful when a 'yes-or-no' answer is needed. The authors feel that only the existence of microabscesses is definite evidence of infection (An et al., 1997). For the histological criteria of experimental prosthetic infection involving a diaphyseal area, one should refer to the work by Petty et al. (1985).

5 SUMMARY

Because of the potential for tragic results and the large number of surgical procedures, implant infection remains a major challenge to orthopaedic surgeons and biomedical engineering researchers. Many animal models of implant infection have been developed with great achievements in understanding the mechanisms of bacterial adhesion and implant infection. Due to the complex nature of bacteria and their sophisticated interaction with biomaterials and host tissues, various animal models are essential and remain the major approach in solving this problem. It is the authors' sincere hope that this chapter has (1) concisely introduced the mechanisms and pathogenesis of orthopaedic implant infection; (2) outlined the animal models that have been established so far; (3) covered the common use of these animal models for studies of the pathogenesis of bacteria, the behaviour of biofilms, the effect of biomaterials on prosthetic infection rate, and the effect of infection on biomaterial surfaces; and (4) helped researchers to design and conduct an animal model of orthopaedic prosthetic infection, including animal selection, implant fabrication, bacterial inoculation, surgical technique, and the methods to be used in evaluating the results.

REFERENCES

An, Y. H., Bradley, J., Powers, D. L. and Friedman, R. J. (1997). The prevention of prosthetic infection using a cross-linked albumin coating in a rabbit model. *Journal of Bone and Joint Surgery – British volume*, 79, 816–19.

An, Y. H., Dickinson, R. B. and Doyle, R. J. (2000). Mechanisms of bacterial adhesion pathogenesis of implant and tissue infections. In *Handbook of Bacterial Adhesion – Principles, Methods, and Applications* (Eds. An, Y. H. and Friedman, R. J.). Humana Press, Totowa, NJ, pp. 1–28.

An, Y. H. and Friedman, R. J. (1996). Prevention of sepsis in total joint arthroplasty. *Journal of Hospital Infection*, 33, 93–108.

An, Y. H. and Friedman, R. J. (1997). Laboratory methods for studies of bacterial adhesion. *Journal of Microbiological Methods*, 30, 141–52.

An, Y. H. and Friedman, R. J. (1998a). Animal models in orthopaedic prosthetic infection. In *Animal Models in Orthopaedic Research* (Eds. An, Y. H. and Friedman, R. J.). CRC Press, Boca Raton, FL, pp. 443–60.

An, Y. H. and Friedman, R. J. (1998b). Animal models of orthopedic implant infection. *Journal of Investigative Surgery*, 11, 139–46.

An, Y. H. and Friedman, R. J. (1998c). Concise review of mechanisms of bacterial adhesion to biomaterial surfaces. *Journal of Biomedical Materials Research*, 43, 338–48.

An, Y. H., McGlohorn, J. B., Bednarski, B. K., Martin, K. L. and Friedman, R. J. (2001). An open channel flow chamber for characterizing biofilm formation on biomaterial surfaces. *Methods in Enzymology*, 337, 79–88.

An, Y. H., Stuart, G. W., McDowell, S. J., McDaniel, S. E., Kang, Q. and Friedman, R. J. (1996). Prevention of bacterial adherence to implant surfaces with a crosslinked albumin coating in vitro. *Journal of Orthopaedic Research*, 14, 846–9.

Andrews, H. J., Arden, G. P., Hart, G. M. and Owen, J. W. (1981). Deep infection after total hip replacement. *Journal of Bone and Joint Surgery – British volume*, 63-B, 53–7.

Arens, S., Schlegel, U., Printzen, G., Ziegler, W. J., Perren, S. M. and Hansis, M. (1996). Influence of materials for fixation implants on local infection. An experimental study of steel versus titanium DCP in rabbits. *Journal of Bone and Joint Surgery – British volume*, 78, 647–51.

Arizono, T., Oga, M. and Sugioka, Y. (1992). Increased resistance of bacteria after adherence to polymethyl methacrylate. An in vitro study. *Acta Orthopaedica Scandinavica*, 63, 661–4.

Bartzokas, C. A., Johnson, R., Jane, M., Martin, M. V., Pearce, P. K. and Saw, Y. (1994). Relation between mouth and haematogenous infection in total joint replacements. *British Medical Journal*, 309, 506–8.

Belmatoug, N., Cremieux, A. C., Bleton, R., Volk, A., Saleh-Mghir, A., Grossin, M., Garry, L. and Carbon, C. (1996). A new model of experimental prosthetic joint infection due to methicillin-resistant *Staphylococcus aureus*: a microbiologic, histopathologic, and magnetic resonance imaging characterization. *Journal of Infectious Diseases*, 174, 414–17.

Benson, M. K. and Hughes, S. P. (1975). Infection following total hip replacement in a general hospital without special orthopaedic facilities. *Acta Orthopaedica Scandinavica*, 46, 968–78.

Blenkinsopp, S. A. and Costerton, W. (1991). Understanding bacterial biofilm. *Tebtech*, 9, 138–42.

Blomgren, G. (1981). Hematogenous infection of total joint replacement. An experimental study in the rabbit. *Acta Orthopaedica Scandinavica Supplement*, 187, 1–64.

Buret, A., Ward, K. H., Olson, M. E. and Costerton, J. W. (1991). An in vivo model to study the pathobiology of infectious biofilms on biomaterial surfaces. *Journal of Biomedical Materials Research*, 25, 865–74.

Canner, G. C., Steinberg, M. E., Heppenstall, R. B. and Balderston, R. (1984). The infected hip after total hip arthroplasty. *Journal of Bone and Joint Surgery – American volume*, 66, 1393–9.

Chadha, H. S., Fitzgerald, R. H., Jr., Wiater, P., Sud, S., Nasser, S. and Wooley, P. H. (1999). Experimental acute hematogenous osteomyelitis in mice. I. Histopathological and immunological findings. *Journal of Orthopaedic Research*, 17, 376–81.

Chang, C. C. and Merritt, K. (1994). Infection at the site of implanted materials with and without preadhered bacteria. *Journal of Orthopaedic Research*, 12, 526–31.

Charnley, J. (1972). Postoperative infection after total hip replacement with special reference to air contamination in the operating room. *Clinical Orthopaedics*, 87, 167–87.

Cheatle, M. D. (1991). The effect of chronic orthopedic infection on quality of life. *Orthopaedic Clinics of North America*, 22, 539–47.

Christensen, G. D., Simpson, W. A., Bisno, A. L. and Beachey, E. H. (1983). Experimental foreign body infections in mice challenged with slime-producing *Staphylococcus epidermidis*. *Infection and Immunity*, 40, 407–10.

Chu, C. C. and Williams, D. F. (1984). Effects of physical configuration and chemical structure of suture materials on bacterial adhesion. A possible link to wound infection. *American Journal of Surgery*, 147, 197–204.

Clasper, J. C., Parker, S. J., Simpson, A. H. and Watkins, P. E. (1999). Contamination of the medullary canal following pin-tract infection. *Journal of Orthopaedic Research*, 17, 947–52.

Clasper, J. C., Stapley, S. A., Bowley, D. M., Kenward, C. E., Taylor, V. and Watkins, P. E. (2001). Spread of infection, in an animal model, after intramedullary nailing of an infected external fixator pin tract. *Journal of Orthopaedic Research*, 19, 155–9.

Collinge, C. A., Goll, G., Seligson, D. and Easley, K. J. (1994). Pin tract infections: silver vs uncoated pins. *Orthopaedics*, 17, 445–8.

Costerton, J. W., Marrie, T. J. and Cheng, K. J. (1985). Phenomena of bacterial adhesion. In *Bacterial Adhesion. Mechanisms and Physiological Significance* (Eds. Savage, D. C. and Fletcher, M.). Plenum, New York, pp. 1–43.

Cremieux, A. C. and Carbon, C. (1997). Experimental models of bone and prosthetic joint infections. *Clinical Infectious Diseases*, 25, 1295–302.

Cremieux, A. C., Mghir, A. S., Bleton, R., Manteau, M., Belmatoug, N., Massias, L., Garry, L., Sales, N., Maziere, B. and Carbon, C. (1996). Efficacy of sparfloxacin and autoradiographic diffusion pattern of [14C]Sparfloxacin in experimental

Staphylococcus aureus joint prosthesis infection. *Antimicrobial Agents and Chemotherapy*, 40, 2111–16.

Curtis, M. J., Brown, P. R., Dick, J. D. and Jinnah, R. H. (1995). Contaminated fractures of the tibia: a comparison of treatment modalities in an animal model. *Journal of Orthopaedic Research*, 13, 286–95.

Darouiche, R. O., Farmer, J., Chaput, C., Mansouri, M., Saleh, G. and Landon, G. C. (1998). Anti-infective efficacy of antiseptic-coated intramedullary nails. *Journal of Bone and Joint Surgery – American volume*, 80, 1336–40.

DeJong, M. E., DeBerardino, M. T., Brooks, D. E., Nelson, M. B., Campbell, A. A., Bottoni, M. C., Pusateri, A. E., Walton, M. R., Guymon, C. H. and McManus, A. T. (2001). Antimicrobial efficacy of external fixator pins coated with a lipid stabilized hydroxyapatite/chlorhexidine complex to prevent pin tract infection in a goat model. *Journal of Trauma*, 50, 1008–14.

DiCarlo, E. F. and Bullough, P. G. (1992). The biologic responses to orthopaedic implants and their wear debris. *Clinical Materials*, 9, 235–60.

Dobbins, J. J., Seligson, D. and Raff, M. J. (1988). Bacterial colonization of orthopedic fixation devices in the absence of clinical infection. *Journal of Infectious Diseases*, 158, 203–5.

Donlan, R. M. (2000). Role of biofilms in antimicrobial resistance. *ASAIO Journal*, 46, S47–S52.

Dougherty, S. H. (1988). Pathobiology of infection in prosthetic devices [see comments]. *Reviews of Infectious Diseases*, 10, 1102–17.

Dougherty, S. H. and Simmons, R. L. (1989). Endogenous factors contributing to prosthetic device infections. *Orthopaedic Clinics of North America*, 3, 199–209.

Eftekhar, N. S., Kiernan, H. A., Jr. and Stinchfield, F. E. (1976). Systemic and local complications following low-friction arthroplasty of the hip joint. A study of 800 consecutive operations. *Archives of Surgery*, 111, 150–5.

Eijer, H., Hauke, C., Arens, S., Printzen, G., Schlegel, U. and Perren, S. M. (2001). PC-Fix and local infection resistance – influence of implant design on postoperative infection development, clinical and experimental results. *Injury*, 32 (Suppl. 2), S-B38–43.

Espersen, F., Frimodt-Moller, N., Corneliussen, L., Thamdrup Rosdahl, V. and Skinhoj, P. (1993). Experimental foreign body infection in mice. *Journal of Antimicrobial Chemotherapy*, 31 (Suppl. D), 103–11.

Fitzgerald, R. H., Jr., Nolan, D. R., Ilstrup, D. M., Van Scoy, R. E., Washington, J. A. D. and Coventry, M. B. (1977). Deep wound sepsis following total hip arthroplasty. *Journal of Bone and Joint Surgery – American volume*, 59, 847–55.

Gallimore, B., Gagnon, R. F., Subang, R. and Richards, G. K. (1991). Natural history of chronic *Staphylococcus epidermidis* foreign body infection in a mouse model. *Journal of Infectious Diseases*, 164, 1220–3.

Garvin, K. L., Miyano, J. A., Robinson, D., Giger, D., Novak, J. and Radio, S. (1994). Polylactide/polyglycolide antibiotic implants in the treatment of osteomyelitis. A canine model. *Journal of Bone and Joint Surgery – American volume*, 76, 1500–6.

Gillespie, W. J. (1990). Infection in total joint replacement. *Orthopaedic Clinics of North America*, 4, 465–84.

Gracia, E., Lacleriga, A., Monzon, M., Leiva, J., Oteiza, C. and Amorena, B. (1998). Application of a rat osteomyelitis model to compare in vivo and in vitro the antibiotic efficacy against bacteria with high capacity to form biofilms. *Journal of Surgical Research*, 79, 146–53.
Gratz, S., Rennen, H. J., Boerman, O. C., Oyen, W. J., Burma, P. and Corstens, F. H. (2001). (99m)Tc-interleukin-8 for imaging acute osteomyelitis. *Journal of Nuclear Medicine*, 42, 1257–64.
Gray, E. D., Peters, G., Verstegen, M. and Regelmann, W. E. (1984). Effect of extracellular slime substance from *Staphylococcus epidermidis* on the human cellular immune response. *Lancet*, 1, 365–7.
Gristina, A. G. and Costerton, J. W. (1985). Bacterial adherence to biomaterials and tissue. The significance of its role in clinical sepsis. *Journal of Bone and Joint Surgery – American volume*, 67, 264–73.
Gristina, A. G., Hobgood, C. D., Webb, L. X. and Myrvik, Q. N. (1987). Adhesive colonization of biomaterials and antibiotic resistance. *Biomaterials*, 8, 423–6.
Gristina, A. G., Jennings, R. A., Naylor, P. T., Myrvik, Q. N. and Webb, L. X. (1989). Comparative in vitro antibiotic resistance of surface-colonizing coagulase-negative staphylococci. *Antimicrobial Agents and Chemotherapy*, 33, 813–16.
Gristina, A. G., Naylor, P. T. and Myrvik, Q. N. (1990). Musculoskeletal infection, microbial adhesion, and antibiotic resistance. *Orthopaedic Clinics of North America*, 4, 391–408.
Gristina, A. G., Naylor, P. T. and Myrvik, Q. N. (1991). Mechanisms of musculoskeletal sepsis. *Orthopaedic Clinics of North America*, 22, 363–71.
Grogan, T. J., Dorey, F., Rollins, J. and Amstutz, H. C. (1986). Deep sepsis following total knee arthroplasty. Ten-year experience at the University of California at Los Angeles Medical Center. *Journal of Bone and Joint Surgery – American volume*, 68, 226–34.
Hebert, C. K., Williams, R. E., Levy, R. S. and Barrack, R. L. (1996). Cost of treating an infected total knee replacement. *Clinical Orthopaedics*, 331, 140–5.
Hughes, S. P. (1988). The role of antibiotics in preventing infections following total hip replacement. *Journal of Hospital Infection*, 11 (Suppl. C), 41–7.
Isiklar, Z. U., Darouiche, R. O., Landon, G. C. and Beck, T. (1996). Efficacy of antibiotics alone for orthopaedic device related infections. *Clinical Orthopaedics*, 332, 184–9.
Itasaka, T., Kawai, A., Sato, T., Mitani, S. and Inoue, H. (2001). Diagnosis of infection after total hip arthroplasty. *Journal of Orthopaedic Science*, 6, 320–6.
James, R. C. and MacLeod, C. J. (1961). Induction of staphylococcal infections in mice with small inocula introduced on sutures. *British Journal of Experimental Pathology*, 42, 266–77.
Jerry, G. J., Jr., Rand, J. A. and Ilstrup, D. (1988). Old sepsis prior to total knee arthroplasty. *Clinical Orthopaedics*, 236, 135–40.
Johansson, A., Lindgren, J. U., Nord, C. E. and Svensson, O. (1999). Local plate infections in a rabbit model. *Injury*, 30, 587–90.
Johansson, A., Lindgren, J. U., Nord, C. E. and Svensson, O. (1999). Material and design in haematogenous implant-associated infections in a rabbit model. *Injury*, 30, 651–7.

Josefsson, G. and Kolmert, L. (1993). Prophylaxis with systematic antibiotics versus gentamicin bone cement in total hip arthroplasty. A ten-year survey of 1,688 hips. *Clinical Orthopaedics,* 292, 210–14.

Kamme, C. and Lindberg, L. (1981). Aerobic and anaerobic bacteria in deep infections after total hip arthroplasty: differential diagnosis between infectious and non-infectious loosening. *Clinical Orthopaedics,* 154, 201–7.

Kieswetter, K., Merritt, K. and Myers, R. (1993). Effects of infection on hydroxyapatite coating. *Transactions of the Society for Biomaterials,* 16, 220.

Kim, Y. Y., Ko, C. U., Ahn, J. Y., Yoon, Y. S. and Kwak, B. M. (1988). Charnley low friction arthroplasty in tuberculosis of the hip. An eight to 13-year follow-up. *Journal of Bone and Joint Surgery – British volume,* 70, 756–60.

Klock, J. C. and Bainton, D. F. (1976). Degranulation and abnormal bactericidal function of granulocytes procured by reversible adhesion to nylon wool. *Blood,* 48, 149–61.

Lidwell, O. M. (1988). Air, antibiotics and sepsis in replacement joints. *Journal of Hospital Infection,* 11 (Suppl. C), 18–40.

Lidwell, O. M., Lowbury, E. J., Whyte, W., Blowers, R., Stanley, S. J. and Lowe, D. (1982). Effect of ultraclean air in operating rooms on deep sepsis in the joint after total hip or knee replacement: a randomised study. *British Medical Journal,* 285, 10–14.

Lowy, F. D. and Hammer, S. M. (1983). *Staphylococcus epidermidis* infections. *Annals of Internal Medicine,* 99, 834–9.

Maderazo, E. G., Judson, S. and Pasternak, H. (1988). Late infections of total joint prostheses. A review and recommendations for prevention. *Clinical Orthopaedics,* 229, 131–42.

Mayberry-Carson, K. J., Tober-Meyer, B., Gill, L. R., Lambe, D. W., Jr. and Hossler, F. E. (1990). Effect of ciprofloxacin on experimental osteomyelitis in the rabbit tibia, induced with a mixed infection of *Staphylococcus epidermidis* and *Bacteroides thetaiotaomicron. Microbios,* 64, 49–66.

Mayberry-Carson, K. J., Tober-Meyer, B., Lambe, D. W., Jr. and Costerton, J. W. (1992). Osteomyelitis experimentally induced with *Bacteroides thetaiotaomicron* and *Staphylococcus epidermidis.* Influence of a foreign-body implant. *Clinical Orthopaedics,* 280, 289–99.

Mayberry-Carson, K. J., Tober-Meyer, B., Smith, J. K., Lambe, D. W., Jr. and Costerton, J. W. (1984). Bacterial adherence and glycocalyx formation in osteomyelitis experimentally induced with *Staphylococcus aureus. Infection and Immunity,* 43, 825–33.

Melcher, G. A., Metzdorf, A., Schlegel, U., Ziegler, W. J., Perren, S. M. and Printzen, G. (1995). Influence of reaming versus nonreaming in intramedullary nailing on local infection rate: experimental investigation in rabbits. *Journal of Trauma,* 39, 1123–8.

Merritt, K., Brown, S. A., Payer, J. H. and Ryerson, D. H. (1991). Influence of bacteria on corrosion of metallic biomaterials. *Transactions of the Society for Biomaterials,* 14, 106.

Merritt, K., Shafer, J. W. and Brown, S. A. (1979). Implant site infection rates with porous and dense materials. *Journal of Biomedical Materials Research,* 13, 101–8.

Monzon, M., Garcia-Alvarez, F., Lacleriga, A., Gracia, E., Leiva, J., Oteiza, C. and Amorena, B. (2001). A simple infection model using pre-colonized implants to reproduce rat chronic *Staphylococcus aureus* osteomyelitis and study antibiotic treatment. *Journal of Orthopaedic Research*, 19, 820–6.
Nakamoto, D. A., Haaga, J. R., Bove, P., Merritt, K. and Rowland, D. Y. (1995). Use of fibrinolytic agents to coat wire implants to decrease infection. An animal model. *Investigative Radiology*, 30, 341–4.
Nelson, J. P., Glassburn, A. R., Jr., Talbott, R. D. and McElhinney, J. P. (1980). The effect of previous surgery, operating room environment, and preventive antibiotics on postoperative infection following total hip arthroplasty. *Clinical Orthopaedics*, 147, 167–9.
Nijhof, M. W., Dhert, W. J., Fleer, A., Vogely, H. C. and Verbout, A. J. (2000). Prophylaxis of implant-related staphylococcal infections using tobramycin-containing bone cement. *Journal of Biomedical Materials Research*, 52, 754–61.
Nijhof, M. W., Fleer, A., Hardus, K., Vogely, H. C., Schouls, L. M., Verbout, A. J. and Dhert, W. J. (2001). Tobramycin-containing bone cement and systemic cefazolin in a one-stage revision. Treatment of infection in a rabbit model. *Journal of Biomedical Materials Research*, 58, 747–53.
Norden, C. W. (1988). Lessons learned from animal models of osteomyelitis. *Reviews of Infectious Diseases*, 10, 103–10.
Norden, C. W., Myerowitz, R. L. and Keleti, E. (1980). Experimental osteomyelitis due to *Staphylococcus aureus* or *Pseudomonas aeruginosa*: a radiographic-pathological correlative analysis. *British Journal of Experimental Pathology*, 61, 451–60.
Norden, C. W., Niederreiter, K. and Shinners, E. M. (1986). Treatment of experimental chronic osteomyelitis due to *Staphylococcus aureus* with teicoplanin. *Infection*, 14, 136–8.
Pascual, A., Ramirez de Arellano, E., Martinez Martinez, L. and Perea, E. J. (1993). Effect of polyurethane catheters and bacterial biofilms on the in-vitro activity of antimicrobials against *Staphylococcus epidermidis*. *Journal of Hospital Infection*, 24, 211–18.
Petty, W., Spanier, S. and Shuster, J. J. (1988). Prevention of infection after total joint replacement. Experiments with a canine model. *Journal of Bone and Joint Surgery – American volume*, 70, 536–9.
Petty, W., Spanier, S., Shuster, J. J. and Silverthorne, C. (1985). The influence of skeletal implants on incidence of infection. Experiments in a canine model. *Journal of Bone and Joint Surgery – American volume*, 67, 1236–44.
Poelstra, K. A., Barekzi, N. A., Grainger, D. W., Gristina, A. G. and Schuler, T. C. (2000). A novel spinal implant infection model in rabbits. *Spine*, 25, 406–10.
Rae, T. (1983). The action of cobalt, nickel and chromium on phagocytosis and bacterial killing by human polymorphonuclear leucocytes; its relevance to infection after total joint arthroplasty. *Biomaterials*, 4, 175–80.
Rediske, A. M., Roeder, B. L., Brown, M. K., Nelson, J. L., Robison, R. L., Draper, D. O., Schaalje, G. B., Robison, R. A. and Pitt, W. G. (1999). Ultrasonic enhancement of antibiotic action on *Escherichia coli* biofilms: an in vivo model. *Antimicrobial Agents and Chemotherapy*, 43, 1211–14.

Rissing, J. P. (1990). Animal models of osteomyelitis. Knowledge, hypothesis, and speculation. *Orthopaedic Clinics of North America*, 4, 377–90.

Saleh Mghir, A., Cremieux, A. C., Bleton, R., Ismael, F., Manteau, M., Dautrey, S., Massias, L., Garry, L., Sales, N., Maziere, B., et al. (1998). Efficacy of teicoplanin and autoradiographic diffusion pattern of [14C]teicoplanin in experimental *Staphylococcus aureus* infection of joint prostheses. *Antimicrobial Agents and Chemotherapy*, 42, 2830–5.

Sanderson, P. J. (1988). The choice between prophylactic agents for orthopaedic surgery. *Journal of Hospital Infection*, 11 (Suppl. C), 57–67.

Sanderson, P. J. (1991). Infection in orthopaedic implants. *Journal of Hospital Infection*, 18 (Suppl. A), 367–75.

Schulz, S., Steinhart, H. and Mutters, R. (2001). Chronic osteomyelitis in a new rabbit model. *Journal of Investigative Surgery*, 14, 121–31.

Shirtliff, M. E., Mader, J. T. and Calhoun, J. (1999). Oral rifampin plus azithromycin or clarithromycin to treat osteomyelitis in rabbits. *Clinical Orthopaedics*, 359, 229–36.

Simpson, W. A., Courtney, H. S. and Ofek, I. (1987). Interactions of fibronectin with streptococci: the role of fibronectin as a receptor for *Streptococcus pyogenes*. *Reviews of Infectious Diseases*, 9 (Suppl. 4), S351–9.

Smeltzer, M. S., Thomas, J. R., Hickmon, S. G., Skinner, R. A., Nelson, C. L., Griffith, D., Parr, T. R., Jr. and Evans, R. P. (1997). Characterization of a rabbit model of staphylococcal osteomyelitis. *Journal of Orthopaedic Research*, 15, 414–21.

Smith, M. M., Vasseur, P. B. and Saunders, H. M. (1989). Bacterial growth associated with metallic implants in dogs. *Journal of the American Veterinary Association*, 195, 765–7.

Southwood, R. T., Rice, J. L., McDonald, P. J., Hakendorf, P. H. and Rozenbilds, M. A. (1985). Infection in experimental hip arthroplasties. *Journal of Bone and Joint Surgery – British volume*, 67, 229–31.

Spangehl, M. J., Masri, B. A., O'Connell, J. X. and Duncan, C. P. (1999). Prospective analysis of preoperative and intraoperative investigations for the diagnosis of infection at the sites of two hundred and two revision total hip arthroplasties. *Journal of Bone and Joint Surgery – American volume*, 81, 672–83.

Sperling, J. W., Kozak, T. K., Hanssen, A. D. and Cofield, R. H. (2001). Infection after shoulder arthroplasty. *Clinical Orthopaedics*, 385, 206–16.

Stocks, G. and Janssen, H. F. (2000). Infection in patients after implantation of an orthopedic device. *ASAIO Journal*, 46, S41–6.

Sugarman, B. and Young, E. J. (1989). Infections associated with prosthetic devices: magnitude of the problem. *Orthopaedic Clinics of North America*, 3, 187–98.

Tollefson, D. F., Bandyk, D. F., Kaebnick, H. W., Seabrook, G. R. and Towne, J. B. (1987). Surface biofilm disruption. Enhanced recovery of microorganisms from vascular prostheses. *Archives of Surgery*, 122, 38–43.

Ueng, W. N., Shih, C. H. and Hseuh, S. (1995). Pulmonary tuberculosis as a source of infection after total hip arthroplasty. A report of two cases. *International Orthopaedics*, 19, 55–9.

Varma, S., Ferguson, H. L., Breen, H. and Lumb, W. V. (1974). Comparison of seven suture materials in infected wounds – an experimental study. *Journal of Surgical Research*, 17, 165–70.

Vaudaux, P. E., Zulian, G., Huggler, E. and Waldvogel, F. A. (1985). Attachment of *Staphylococcus aureus* to polymethylmethacrylate increases its resistance to phagocytosis in foreign body infection. *Infection and Immunity*, 50, 472–7.

Vergères, P. and Blaser, J. (1992). Amikacin, ceftazidime and flucloxacillin against suspended and adherent *Pseudomonas aeruginosa* and *Staphylococcus epidermidis* in an *in vitro* model of infection. *Journal of Infectious Diseases*, 165, 281–9.

Verheyen, C. C., Dhert, W. J., Petit, P. L., Rozing, P. M. and de Groot, K. (1993). In vitro study on the integrity of a hydroxylapatite coating when challenged with staphylococci. *Journal of Biomedical Materials Research*, 27, 775–81.

Vogely, H. C., Oosterbos, C. J., Puts, E. W., Nijhof, M. W., Nikkels, P. G., Fleer, A., Tonino, A. J., Dhert, W. J. and Verbout, A. J. (2000). Effects of hydrosyapatite coating on Ti-6A1-4V implant-site infection in a rabbit tibial model. *Journal of Orthopaedic Research*, 18, 485–93.

Voos, K., Rosenberg, B., Fagrhi, M. and Seligson, D. (1999). Use of a tobramycin-impregnated polymethylmethacrylate pin sleeve for the prevention of pin-tract infection in goats. *Journal of Orthopaedic Trauma*, 13, 98–101.

Wimmer, C. and Gluch, H. (1996). Management of postoperative wound infection in posterior spinal fusion with instrumentation. *Journal of Spinal Disorders*, 9, 505–8.

Zimmerli, W. (1993). Experimental models in the investigation of device-related infections. *Journal of Antimicrobial Chemotherapy*, 31 (Suppl. D), 97–102.

Zimmerli, W., Waldvogel, F. A., Vaudaux, P. and Nydegger, U. E. (1982). Pathogenesis of foreign body infection: description and characteristics of an animal model. *Journal of Infectious Diseases*, 146, 487–97.

CHAPTER FIVE

Intravascular-Catheter-Related Infections

Hend A. Hanna and Issam Raad

1 INTRODUCTION

The past few decades have witnessed the advent of many medical advancements, which undoubtedly have aided health care professionals and ameliorated the suffering of many critically ill patients. Among these advancements are intravascular catheters, which have become indispensable for the administration of medications, especially chemotherapeutic agents, intravenous antibiotics, blood products, parenteral nutritional solutions, and fluids; for monitoring critically ill patients haemodynamically; and for providing access for haemodialysis (Flowers et al., 1989). The use of such devices is associated with an increased risk for catheter-related infections, such as local catheter site infection, thrombophlebitis, catheter-related bloodstream infections (CR-BSI), endocarditis, metastatic infections such as osteomyelitis, arthritis, endophthalmitis, and distant organ abscesses (Arnow, Quimosing, and Beach, 1993). Our understanding of the process that leads to CR-BSI has evolved over the years, and now the role of bacterial biofilms in this process is certainly undeniable.

Bacterial biofilms have always existed ubiquitously in the world around us. Costerton and Stewart describe some examples: 'the slippery coating on a rock in a stream, dental plaque (which most of us confront daily), and the slime that inevitably materializes inside a flower vase after two or three days' (Costerton and Stewart, 2001). In addressing the medical implications of biofilms as they relate to intravascular-catheter-related infections, we will begin by defining the magnitude of the problem of catheter-related infections, followed by addressing the role of biofilms in the pathogenesis of CR-BSI.

Finally, we will review some strategies to prevent biofilm formation on central venous catheters (CVC).

2 EPIDEMIOLOGY

The majority of intravascular devices used in the United States are peripheral venous catheters. It is estimated that health care facilities in the United States purchase annually more than 150 million intravascular devices, of which more than 5 million CVCs are inserted annually (Maki and Mermel, 1998). It is estimated that each year more than 200,000 nosocomial bloodstream infections occur in the United States. Non-tunneled intravascular devices are responsible for the majority of these infections (Jarvis et al., 1991). Using results generated by a computer model based on American Hospital Association data, Mermel estimated the magnitude of the problem among intensive care units (ICU) patients in the United States to be an average of 5.3 CR-BSI per 1,000 catheter-days with approximately 80,000 CR-BSI occurring annually (Mermel, 2000a,b) and approximately 2,400 to 20,000 deaths due to central line-associated bloodstream infections. The estimated annual cost of managing patients with CR-BSI in the ICU could range from $296 million to $2.3 billion (Mermel, 2000a,b). In addition, many non-ICU inpatients also have CVCs and oftentimes are discharged with the catheters still in place for future use on an outpatient basis. Such patients are also at risk for catheter-related complications and infections. Considering all of the hospital services, not only the ICUs, it is estimated that a total of 200,000 cases of CR-BSI occur annually (Raad, 1998), with an attributable mortality of 10–25 per cent and an economic burden ranging from $4,869 to $19,476 per episode of CR-BSI (Veenstra, Saint, and Sullivan, 1999b).

Estimating the exact infection rates for CR-BSI is a complicated process because these rates could differ according to a number of factors. The risk of CR-BSI could vary according to:

1. Duration of catheter placement: short term (less than 10 days) versus long term (more than 10 days)
2. Site of insertion: CVC inserted through a peripheral vein such as basilic or cephalic versus CVC inserted directly into a central vein
3. Anatomic location of vein: internal jugular vein, subclavian vein, or femoral vein
4. Technique in inserting the catheter: surgically placed, as in tunnelling or inserting a port, versus direct insertion into a vein, which does not require surgery

5. Type of catheter material: polyurethane, polyvinylchloride, Teflon, or silicone
6. Number of catheter lumens: single, double, or triple lumen
7. Other catheter-related factors: length of CVC, presence or absence of impregnation or cuff
8. Conditions under which insertion takes place: emergency versus scheduled insertion, presence or absence of maximal sterile barrier precautions

Infection rates associated with peripheral venous catheters have been reported to be less than 1 per 1,000 catheter-days (Maki and Ringer, 1987). Tunnelled CVCs are used more frequently in cancer patients, where they can remain for years. These catheters are tunnelled into the subcutaneous tissue on their way to a central vein. Subcutaneous tunnelling of short-term CVCs inserted in the jugular vein has been shown to reduce the incidence of catheter-related infections in critically ill patients (Timsit et al., 1999). However, there are no large prospective randomised trials to demonstrate the role of tunnelling in preventing CR-BSIs occurring with long-term CVCs. In a study comparing infection rates of 707 ports and 923 tunnelled CVCs, ports were associated with a twelvefold lower rate of infection (Groeger et al., 1993b). The average infection rate for long-term non-tunnelled CVCs in cancer patients is reported in the literature to range from one to two episodes per 1,000 catheter-days (Howell et al., 1995). Saint, Veenstra, and Lipsky estimated the clinical impact of short-term CVC-related infections by reviewing and synthesising the available relevant literature (Saint, Veenstra, and Lipsky, 2000). After pooling the data statistically, they found the incidence rate of CR-BSI for short-term CVCs to range from 4.9 to 8.2 infections per 1,000 catheter-days.

Although studies may differ in the calculations of the exact numbers describing morbidity, mortality, and the economic burden associated with intravascular device-related infections, the majority of such studies succeed in highlighting the problem of catheter-related infections. A thorough understanding of the mechanisms by which device-related infections occur is an imperative prerequisite to outlining successful strategies for both prevention and management of these infections.

3 BIOFILMS AS THEY RELATE TO CR-BSI

Since intravascular catheters are inserted through the skin and into a blood vessel, they connect a non-sterile surface, namely, the skin, with the blood's sterile milieu. Many factors play a role in the process that eventually leads to catheter colonisation and infection, including the process of biofilm formation. This process is the byproduct of the interaction between the microorganisms, the host, the catheter material, and possibly other factors (Raad and Hanna, 1999).

Catheter colonisation is necessary, but not sufficient alone, in leading to catheter-related infection (Raad et al., 1993). Microbial biofilms start forming when microorganisms adhere to the indwelling catheter surface. Since uncovering the existence of biofilms, scientists have offered many definitions of the term 'biofilm'. In one definition, a biofilm is said to be 'cells immobilized at a substratum and frequently embedded in an organic polymer of microbial origin' (Characklis and Marshall, 1990). Costerton et al. defined a biofilm as 'a matrix-enclosed microbial population adherent to each other and/or surfaces or interfaces' (Costerton et al., 1995). During the last few years, researchers have been probing deeper into the essence of microbial biofilms, trying to gain a better understanding. Among areas of current research are gene expressions of sessile organisms in biofilms, bacterial surface proteins and their role in the initial process of attachment during the formation of a biofilm, and subtle phenotypic differences among sessile bacteria that do not involve changes in genotype. New technologies, such as confocal laser scanning microscopy (CLSM) and fluorescent *in situ* hybridisation (FISH) (Gieseke et al., 2001), offer the ability to examine biofilms in more comprehensive manners than both scanning electron microscopy (SEM) and transmission electron microscopy (TEM). Definitions of biofilms are more likely to continue to change, reflecting our current understanding of the issue.

The process of biofilm formation involves (1) a microbial factor, (2) a host factor, and (3) a material or the CVC factor. Biofilm formation is the byproduct of the interaction of these factors with each other (Figure 5.1).

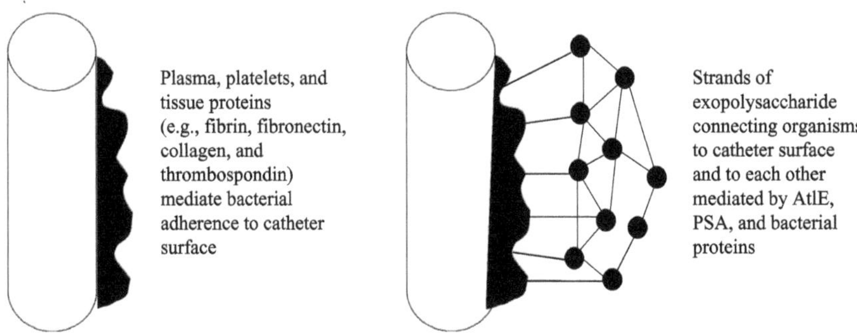

Figure 5.1: Biofilm formation.

When a CVC is inserted, microorganisms may migrate from the colonised skin at the catheter insertion site onto the catheter surface. Bacteria have the ability to attach themselves to the catheter surface and synthesise or produce a hydrated extracellular polymeric exopolysaccharide to create a complex matrix in which they aggregate, forming a biofilm (Costerton, Stewart, and Greenberg, 1999). These exopolysaccharides are made up of either thin strands connecting the microorganisms to the catheter surface and to one another or amorphous material shaped as a sheet on the catheter surface (Characklis, 1990). In earlier studies, proteinacious autolysin (AtlE) was found to be associated with initial attachment of *Staphylococcus epidermidis* to plastic (Heilmann et al., 1996a, 1997). Polysaccharide adhesin (PSA), also known as polysaccharide intercellular adhesin (PIA), was shown to mediate cell-to-cell adherence and biofilm accumulation and also haemaglutination and was found to be essential in the pathogenesis of foreign body infections in *in vivo* models (Fey et al., 1999; Mack et al., 1994, 1996; Mack, Siemssen, and Laufs, 1992; Rupp et al., 1999a,b).

Rupp et al. used a rat CVC infection model to study the pathogenesis of CVC-associated infection due to *S. epidermidis*, as it relates to AtlE and to PSA (Rupp et al., 2001). In the rat model, CVCs were inserted through the rats' jugular veins and into the superior vena cava, then the rats were divided into three groups and a different strain of *S. epidermidis* was injected into each group's CVCs. The first group was injected with a clinical strain of *S. epidermidis* that was AtlE and PSA positive. The second group was injected with a strain of *S. epidermidis* that was deficient in AtlE production. The last group was injected with a *S. epidermidis* strain that was deficient in PSA. Rupp and colleagues observed that the strain of *S. epidermidis* that was AtlE and PSA positive was more likely to cause a CVC-associated infection with bacteraemia than were the strains deficient in either AtlE or PSA. This study demonstrated that adhesion and aggregation are essentially equally important in the pathogenesis of CVC-associated infection in the rat model (Rupp et al., 2001).

Bacterial adherence to the catheter surface, according to Gristina (1987), can be divided into three phases:

1. Bacterial attachment to the catheter surface
2. Bacterial adhesion to the surface
3. Bacterial aggregation

Bacterial attachment to the catheter surface is believed to be due to factors such as catheter surface charge and hydrophobicity (Rupp et al., 2001).

However, adhesion of bacteria to the surface of the catheter depends on more specific factors such as bacterial proteins (e.g., the 60-kDa AtlE protein of *S. epidermidis*) that are essential for the adhesion process (Heilmann et al., 1996a, 1997). In addition, PSA was also found to be important in the process of adhesion to plastics as well as in the aggregation phase (Timmerman et al., 1991; Veenstra et al., 1996). Bacterial aggregation involves cell-to-cell adherence and is facilitated by PSA (Mack et al., 1992). Bacterial strains that are deficient in PSA will adhere, but will not be able to aggregate and form microcolonies (Mack et al., 1994).

The host also contributes to the formation of the biofilm by coating the surface of the CVC with platelets, plasma, and tissue proteins such as fibrin, fibronectin, collagen, thrombospondin, and laminin, which act as mediators of bacterial adherence to the surface of the CVC (Hawiger et al., 1982; Herrmann et al., 1991; Kuusela, 1978; Lopes, Dos Reis, and Brentani, 1985; Raad, 1998; Vaudaux et al., 1989). Different organisms adhere differently to these proteins. For example, coagulase-negative staphylococci such as *S. epidermidis* adhere only to fibrin, whereas *S. aureus* adheres strongly to fibronectin, fibrin and laminin (Hawiger et al., 1982; Kuusela, 1978). *Candida albicans* has the ability to bind to fibrin (Bouali et al., 1987). *S. aureus* has also been found to adhere to immobilised platelets in the presence of plasma, which further promotes its adherence to the catheter surface (Herrmann et al., 1993). Adherence of both streptococci and staphylococci to a fibrin matrix is greatly enhanced in the presence of platelets in the matrix, even in the presence of a small number of platelets (Herrmann et al., 1991).

This complex structure of the biofilm provides a protected mode of growth that allows microorganisms to survive in a hostile environment. Bacteria inhabiting the biofilm itself are called sessile bacteria (Costerton, Geesey, and Cheng, 1978). The structure of the biofilm contains channels through which nutrients can flow to the sessile bacteria within the biofilm to ensure the maintenance of their growth (DeBeer, Stoodley, and Lewandowski, 1994). Biofilm sessile communities of bacteria can give rise to non-sessile bacteria, called planktonic bacteria, or free-floating bacteria, which can rapidly multiply and spread from the biofilm matrix into the bloodstream. The process of biofilm formation and release of microorganisms is a slow one and could be described as involving three components (Figure 5.2):

1. Bacterial attachment, adhesion, and aggregation (as referred to earlier)
2. Bacterial growth on the surface
3. Detachment of bacteria from the surface

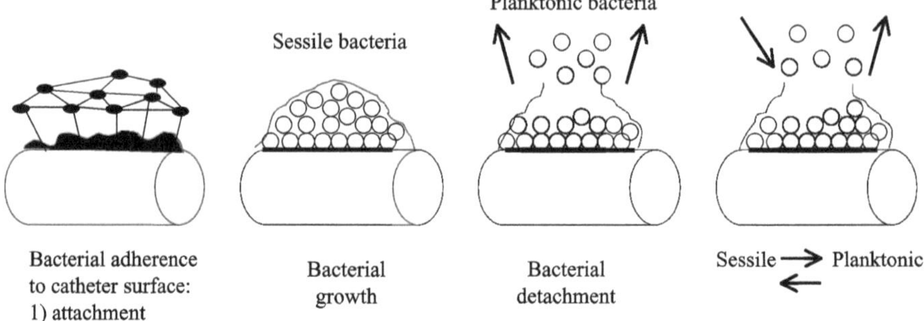

Figure 5.2: Stages of biofilm formation.

Lappin-Scott and Bass (2001) developed a method to monitor these steps in the formation of biofilms in real time by the use of digital time-lapse microscopy imaging. They noted that bacteria, once on the surface, form an anchor to hold them in place. After attachment, some bacteria grow on the surface and become sessile, some detach from the surface and become planktonic, whereas others join the surface from the planktonic phase to become sessile. They also observed that during the initial stages of bacterial colonisation of the surface, the rate of bacterial attachment to the surface was greater than the rate of bacterial growth or detachment from the surface. Their work also demonstrated that slow-growing anaerobic bacteria colonise the surface at a much slower rate than fast-growing aerobic bacteria (Lappin-Scott and Bass, 2001).

While in the matrix of the biofilm, sessile bacterial cells release antigens, which stimulate the release of antibodies from the host. However, these antibodies, as well as phagocytes and antibiotics, are not effective in killing sessile bacteria within the substance of the biofilm (Cochrane et al., 1988). Therefore, in treating a biofilm-associated infection with antibiotics, symptoms caused by planktonic bacteria released from the biofilm will often disappear, because planktonic bacteria can be killed by the antibiotic therapy as well as by the host's own immune system through antibodies and phagocytes. However, all these measures fail to eradicate the biofilm and its sessile bacteria.

4 MICROORGANISMS IN THE BIOFILM

Skin commensals are the most common organisms responsible for causing CR-BSI. These organisms arise from the patient's own skin flora, the contaminated hands of health care workers, or less often the contaminated infusate. Since the use of intravascular catheters and other indwelling devices has increased over the years, nosocomial bloodstream infections caused by skin commensals, especially *S. epidermidis*, have also increased. Bacteria commonly isolated from indwelling CVCs include Gram-positive, coagulase-negative staphylococci (such as *S. epidermidis*); *S. aureus*; or *Enterococcus faecalis*. During the 1990s, coagulase-negative staphylococci accounted for 37.3 per cent of nosocomial bloodstream infections, according to the National Nosocomial Infection Surveillance system report by the Centers for Disease Control and Prevention (1999). Gram-negative microorganisms are the second most common group of organisms causing CR-BSI, especially in ICU and haemodialysis units and among neutropenic patients (Kolls and Brown, 1993). These include organisms such as *Pseudomonas* spp., *Acinetobacter* spp., *Xanthomonas maltophilia, Escherichia coli*, and *Proteus mirabilis. Candida* spp., such as *C. albicans* and *C. parapsilosis*, and various other fungi, such as *Malassezia furfur* and *Fusarium oxysporum* as well as rapidly growing mycobacteria, have also been associated with CVC infections (Maki and Mermel, 1998; Raad, 1998). Biofilms may contain a single microbial species or several different species (Donlan, 2001). During the past decade, among the pathogens responsible for causing CR-BSI, there has been an increase in the frequency of organisms resistant to the most commonly used antibiotics in hospitals, such as methicillin-resistant, coagulase-negative staphylococci; vancomycin-resistant enterococci; methicillin-resistant *S. aureus; Candida;* and resistant Gram-negative bacilli such as *Enterobacter, Serratia*, and *Klebsiella* (Farr, 1995).

5 BIOFILMS AS A PREDICTOR OF SOURCE OF INFECTION

The skin surrounding the catheter insertion site has been implicated as the most common source of CVC colonisation. Bacteria colonising the skin surface migrate from the catheter insertion site along the external catheter surface into the subcutaneous segment of the catheter, forming a biofilm and ultimately causing catheter-related infections (Maki, 1989). This external catheter colonisation, manifested as a microbial biofilm, was demonstrated by conventional light microscopy (Cooper and Hopkins, 1985) and is the basis upon which the semiquantitative roll-plate culturing technique was devised (Maki, Weise, and Sarafin, 1977). According to the roll-plate technique, the catheter segment is rolled over the surface of an agar plate back and forth. Then the isolated growing microorganisms are quantified. This technique allows for the retrieval of bacteria colonising only the outer surface of the catheter through the extraluminal route.

On the other hand, Sitges-Serra and colleagues emphasised the catheter hub rather than the skin surface as the most common source of catheter-related infections (Linares et al., 1985; Stewart, 1996; Stiges-Serra et al., 1985). A study by Raad et al. (1993) reconciled these differences by showing that external catheter surface colonisation, probably originating from the contaminated skin surface, is the predominant source of contamination for short-term catheters placed for up to 10 days. However, luminal colonisation and biofilm formation originate from a contaminated hub, which is the predominant source of contamination for long-term catheters placed for more than 30 days (Raad et al., 1993). In this study, Raad and colleagues showed through SEM and TEM that almost all indwelling CVCs had visible adherent microorganisms on their surfaces and concluded that colonisation of indwelling catheters is universal for all indwelling CVCs regardless of culture results (Raad et al., 1993).

It has been estimated that 50 per cent of catheter-related infections originate from the skin, 40 per cent from the contaminated hub, and 10 per cent from other pathways including haematogenous seeding through bacteraemia arising from a distant focus or from contaminated fluids or parenteral nutritional solutions (Capdevila, 1998).

6 BIOFILMS AS A RESISTANCE FACTOR IN CR-BSI

A biofilm is formed when bacteria attach themselves to the surface of the CVC and aggregate inside a protective exopolysaccharide matrix. Infected CVCs may respond to antibiotic therapy without the removal of the catheter, especially in cases of uncomplicated infections caused by coagulase-negative staphylococci. However, in many cases, where the CVC is not removed, catheter-related infections may respond initially to conventional antibiotic treatment, as evident by the temporary resolution of the symptoms and signs of the infection, only to recur and necessitate the removal of the infected intravascular device. This clinical observation is explained through the aid of TEM, which revealed in one study that microbial cells in the inner biofilm layers tend to remain intact even after antibiotic treatment (Amorena et al., 1999). In this *in vitro* study, eleven antibiotics in three concentrations (4 × MBC, 100 and 500 mg/L) were used to conduct antibiotic susceptibility assays on *Staphylococcus aureus* isolates. Two biofilms of the slime-producing *S. aureus*, aged 6 and 48 hours, were tested for susceptibility. A non-slime producing type of *S. aureus* was also tested for susceptibility. The study showed that various antibiotics had a greater effect on the viability of microorganisms when used at concentrations much higher than the minimal bactericidal concentration (MBC) and ≥100 mg/L. Antibiotics were also more effective when used on younger biofilms (6 hours). The density of the microbial cells was enhanced as the biofilm aged (Amorena et al., 1999). Although bacteria in the biofilm required higher concentrations of antibiotics than bacteria that did not produce biofilms, the susceptibilities of both the dispersed slime-producing *S. aureus* and the non-slime-producing *S. aureus* were similar as judged by their minimal inhibitory concentration (MIC) and MBC. In another *in vitro* study, the penetration of antibiotics into biofilms of *S. epidermidis* was evaluated. The study showed that extremely high vancomycin concentrations of 1,184 μg/mL, which was 74 times the MBC for *S. epidermidis*, did not eradicate bacteria embedded in biofilms, although the high concentrations of vancomycin in the biofilm exceeded those in the surrounding medium (Darouiche et al., 1994). The study also showed that sessile *S. epidermidis* organisms tended to remain viable within the biofilm despite treatment with antibiotics at much higher concentrations than those needed to kill corresponding planktonic bacteria (Darouiche et al., 1994). Consistent with the data published by Amorena et al. (1999), Darouiche et al. had earlier found that dispersed bacteria originating from biofilms were as sensitive to a particular antibiotic as were their planktonic counterparts. Hence, any bacterial resistance to antibiotics is usually lost when organisms are dislodged from the biofilm (Darouiche et al., 1994).

A microorganism is said to be resistant to an antimicrobial agent if it can survive and grow in the presence of a level of the antimicrobial agent that is higher than its MIC and MBC. Because *in vitro* testing shows that microbial cells embedded in the biofilm require higher concentrations of antibiotics, one may assume that these sessile microorganisms are more resistant to antibiotics than planktonic free-floating bacteria only when sheltered inside the biofilm matrix. To understand this paradox of relative resistance occurring only in the setting of the biofilm, it is important to know how bacterial cells survive inside a biofilm and what are the elements that may influence the biofilm resistance. This area is not fully understood, and ongoing research is trying to solve its mystery (see also Chapter 3). Biofilm resistance could be attributable to many factors (Table 5.1), each of which may act alone or in concert with each other and could be summarised as:

1. Decreased and delayed infiltration or diffusion of antimicrobial agents into the biofilm
2. Decreased bacterial growth in the biofilm
3. Possible expression of resistance genes by the microorganisms in the biofilm

Table 5.1: Mechanisms of biofilm resistance

- Biofilm as a barrier:
  a. Decreases diffusion of antimicrobial proteins, lysozymes, and defensins
  b. Decreases diffusion of antibiotics
  c. Shields bacteria from phagocytes
- Biofilm concentrates bacterial enzymes that degrade antibiotics, e.g., β-lactamases
- Altered growth of bacteria in the biofilm (due to limited nutrients and oxygen and accumulation of metabolic wastes)
- Expression of resistance genes by bacteria in the biofilm

6.1 Decreased Infiltration of Antimicrobial Agents into Biofilms

The exopolysaccharide matrix or glycocalyx of the biofilm may act to delay and restrict the diffusion of antibiotics through the full thickness of the biofilm (Ishida et al., 1998). *Pseudomonas aeruginosa* is used extensively in *in vitro* models to study the resistance of biofilms. In one study, it was shown through the use of infrared spectroscopy that ciprofloxacin was transported to a surface colonised with *P. aeruginosa* at a much slower rate than to a sterile surface (Suci et al., 1994). The study suggested that the slow transportation is due to the binding of ciprofloxacin to the biofilm components. The biofilm acts as a barrier against the diffusion of large molecules, for example, antimicrobial proteins such as lysozymes (Lewis, 2001). Biofilms also act as a barrier against the diffusion of the smaller defensins, their analogues, and other antimicrobial peptides. The exopolysaccharides in the biofilm are negatively charged, a fact that renders them extremely effective in interfering with the diffusion of the positively charged aminoglycosides antibiotics (Lewis, 2001). In addition to the retardation of diffusion of the antibiotic into the biofilm, sessile *P. aeruginosa* in the biofilm produce enzymes, such as β-lactamase, to degrade diffused antibiotics (Giwercman et al., 1991). The retarded diffusion of antibiotics, aided by the destruction of the diffused antibiotics by an enzyme such as β-lactamase, attributes to the observed resistance of the biofilm. This synergy between delayed antibiotic diffusion into the biofilm and antibiotic degradation by bacterial enzymes is the basis of a theoretical mathematical model that was used to study antibiotic penetration of the biofilm through the processes of diffusion, sorption, and reaction. Applying this model, Stewart concluded that the extent of retardation of antibiotic diffusion due to sorption alone does not seem to be sufficient in explaining the reduced biofilm susceptibility to antimicrobial agents (Stewart, 1996).

Anderl, Franklin, and Stewart (2000) evaluated the penetration of two antibiotics, ampicillin and ciprofloxacin, through biofilms of wild-type *K. pneumoniae* as measured by assaying the concentration of the antibiotic that diffused through the biofilm to an overlying filter disk. Simultaneously, the same experiments were done on a mutant strain of *K. pneumoniae* in which β-lactamase activity was eliminated. Ampicillin was unable to penetrate the biofilm of the wild-type *K. pneumoniae*, whereas ciprofloxacin and a non-reactive tracer of chloride ions were able to penetrate the biofilm quickly. However, the biofilm was still resistant to ciprofloxacin. On the other hand, ampicillin was able to penetrate the biofilm of the mutant strain of *K. pneumoniae*, and similarly, the biofilm was still resistant to the ampicillin in spite of the absence of the ampicillin-degrading enzyme, β-lactamase. In this study, poor penetration contributed to the resistance of the wild-type *K. pneumoniae* biofilm to ampicillin, a fact that does not explain its resistance to ciprofloxacin. The study concluded that the resistance exhibited by the wild-type strain to ciprofloxacin and the mutant strain to ampicillin and ciprofloxacin could not be explained only by poor diffusion since these antibiotics diffused through the biofilm (Anderl et al., 2000). So, it is evident that the exopolysaccharide matrix could be penetrated by some antimicrobial agents and that other mechanisms must be involved to further explain the biofilm resistance.

6.2 Decreased Bacterial Growth in Biofilms

Antimicrobial agents kill rapidly growing bacterial cells more effectively than slower growing ones (Costerton et al., 1999; Lewis, 2001). Some antibiotics, in order to be active, require certain rates of bacterial growth, for example, penicillin and ampicillin fail to kill non-growing bacterial cells. Other antibiotics, such as the newer β-lactams, aminoglycosides, and fluoroquinolones, can kill non-growing bacterial cells, but they are more effective at killing fast growing cells (Lewis, 2001). Bacterial cells inside the biofilm experience limited diffusion of nutrients and oxygen and also accumulation of metabolic waste. These conditions favour anaerobic metabolism which produces acidic compounds (König, Schwank, and Blaser, 2001). Acidic pH and hypoxia act in the deeper regions of the biofilm to further reduce the activity of certain antibiotics, such as aminoglycosides and quinolones (König, Simmen, and Blaser, 1993).

Wentland et al. (1996) used acridine orange to visualise and quantify spatial variations in growth rate within *K. pneumoniae* colonies and biofilms. Confocal microscopy and image analysis of stained cross sections of bacterial colonies were used to study the spatial patterns of fluorescent colour and intensity with depth in the colony. The profiles obtained in this study showed that orange staining, which corresponds to rapid growth, occurred preferentially near the air and close to the surface of the colony and the biofilm. Green staining, which corresponds to slow growth, was predominantly in the interior area of the colony and the biofilm interior (Wentland et al., 1996).

6.3 Expression of Resistance Genes in the Biofilm and Other Factors

When bacteria adhere to the surface of a CVC and start forming biofilms, they produce proteins that are not found in free-floating bacterial cells. The ability of *S. epidermidis* to produce biofilms on a catheter surface depends on the production of the PSA. This adhesin controls the interaction and contact of bacterial cells with each other, which result in the formation of multilayered biofilms (Mack et al., 1996). The enzymes that are involved in the synthesis of PSA are encoded by the *ica* operon, which contains the genes *icaA, icaD, icaB,* and *icaC* (Heilmann et al., 1996b). Rachid et al. showed that *S. epidermidis* biofilm formation and the enhancement of *ica* expression are induced by external stress, such as high temperature and osmolarity (Rachid et al., 2000). They also showed that when *S. epidermidis* was exposed during growth and development of biofilms to subinhibitory concentrations ($1/_{70}$ to $^1/_2$ of the MIC) of tetracycline and the semisynthetic streptogramin antibiotic quinopristin-dalfopristin, *ica* expression was enhanced nine to elevenfold, and, consequently, biofilm formation was enhanced strongly (Rachid et al., 2000). On the other hand, *ica* expression was not induced and bacterial growth was inhibited when the antibiotic concentrations were increased to more than half of the MIC. Researchers studied biofilms of *Escherichia coli* and found that the multiple antibiotic resistance operon *mar* and the multidrug efflux pump *acrAB* play a role in biofilm resistance at low antibiotic concentrations (Maira-Litrán, Allison, and Gilbert, 2000). However, they concluded that biofilm resistance is likely to be related, in addition, to other aspects of the biofilm phenotype. Whiteley and colleagues studied *P. aeruginosa* biofilms and were able to identify seventy-three genes that were differentially expressed in the biofilms (Whiteley et al., 2001). They showed that exposure of biofilms to high levels of tobramycin caused differential expression of twenty genes.

Brooun and co-workers studied the dose-response elimination of *P. aeruginosa* by the quinolones, ofloxacin and ciprofloxacin (Brooun, Liu, and Lewis, 2000). They showed that the bulk of bacterial cells (~99.9 per cent) were eliminated by clinically achievable concentrations of the antibiotics

(1–5 μg/mL) similar to what occurred among planktonic bacterial cells. However, when the levels of antibiotics were increased, it had no further effect on the elimination of the surviving bacterial cells. The study showed the presence of a small super-resistant cell fraction that is responsible for the very high resistance of *P. aeruginosa* to quinolones. Therefore, in treating a CR-BSI, if the concentration of the antibiotic temporarily drops or if therapy is stopped altogether due to the resolution of signs and symptoms caused by the eradication of the planktonic cells, the persister bacterial cells will reform the biofilm and new planktonic bacterial cells will be shed (Lewis, 2001).

7 PREVENTION OF BIOFILM FORMATION

Once a biofilm forms on the surface of the intravascular device, due to the resistance factors discussed previously, it is not a simple task to eradicate it. Currently, preventing the formation of biofilms remains the best defensive mechanism against CR-BSI. Effective preventive measures should aim at decreasing the likelihood of skin colonisation around the catheter exit site and protecting the catheter surface against being colonised both internally and externally. This Section describes approaches to some preventive measures. It is important to realise that some of these measures do not necessarily decrease biofilm formation, per se, but decrease catheter colonisation.

7.1 Aseptic CVC Insertion

Since skin is a major source of intravascular contamination (Centers for Disease Control and Prevention, 1999), assuring sterile conditions during CVC insertion and dressing changes is vital to avoiding CVC colonisation that often leads to biofilm formation and CR-BSI. Catheter insertion and maintenance by a trained and skilled infusion therapy team were shown to contribute to lower rates of catheter-related infections (Faubion et al., 1986; Nelson et al., 1986). Also, inserting catheters under maximal sterile barrier conditions proved to lower CR-BSI more than sixfold (Raad et al., 1994). The maximal sterile barrier measures used were sterile gloves, gowns, masks, caps, and the use of large sterile drapes.

7.2 Topical Disinfecting and Antimicrobial Ointments

For a biofilm to form on the catheter surface, bacteria colonising the skin around the catheter exit site must migrate onto the catheter, either through the external surface or the internal lumen, to eventually colonise it and start the process of biofilm formation. Protecting the skin at the CVC exit site from being colonised may reduce CR-BSI. Topical polyantibiotic ointment applied at the catheter exit site, during CVC insertion and once every 48 hours thereafter, was shown to decrease catheter tip colonisation when compared to no ointment and to iodophor ointment applied every 48 hours (Maki and Band, 1981). In another prospective randomised study, 2 per cent chlorhexidine gluconate applied at the catheter exit site lowered CR-BSI fourfold when compared to 10 per cent povidone iodine and 70 per cent alcohol (Maki, Ringer, and Alvarado, 1991).

7.3 Anticoagulant and Antimicrobial Flush Solutions

This technique is particularly useful for long-term catheters where hub contamination is the main source of catheter colonisation and, ultimately, CR-BSI (Salzman et al., 1993; Stiges-Serra et al., 1985). According to this method, the catheter lumen is flushed and then filled with an anticoagulant in combination with an antimicrobial agent. Heparin as an anticoagulant, in combination with vancomycin, was used to flush the CVC lumen. Heparin interferes with the thrombin sheath formation on the catheter surface and was shown to reduce catheter tip colonisation when combined with vancomycin (Bailey, 1979; Schwartz et al., 1990). However, to avoid the increase in emergence of vancomycin-resistant organisms, using vancomycin prophylactically is discouraged. Raad et al. used a lock solution of 3 mg minocycline (M) and 30 mg ethylenediaminetetraacetate (EDTA) in water as prophylaxis in three patients (Raad et al., 1997). EDTA has an anticoagulant activity and is able to interfere with biofilm formation (Hanna, Darouiche, and Raad, 2001). The combination of M-EDTA showed synergistic activity against methicillin-resistant staphylococci, Gram-negative bacilli, and *C. albicans* and a cidal effect on bacteria in the biofilm (Raad et al., 1997). Contrary to vancomycin, neither M nor EDTA is used therapeutically in the treatment of bloodstream infections, and hence, the emergence of resistant organisms because of the use of this combination is unlikely. Solutions similar to M-EDTA may be useful in keeping the CVC patent and in interfering with biofilm formation. Recently, the Infectious Diseases Society of America published guidelines highlighting the role of antibiotic flush solutions as adjunctive treatment of CR-BSI without removal of the catheter (Mermel et al., 2001). It is important to determine the impact of flush solutions on bacteria embedded in biofilms. Salicylic acid inhibits biofilm exopolymer production. Theoretically, salicylic acid could be used as a component of flush solutions. However, the concentration required to inhibit polymer production far exceeds the normal human dosing (Teichberg et al., 1993).

7.4 Ionotophoretic Catheters and Electric Current

A collagen cuff impregnated with silver and connected to the CVC is placed subcutaneously to act as a mechanical barrier against migrating organisms and also to exert antimicrobial activity due to the presence of the silver. The device failed to reduce the rates of CR-BSI when used with long-term catheters (Clementi et al., 1991; Groeger et al., 1993a), possibly due to the biodegradable nature of the collagen since it degrades within a week. Another novel silver ionotophoretic catheter works when low-voltage current runs through silver wires to release the silver ions. This device remains efficacious for up to 5 months and has inhibitory activities against Gram-positive and Gram-negative bacteria as well as *Candida* spp. (Raad et al., 1996a,b). Novel technologies should be validated for clinical safety before they are incorporated into routine clinical practice. Several investigators have shown that antibiotics alone cannot eradicate biofilms unless they are used simultaneously with a low DC electric current (15–64 $\mu A\ cm^{-2}$) (Benson et al., 1994; Costerton et al., 1994; Khoury et al., 1992). This phenomenon has been referred to as the bioelectric effect. The bioelectric effect could have contributed to the efficacy of silver ionotophoretic catheters.

7.5 Coating or Impregnation of Catheters with Antimicrobial Agents

This technology uses both antiseptics and antibiotics for the impregnation of CVCs. Catheters impregnated externally with chlorhexidine-silver sulphadiazine were found in a controlled, randomised, double-blind prospective study to be more efficacious than non-impregnated catheters in reducing both catheter colonisation and CR-BSI (Maki et al., 1997). Veenstra et al. (1999a) used meta-analysis to evaluate the efficacy of these catheters and came to the same conclusion. Mermel (2000a) used the Mantel-Haenzel method to estimate a summary measure of the effect of this device on CR-BSI. His review showed that this catheter reduces the risk for CR-BSI in short-term catheterisation, but it is not effective for catheters placed for an average of 3 weeks, probably due to the reduction of its antimicrobial activity over time. Mermel also attributed the lack of efficacy with long-term use to the fact that only the outer surface of the catheter was impregnated with chlorhexidine-silver sulphadiazine, hence not providing protection against luminal colonisation. Catheters impregnated on both surfaces are now available. There have been reports, mostly in Japan, of anaphylactic reactions associated with the chlorhexidine gluconate in these catheters.

Commercially available catheters impregnated intraluminally and extraluminally with M and rifampin were more efficacious when compared, in a prospective, randomised, double-blind controlled study, with catheters impregnated with chlorhexidine-silver sulphadiazine (Darouiche et al., 1999). Catheters impregnated with M and rifampin were twelvefold less likely to be associated with CR-BSI (relative risk 0.1; 95 per cent CI 0.0–0.6) and threefold less likely to be colonised. Using quantitative SEM, Raad et al. studied biofilm formation on CVCs coated with M and rifampin and on uncoated catheters (Raad et al., 1998). They found that coated catheters had a decreased risk of microbial colonisation. An *in vitro* study suggested that the use of these catheters might lead to resistance developing against M and rifampin. Clinical studies, however, did not demonstrate resistance to M and rifampin (Darouiche et al., 1999). Also, the use of these catheters may lead to reduction in the use of systemic antibiotics and, therefore, may actually reduce the emergence of resistant organisms. In a review, Wenzel and Edmond estimated that 4,745 to 9,450 lives would be saved with the use of catheters impregnated with M and rifampin according to varying attributable mortality rate estimates (Wenzel and Edmond, 2001). Catheters impregnated with either antiseptic agents or antibiotic agents should be used in patients at high risk for developing CR-BSI.

8 WHAT DOES THE FUTURE HOLD?

Scientists and researchers have unlocked many of the mysteries surrounding biofilm formation and its role in CR-BSI. We understand most of its complexity as a microbial society with inhabitants armed by multiple defence mechanisms. However, this field of studying and understanding biofilms and their impact on device-related infections can certainly accommodate more effort. Areas that provide future challenges may include reliable techniques for collecting and measuring biofilms, the role of biofilms in developing resistance to antimicrobial agents, identifying genes responsible for bacterial persistence, and developing newer drugs to be used alone or in combination with conventional antibiotics to target these genes.

REFERENCES

Amorena, B., Gracia, E., Monzón, M., Leiva, J., Oteiza, C., Pérez, M., Alabart, J. L. and Hernández-Yago, J. (1999). Antibiotic susceptibility assay for *Staphylococcus aureus* in biofilms developed *in vitro*. *Journal of Antimicrobial Chemotherapy*, 44, 43–55.

Anderl, J. N., Franklin M. J. and Stewart, P. S. (2000). Role of antibiotic penetration limitation in *Klebsiella pneumoniae* biofilm resistance to ampicillin and ciprofloxacin. *Antimicrobial Agents and Chemotherapy,* 44, 1818–1824.

Arnow, P. M., Quimosing, E. M. and Beach, M. (1993). Consequences of intravascular catheter sepsis. *Clinical Infectious Diseases,* 16, 778–784.

Bailey, M. J. (1979). Reduction of catheter-associated sepsis in parenteral nutrition using low-dose intravenous heparin. *British Medical Journal,* 1, 1671–1673.

Benson, D. E., Grissom, C. B., Burns, G. L. and Hohammad, S. F. (1994). Magnetic field enhancement of antibiotic activity in biofilm forming *Pseudomonas aeruginosa. ASAIO Journal,* 40, M371–M376.

Bouali, A., Robert, R., Tronchin, G. and Senet, J. M. (1987). Characterization of binding of human fibrinogen to the surface of germ-tubes and mycelium of *Candida albicans. Journal of General Microbiology,* 133, 454–551.

Brooun, A., Liu, S. and Lewis, K. (2000). A dose-response study of antibiotic resistance in *Pseudomonas aeruginosa* biofilm. *Antimicrobial Agents and Chemotherapy,* 44, 640–646.

Capdevila, J. A. (1998). Catheter-related infection: an update on diagnosis, treatment and prevention. *International Journal of Infectious Diseases,* 2, 230–236.

Centers for Disease Control and Prevention. (1999). National Nosocomial Infection Surveillance (NNIS) system report: data summary from January 1990–May 1999. *American Journal of Infection Control,* 27, 520–532.

Characklis, W. G. (1990). Biofilm processes. In *Biofilms,* eds. W. G. Characklis and K. C. Marshall, pp. 195–233. New York: Wiley.

Characklis, W. G. and Marshall, K. (1990). Biofilms: a basis for an interdisciplinary approach. In *Biofilms,* eds. W. G. Characklis and K. C. Marshall, pp. 3–17. New York: Wiley.

Clementi, E., Mario, O., Arlet, G., Villers, S., Boudaoud, S. and Falkman, H. (1991). Usefulness of an attachable silver-impregnated cuff for the prevention of catheter-related sepsis (CRS)? In Programs and Abstracts of the 31st Interscience Conference on Antimicrobial Agents and Chemotherapy, September 29–October 2, 1991. Chicago, IL. Washington, DC: American Society for Microbiology, Abstract #460.

Cochrane, D. M. G., Brown, M. R. W., Anwar, H., Weller, P. H., Lam, K. and Costerton, J. W. (1988). Antibody response to *Pseudomonas aeruginosa* surface protein antigens in a rat model of chronic lung infection. *Journal of Medical Microbiology,* 27, 255–261.

Cooper, G. L. and Hopkins, C. C. (1985). Rapid diagnosis of intravascular catheter-associated infection by direct Gram staining of catheter segments. *New England Journal of Medicine,* 312, 1142–1147.

Costerton, J. W., Ellis, B., Lam, K., Johnson, F. and Khoury, A. E. (1994). Mechanism of electrical enhancement of efficacy of antibiotics in killing biofilm bacteria. *Antimicrobial Agents and Chemotherapy,* 38, 2803–2809.

Costerton, J. W., Geesey, G. G. and Cheng, G. K. (1978). How bacteria stick. *Scientific American,* 238, 86–95.

Costerton, J. W., Lewandowski, Z., Caldwell, D. E., Korber, D. R. and Lappin-Scott, H. M. (1995). Microbial biofilms. *Annual Review of Microbiology,* 49, 711–745.

Costerton, J. W. and Stewart, P. S. (2001). Battling biofilms. *Scientific American*, 285, 74–81.

Costerton, J. W., Stewart, P. S. and Greenberg, E. P. (1999). Bacterial biofilms: a common cause of persistent infections. *Science*, 284, 1318–1322.

Darouiche, R. O., Dahir, A., Miller, A. J., Landon, G. C., Raad, I. I. and Musher, D. M. (1994). Vancomycin penetration into biofilm covering infected prostheses and effect on bacteria. *Journal of Infectious Diseases*, 170, 720–723.

Darouiche, R. O., Raad, I. I., Heard, S. O., Thornby, J. I., Wenker, O. C., Gabrielli, A., Berg, J., Kahrdori, N., Hanna, H., Hachem, R., Harris, R. L. and Mayhall, G. A. (1999). Comparison of two anti-microbial impregnated central venous catheters. *New England Journal of Medicine*, 340, 1–8.

DeBeer, D., Stoodley, P. and Lewandowski, Z. (1994). Liquid flow in heterogenous biofilms. *Biotechical Bioengineering*, 44, 636–641.

Donlan, R. (2001). Biofilms and device-associated infections. *Emerging Infectious Diseases*, 7, 277–281.

Farr, B. M. (1995). Vascular catheter related infections in cancer patients. *Surgical Oncology Clinics of North America*, 4, 493–503.

Faubion, W. C., Wesley, J. R., Khalidi, N. and Silva, J. (1986). Total parenteral nutrition catheter sepsis: impact of the team approach. *Journal of Parenteral Enteral Nutrition* , 10, 642–645.

Fey, P. D., Ulphani, J. S., Götz, F., Heilmann, C., Mack, D. and Rupp, M. E. (1999). Characterization of the relationship between polysaccharide intercellular adhesion and hemagglutination in *Staphylococcus epidermidis*. *Journal of Infectious Diseases*, 179, 1561–1564.

Flowers, R. H., Schwenzer, K. J., Kopel, R. F., Fisch, M. J., Tucker, S. I. and Farr, B. M. (1989). Efficacy of an attachable subcutaneous cuff for the prevention of intravascular catheter-related infection. *Journal of the American Medical Association*, 261, 878–883.

Gieseke, A., Purkhold, U., Amann, R. and Schamm, A. (2001). Community structure and activity dynamics of nitrifying bacteria in a phosphate-removing biofilm. *Applied and Environmental Microbiology*, 67(3), 1351–1362.

Giwercman, B., Jensen, E. T., Hoiby, N., Kharazmi, A. and Costerton, J. W. (1991). Induction of beta-lactamase production in *Pseudomonas aeruginosa* biofilm. *Antimicrobial Agents and Chemotherapy*, 35, 1008–1010.

Gristina, A. G. (1987). Biomaterial-centered infection: microbial adhesion versus tissue integration. *Science*, 237, 1588–1595.

Groeger, J. S., Lucas, A. B., Coit, D., LaQuaglia, M., Brown, A. E., Turnbull, A. and Exelby, P. (1993a). A prospective randomized evaluation of silver-impregnated subcutaneous cuffs for preventing tunneled chronic venous access catheter infections. *Annals of Surgery*, 218, 206–210.

Groeger, J. S., Lucas, A. B., Thaler, H. T., Friedlander-Klar, H., Brown, A. E., Kiehn, T. E. and Armstrong, D. (1993b). Infectious morbidity associated with long-term use of venous access devices in patients with cancer. *Annals of Internal Medicine*, 119, 1168–1174.

Hanna, H., Darouiche, R. and Raad, I. (2001). New approaches for prevention of intravascular catheter-related infections. *Infections in Medicine*, 18, 38–48.

Hawiger, J., Timmons, S., Strong, D. D., Cottrell, B. A., Riley, M. and Doolittle, R. F. (1982). Identification of a region of human fibrinogen interacting with staphylococcal clumping factor. *Biochemistry*, 21, 1407–1413.

Heilmann, C., Gerke, C., Perdreau-Remington, F. and Götz, F. (1996a). Characterization of Tn*917* insertion mutants of *Staphylococcus epidermidis* affected in biofilm formation. *Infection and Immunity*, 64, 277–282.

Heilmann, C., Hussain, M., Peters, G. and Götz, F. (1997). Evidence for autolysin-mediated primary attachment of *Staphylococcus epidermidis* to a polystyrene surface. *Molecular Microbiology*, 24, 1013–1024.

Heilmann, C., Schweitzer, O., Gerke, C., Vanittanakom, N., Mack, D. and Götz, F. (1996b). Molecular basis of intercellular adhesion in the biofilm-forming *Staphylococcus epidermidis. Molecular Microbiology*, 20, 1083–1091.

Herrmann, M., Lai, Q. J., Albrecht, R. M., Mosher, D. F. and Proctor, R. A. (1993). Adhesion of *Staphylococcus aureus* to surface-bound platelets: role of fibrinogen/fibrin and platelet integrins. *Journal of Infectious Diseases*, 167, 312–322.

Herrmann, M., Suchard, S. J., Boxer, L. A., Waldvogel, F. A. and Lew, P. D. (1991). Thrombospondin binds to *Staphylococcus aureus* and promotes staphylococcal adherence to surfaces. *Infection and Immunity*, 59, 279–288.

Howell, P. B., Walters, P. E., Donowitz, G. R. and Farr, B. M. (1995). Risk factors for infection of tunneled central venous catheters in adult cancer patients. *Cancer*, 75(6), 1367–1374.

Ishida, H., Ishida, Y., Kurosaka, T., Otani, K., Sato, K. and Kobayashi, H. (1998). In-vitro and in-vivo activities of levofloxacin against biofilm-producing *Pseudomonas aeruginosa. Antimicobrial Agents and Chemotherapy*, 42, 1641–1645.

Jarvis, W. R., Edwards, J. E., Culver, D. H., Hughes, J. M., Horan, T., Emori, G. G., Banerjee, S., Tolson, J., Henderson, T. and Gaynes, R. P. (1991). Nosocomial infection rates in adult and pediatric intensive care units in the United States. *American Journal of Medicine*, 91(Suppl. 3B), 185S-191S.

Khoury, A. E., Lam, K., Ellis, B. and Costerton, J. W. (1992). Prevention and control of bacterial infections associated with medical devices. *ASAIO Journal*, 38, M174–M178.

Kolls, S. and Brown, A. E. (1993). The changing epidemiology of infections at cancer hospitals. *Clinical Infectious Diseases*, 17, S322–S328.

König, C., Schwank, S. and Blaser, J. (2001). Factors compromising antibiotic activity against biofilms of *Staphylococcus epidermidis. European Journal of Clinical Microbiology and Infectious Diseases*, 20, 21–26.

König, C., Simmen, H. P. and Blaser, J. (1993). Effect of pathological changes of pH, pO_2 and pCO_2 on the activity of antimicrobial agents in vitro. *European Journal of Clinical Microbiology and Infectious Diseases*, 12, 519–526.

Kuusela, P. (1978). Fibronectin binds to *Staphylococcus aureus. Nature*, 276, 718–720.

Lappin-Scott, H. M. and Bass, C. B. (2001). Biofilm formation: attachment, growth and detachment of microbes from surfaces. *American Journal of Infection Control*, 29, 250–251.

Lewis, K. (2001). Riddle of biofilm resistance. *Antimicrobial Agents and Chemotherapy*, 45, 999–1007.

Linares, J., Sitges-Serra, A., Garau, J., Perez, J. L. and Martin, R. (1985). Pathogenesis of catheter sepsis: a prospective study with quantitative and semiquantitative cultures of catheter hub and segments. *Journal of Clinical Microbiology*, 21, 357–360.

Lopes, J. D., Dos Reis, M. and Brentani, R. R. (1985). Presence of laminin receptors in *Staphylococcus aureus*. *Science*, 229, 275–277.

Mack, D., Fischer, W., Krokotsch, A., Leopold, K., Hartmann, R., Egge, H. and Laufs, R. (1996). The intercellular adhesion involved in biofilm accumulation of *Staphylococcus epidermidis* is a linear β-1, 6-linked glucosamineoglycan: purification and structural analysis. *Journal of Bacteriology*, 178, 175–183.

Mack, D., Nedelmann, M., Krokotsch, A., Schwarzkopf, A., Heesemann, J. and Laufs, R. (1994). Characterization of transposon mutants of biofilm-producing *Staphylococcus epidermidis* impaired in the accumulative phase of biofilm production: genetic identification of a hexosamine-containing polysaccharide intercellular adhesion. *Infection and Immunology*, 62, 3244–3253.

Mack, D., Siemssen, N. and Laufs, R. (1992). Parallel induction by glucose of adherence and a polysaccharide antigen specific for plastic-adherent *Staphylococcus epidermidis*: evidence for functional relation to intercellular adhesion. *Infection and Immunology*, 60, 2048–2057.

Maira-Litrán, T., Allison, D. G. and Gilbert, P. (2000). An evaluation of the potential of the multiple antibiotic resistance operon (*mar*) and the multidrug efflux pump *acr*AB to moderate resistance towards ciprofloxacin in *Escherichia coli* biofilms. *Journal of Antimicrobial Chemotherapy*, 45, 789–795.

Maki, D. G. (1989). Pathogenesis, prevention, and management of infections due to intravascular devices used for infusion therapy. In *Infections Associated with Indwelling Medical Devices*, eds. S. L. Bisno and F. A. Waldvogal, pp. 161–177. Washington, DC: American Society for Microbiology.

Maki, D. G. and Band, J. D. (1981). A comparative study of polyantibiotic and iodophor ointments in prevention of vascular catheter-related infection. *American Journal of Medicine*, 70, 739–744.

Maki, D. G. and Mermel, L. A. (1998). Infections due to infusion therapy. In *Hospital Infections*, ed. J. V. Bennett and P. S. Brachman, pp. 689–724. Philadelphia, PA: Lippincott-Raven.

Maki, D. G. and Ringer, M. (1987). Evaluation of dressing regimens for prevention of infection with peripheral intravenous catheters gauze, a transparent polyurethane dressing, and an iodophor-transparent dressing. *Journal of the American Medical Society*, 258, 2396–2403.

Maki, D. G., Ringer, M. and Alvarado, C. J. (1991). Prospective randomized trial of povidone-iodine, alcohol, and chlorhexidine for prevention of infection associated with central venous and arterial catheters. *Lancet*, 338, 339–343.

Maki, D. G., Stolz, S. M., Wheeler, S. and Mermel, L. A. (1997). Prevention of central venous catheter-related bloodstream infection by use of an antiseptic-impregnated catheter: a randomized controlled trial. *Annals of Internal Medicine*, 127, 257–266.

Maki, D. G., Weise, C. E. and Sarafin, H. W. (1977). A semiquantitative culture method for identifying intravenous catheter-related infections. *New England Journal of Medicine*, 296, 1305–1309.

Mermel, L. A. (2000a). Prevention of intravascular catheter-related infections. *Annals of Internal Medicine*, 132, 391–402.
Mermel, L. A. (2000b). Correction: catheter-related bloodstream infections. *Annals of Internal Medicine*, 133, 395.
Mermel, L. A., Farr, B. M., Sherertz, R. J., Raad, I. I, O'Grady, N., Harris, J. A. S. and Craven, D. E. (2001). Guidelines for the management of intravascular catheter-related infections. *Clinical Infectious Diseases*, 32, 1249–1272.
Nelson, D. B., Kien, C. L., Mohr, B., Frank, S. and Davis, S. D. (1986). Dressing changes by specialized personnel reduce infection rates in patients receiving central venous parenteral nutrition. *Journal of Parenteral Enteral Nutrition*, 10, 220–222.
Raad, I. (1998). Intravascular-catheter-related infections. *Lancet*, 351, 893–898.
Raad, I., Buzaid, A., Rhyne, J., Hachem, R., Darouiche, R., Safar, H., Albitar, M. and Sherertz, R. (1997). Minocycline and ethlenediaminetetraacetate for the prevention of recurrent vascular catheter infections. *Clinical Infectious Diseases*, 25, 149–151.
Raad, I., Darouiche, R.O., Hachem, R., Abi-Said, D., Safar, H., Darnule, T., Mansouri, M. and Morck, D. (1998). Antimicrobial durability and rare ultrastructural colonization of indwelling central catheters coated with minocycline and rifampin. *Critical Care in Medicine*, 26, 219–224.
Raad, I., Costerton, W., Sabharwal, U., Sacilowski, M., Anaissie, E. and Bodey, G. P. (1993). Ultrastructural analysis of indwelling vascular catheters: a quantitative relationship between luminal colonization and duration of placement. *Journal of Infectious Diseases*, 168, 400–407.
Raad, I., Hachem, R., Zermeno, A., Dumo, M. and Bodey, G. P. (1996a). In vitro antimicrobial efficacy of silver iontophoretic catheter. *Biomaterials*, 17, 1055–1059.
Raad, I., Hachem, R., Zermeno, A., Stephens, L. C. and Bodey, G. P. (1996b). Silver iontophoretic catheter: a prototype of a long-term antiinfective vascular access device. *Journal of Infectious Diseases*, 173, 495–498.
Raad, I. I. and Hanna, H. (1999). Nosocomial infections related to use of intravascular devices inserted for long-term vascular access. In *Hospital Epidemiology and Infection Control*, ed. C. G. Mayhall, pp. 165–171. Philadelphia, PA: Lippincott Williams & Wilkins.
Raad, I. I., Hohn, D. C., Gilbreath, B., Suleiman, N., Hill, L. A., Bruso, P. A., Marts, K., Mansfield, P. F. and Bodey, G. P. (1994). Prevention of central venous catheter-related infections using maximal sterile barrier precautions during insertion. *Infection Control and Hospital Epidemiology*, 15, 231–238.
Rachid, S., Ohlsen, K., Wite, W., Hacker, J. and Ziebuhr, W. (2000). Effect of subinhibitory concentrations on polysaccharide intercellular adhesion expression in biofilm-forming *Staphylococcus epidermidis*. *Antimicrobial Agents and Chemotherapy*, 44, 3357–3363.
Rupp, M. E., Ulphani, J. S., Fey, P. D. and Mack, D. (1999a). Characterization of *Staphylococcus epidermidis* polysaccharide intercellular adhesion/hemagglutinin in the pathogenesis of intravascular catheter-associated infection in a rat model. *Infection and Immunity*, 67, 2656–2659.
Rupp, M. E., Fey, P. D., Heilmann, C. and Götz, F. (2001). Characterization of the importance of *Staphylococcus epidermidis* autolysin and polysaccharide

intercellular adhesin in the pathogenesis of intravascular catheter-associated infection in a rat model. *Journal of Infectious Diseases*, 183, 1038–1042.
Rupp, M. E., Ulphani, J. S., Fey, P. D., Bartscht, K. and Mack, D. (1999b). Characterization of the importance of polysacchride intercellular adhesin/hemagglutinin of *Staphylococcus epidermidis* in the pathogenesis of biomaterial-based infection in a mouse foreign body infection model. *Infection and Immunity*, 67, 2627–2632.
Saint, S., Veenstra, D. L. and Lipsky, B. A. (2000). The clinical and economic consequences of nosocomial central venous catheter-related infection: are antimicrobial catheters useful? *Infection Control and Hospital Epidemiology*, 21, 375–380.
Salzman, M. B., Isenberg, H. D., Shapiro, J. F., Lipsitz, P. J. and Rubin, L. G. (1993). A prospective study of the catheter hub as the portal of entry for microorganisms causing catheter-related sepsis in neonates. *Journal of Infectious Diseases*, 167, 487–490.
Schwartz, C., Henrickson, K. J., Roghmann, K. and Powell, K. (1990). Prevention of bacteremia attributed to luminal colonization of tunneled central venous catheters with vancomycin-susceptible organisms. *Journal of Clinical Oncology*, 8, 591–597.
Stewart, P. S. (1996). Theoretical aspects of antibiotic diffusion into microbial biofilm. *Antimicrobial Agents and Chemotherapy*, 40, 2517–2522.
Stiges-Serra, A., Linares, J., Perez, J. L., Jaurrieta, E. and Lorente, L. (1985). A randomized trial on the effect of tubing changes on hub contamination and catheter sepsis during parenteral nutrition. *Journal of Parenteral Enteral Nutrition*, 9, 322–325.
Stiges-Serra, A., Puig, P., Linares, J., Perez, J. L., Farrero, N., Jaurrieta, E. and Garau, J. (1984). Hub colonization as the initial step in an outbreak of catheter-related sepsis due to coagulase negative staphylococci during parenteral nutrition. *Journal of Parenteral Enteral Nutrition*, 8, 668–672.
Suci, P. A., Mittelman, M. W., Yu, F. P. and Geesey, G. G. (1994). Investigation of ciprofloxacin penetration into *Pseudomonas aeruginosa* biofilm. *Antimicrobial Agents and Chemotherapy*, 38, 2125–2133.
Teichberg, S., Farber, B. F., Wolff, A. G. and Roberts, B. (1993). Salicylic acid decreases extracellular biofilm production by *Staphylococcus epidermidis* electron microscopic analysis. *Journal of Infectious Diseases*, 167, 1501–1503.
Timmerman, C. P., Fleer, A., Besnier, J. M., De Graff, L., Cremers, F. and Verhoef, J. (1991). Characterization of a proteinacious adhesion of *Staphylococcus epidermidis* which mediates attachment to polystyrene. *Infection and Immunity*, 59, 4187–4192.
Timsit, J. F., Bruneel, F., Cheval, C., Mamzer, M. F., Garrouste-Oregeas, M., Wolff, M., Misset, B., Chevret, S., Regnier, B. and Carlet, J. (1999). Use of tunneled femoral catheters to prevent catheter-related infection: a randomized, controlled trial. *Annals of Internal Medicine*, 130, 729–735.
Vaudaux, P., Pittet, D., Haeberli, A., Huggler, E., Nydegger, U. E. and Lew, D. P. (1989). Host factors selectively increase staphylococcal adherence on inserted catheters: a role for fibronectin and fibrinogen or fibrin. *Journal of Infectious Diseases*, 160, 865–875.
Veenstra, G. J., Cremers, F. M., Van Dijk, H. and Fleer, A. (1996). Ultrastructural organization and regulation of a biomaterial adhesion of *Staphylococcus epidermidis*. *Journal of Bacteriology*, 178, 537–541.

Veenstra, D. L., Saint, S., Saha, S., Lumley, T. and Sullivan, S. D. (1999a). Efficacy of antiseptic-impregnated central venous catheters in preventing catheter-related bloodstream infection: a meta analysis. *Journal of the American Medical Association*, 281, 261–267.

Veenstra, D. L., Saint, S. and Sullivan, S. D. (1999b). Cost-effectiveness of antiseptic-impregnated central venous catheters for the prevention of catheter-related bloodstream infection. *Journal of the American Medical Association*, 28, 554–560.

Wentland, E. J., Stewart, P. S., Huang, C. T. and McFeters, G. A. (1996). Spatial variations in growth rate within *Klebsiella pneumoniae* colonies and biofilm. *Biotechnological Programs*, 12, 316–321.

Wenzel, R. P. and Edmond, M. B. (2001). The impact of hospital-acquired bloodstream infections. *Emerging Infectious Diseases*, 7, 174–177.

Whiteley, M., Bangera, M. G., Bumgarner, R. E., Parsek, M. R., Teltzel, G. M., Lory, S. and Greenberg, E. P. (2001). Gene expression in *Pseudomonas aeruginosa* biofilms. *Nature*, 413, 860–864.

CHAPTER SIX

Molecular Basis of Biofilm Formation by *Staphylococcus epidermidis*

Christine Heilmann

1 INTRODUCTION

Staphylococcus epidermidis is a ubiquitous commensal of human skin. In the past 20 years, this organism has emerged as a serious pathogen, causing frequent nosocomial infections. In contrast to *S. aureus* and except for rare cases of native valve endocarditis, *S. epidermidis* needs a predisposed host to cause infection. Individuals who are especially susceptible to *S. epidermidis* infections are immunocompromised patients, intravenous drug abusers, and, most commonly, patients with indwelling medical devices (Heilmann and Peters, 2000). Immunocompromised patients include those with malignant disease who are undergoing cytostatic and/or immunosuppressive therapy and premature newborns. In premature newborns, *S. epidermidis* is the most frequent cause of septicaemia with an onset later than 48 hours after birth. In drug abusers, continual injections of heroin containing crystals are believed to cause microlesions in the valvular endothelium leading to right-sided endocarditis; the damaged tissue is then infected with *S. epidermidis* that is introduced due to non-sterile conditions. By far, the largest number of infections caused by *S. epidermidis* are those associated with implanted foreign bodies such as orthopaedic devices, intravascular catheters, cerebrospinal fluid shunts, prosthetic heart valves, continuous ambulatory peritoneal dialysis (CAPD) catheters, and cardiac pacemakers (George, Leibrock, and Epstein, 1979; Karchmer, Archer, and Dismukes, 1983; Peters, Locci, and Pulverer, 1982; Peters et al., 1984; West et al., 1986). These devices are manufactured from various polymers, most commonly from polyvinylchloride, polyethylene, silicon rubber, and polyurethane.

Colonisation of the polymer surface by *S. epidermidis* is the most critical step in the pathogenesis of foreign-body-associated infections. This most likely occurs during the surgical implantation of the device due to contamination by very small numbers of the organism from the patient's skin or mucous membranes. Occasionally, the organism is acquired from the surgical or clinical staff. The bacteria proliferate on the polymer surface and form multilayered cell clusters, which are embedded in an amorphous extracellular material (Peters et al., 1982). The bacteria together with the extracellular material, which is mainly composed of cell wall teichoic acids (Hussain, Wilcox, and White, 1993) and host products, are referred to as biofilm. The presence of such large adherent biofilms on polymer surfaces has been demonstrated by scanning electron microscopy (Figure 6.1) (Franson et al., 1984; Peters et al., 1984; Peters, Locci, and Pulverer, 1981).

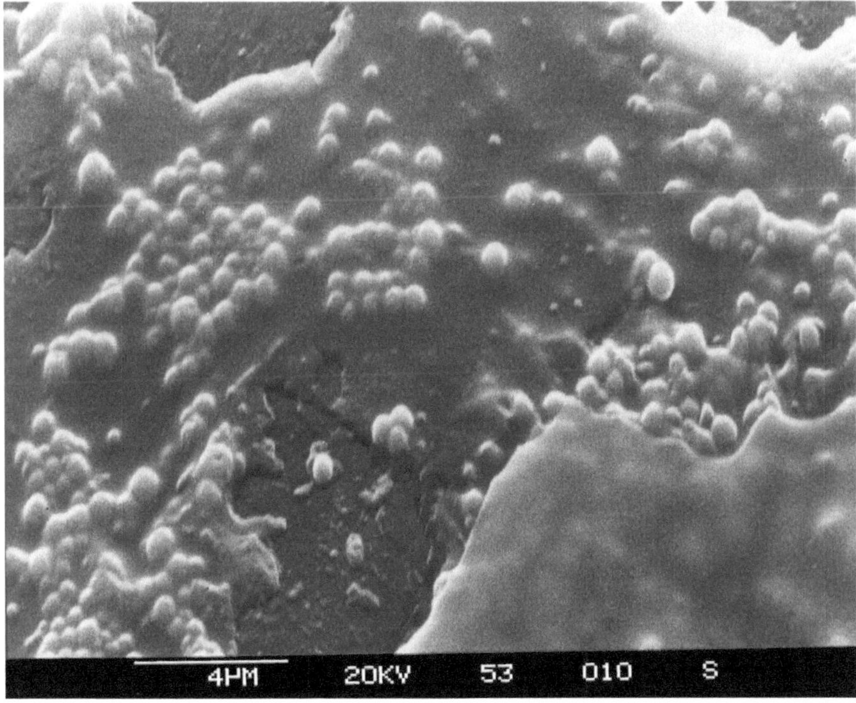

Figure 6.1: Scanning electron micrograph of a *S. epidermidis* biofilm on the surface of the plastic sheath of a pacemaker lead. (Reprinted with permission from Peters et al., *American Heart Journal*, 108, 362.)

2 BIOFILM FORMATION

Biofilm formation takes place in two stages. First, the bacteria rapidly attach to the polymer material and then multiply to form multilayered cell clusters on the polymer surface. The second, more prolonged accumulation phase involves intercellular adhesion leading to the development of a confluent biofilm. In the past 5 years, substantial progress has been made in characterising the molecular mechanisms that participate in *S. epidermidis* biofilm formation (summarised in Figure 6.2).

2.1 Attachment to Unmodified Polymer Surfaces

Microbial adherence to biomaterials largely depends on bacterial cell surface characteristics and on the nature of the polymer material. The initial interactions are believed to occur via non-specific physicochemical forces such as polarity, van der Waal's forces, and hydrophobic interactions. Recently, the net charge of teichoic acid has been found to play a key role in *S. aureus* colonisation of artificial surfaces (Gross et al., 2001). *S. aureus* teichoic acids are highly charged cell wall polymers composed of alternating phosphate and ribitol (wall teichoic acids) or glycerol (lipoteichoic acids) groups, which are substituted with D-alanine and *N*-acetylglucosamine. *dltA* encodes the

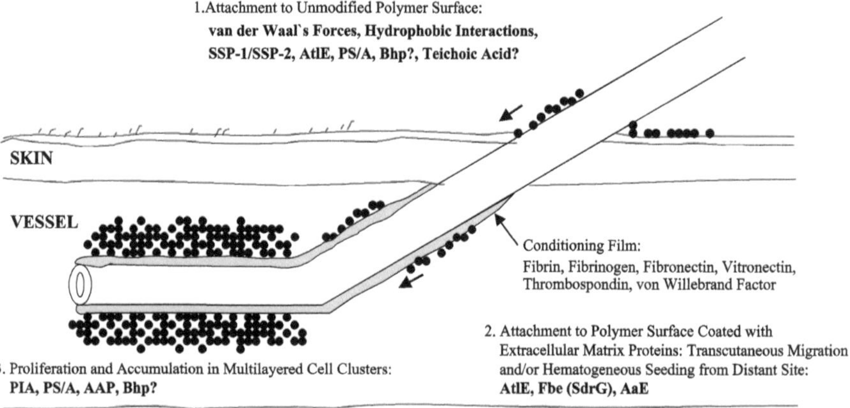

Figure 6.2: Model of the phases involved in *S. epidermidis* biofilm formation showing the bacterial factors whose participation in the process is proven or very likely. Factors whose homologous counterparts in *S. aureus* have been shown to be involved in *S. aureus* biofilm formation are labelled with a question mark. Abbreviations: SSP-1/SSP-2, staphylococcal surface proteins; AtlE, autolysin; PS/A, polysaccharide/adhesin; Bhp, biofilm-associated (Bap)-homologous protein; Fbe, fibrinogen-binding protein; AaE, autolysin/adhesin; SdrG, SD (serine-aspartate)-repeat-containing protein G; PIA, polysaccharide intercellular adhesin; AAP, accumulation-associated protein.

D-alanine-D-alanyl carrier protein ligase DltA that mediates the incorporation of the substituent D-alanine into the teichoic acid. A *dltA* mutant has a biofilm-negative phenotype. Without the ester-linked D-alanine in its teichoic acid, the *dltA* mutant has a higher negative surface charge, resulting in the inhibition of initial attachment to polystyrene as well as to glass, which is negatively charged. The main difference between the wall teichoic acids of *S. epidermidis* and *S. aureus* is the presence of glycerol rather than ribitol. The genome of *S. epidermidis* RP62A contains a gene encoding a protein which is 75 per cent identical to DltA from *S. aureus* (preliminary sequence data was obtained from The Institute for Genomic Research website at http://www.tigr.org). Therefore, it may be hypothesised that analogously to *S. aureus*, teichoic acid contributes to the surface charge of *S. epidermidis* and thus participates in its colonisation of abiotic surfaces.

Cell surface hydrophobicity and initial adherence have been ascribed to bacterial surface-associated proteins. Using monoclonal antibodies to block adherence, the two antigenically related staphylococcal surface proteins, SSP-1 (280 kDa) and SSP-2 (250 kDa), have been found to contribute to the adherence of *S. epidermidis* strain 354 to polystyrene (Veenstra et al., 1996). SSP-1 and SSP-2 appear to be fimbria-like polymers; however, the genes encoding these proteins have not been identified.

Other proteins also seem to be associated with the initial attachment phase of biofilm formation. Using Tn*917* mutagenesis, we isolated mutants from the biofilm-forming clinical strain *S. epidermidis* O-47 (Heilmann et al., 1996a). *S. epidermidis* mut1 was deficient in initial adherence to polystyrene and lacked surface-associated proteins. Cloning and sequencing of the DNA region that was inactivated in mut1 and complementation of mut1 revealed that the surface-associated autolysin AtlE is involved in primary attachment of the cells to the polymer surface (Heilmann et al., 1997). AtlE shows high similarity to the *S. aureus* autolysin Atl (61 per cent identical amino acids). It is proteolytically cleaved from a 148-kDa precursor into two bacteriolytically active polypeptides; the N-terminal portion yields a 60-kDa amidase and the C-terminal portion yields a 52-kDa glucosaminidase (Figure 6.3). In the AtlE precursor, the enzymatically active domains are separated from each other by three repetitive sequences (R1 [167 amino acids], R2 [165 amino acids], and R3 [171 amino acids]), with the repeats R1 and R2 being located in the C-terminal region of the amidase and the repeat R3 being located in the N-terminal region of the glucosaminidase.

Recently, another protein, the biofilm-associated protein Bap, was reported to be involved in *S. aureus* biofilm formation (Cucarella et al., 2001). Tn*917*

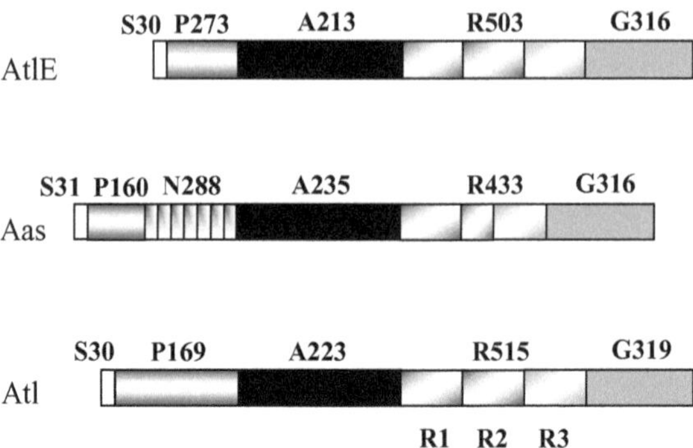

Figure 6.3: Model of the structural organisation of the autolysins AtlE (1,335 amino acids), Aas (1,463 amino acids), and Atl (1,256 amino acids) from *S. epidermidis, S. saprophyticus,* and *S. aureus,* respectively (modified from Heilmann et al., 1997 and Hell et al., 1998). The relative positions and sizes (number of amino acids) of their signal sequences (S); propeptides (P); N-terminal repeats (N, Aas only); amidase domains (A); repeats (R) R1, R2, R3; and glucosaminidase domains (G) are indicated.

insertion mutants of *S. aureus* that were deficient in Bap showed markedly decreased attachment to a polystyrene surface, intercellular adhesion, and biofilm formation. Thus, Bap seems to be involved in both phases of biofilm formation. The *bap* mutants also had reduced persistence of infection in a mouse foreign body infection model as compared to the wild type. *bap* encodes a novel cell surface-associated protein with 2,276 amino acids and a predicted molecular mass of 238.7 kDa. The structural features of Bap correspond to those of other Gram-positive surface proteins, such as a signal sequence at the N-terminus, multiple repeat regions, and a C-terminal cell wall anchor region. The cell wall anchor region is composed of an LPXTG motif preceding a hydrophobic membrane-spanning region and a positively charged tail, which is thought to act as a stop-secretion signal at the C-terminus (Schneewind, Mihaylova-Petkov, and Model, 1993). Bap has been found to be present in only 5 per cent of 350 bovine mastitis *S. aureus* isolates tested and was absent in all 75 clinical human *S. aureus* isolates tested. However, a gene encoding a Bap-homologous protein, Bhp (Tormo et al., unpublished; EMBL database accession number: AAK29746), was found to be present in *S. epidermidis* strain RP62A – a clinical human isolate associated with foreign body infection and a strong biofilm producer (Christensen et al., 1982). Bhp consists of 2,402 amino acids with a predicted molecular mass of 258 kDa and was assumed to promote biofilm formation.

Aside from proteins, a polysaccharide, capsular polysaccharide/adhesin (PS/A), has been associated with initial adherence and slime production

(Muller et al., 1993). Tn*917* mutants deficient in PS/A and initial adherence do not cause endocarditis in a rabbit model in contrast to the isogenic parent (Shiro et al., 1994). Furthermore, immunisation with PS/A results in protection against infection. The relationship of PS/A to other polysaccharides of *S. epidermidis* is discussed in Section 2.2.

2.2 Attachment via Interaction with Host Extracellular Matrix Proteins

The direct interaction between bacteria and the unmodified polymer surface plays a crucial role in the early stages of the adherence process *in vitro* and probably also *in vivo*. After insertion of the medical device, the implanted material rapidly becomes coated with plasma and extracellular matrix proteins such as fibrinogen (Fg), fibronectin (Fn), thrombospondin, vitronectin (Vn), and von Willebrand factor (Cottonaro et al., 1981; Dickinson and Bisno, 1989; Kochwa et al., 1977). Therefore, in later stages of infection, plasma and extracellular matrix proteins deposited on the polymer surface could serve as specific receptors for colonising pathogens, for example, *S. aureus* (Chhatwal et al., 1987; Hartleib et al., 2000; Herrmann et al., 1997; Lopes, dos Reis, and Brentani, 1985). *S. epidermidis* also appears to possess surface proteins which may be involved in the interaction with these host factors and thus in biofilm formation on protein-coated implants. Adherence of clinical isolates of coagulase-positive and -negative staphylococci to biomaterials is markedly enhanced by surface-bound Fn (Herrmann et al., 1988). Adherence of all *S. aureus* strains tested is significantly promoted by immobilised Fg, whereas adherence of *S. epidermidis* strains to Fg varies significantly among strains.

Numerous genes for host-factor-binding proteins from *S. aureus* have been cloned and sequenced, such as Fn-binding proteins and Fg-binding proteins. Much less is known about host-factor-binding proteins from *S. epidermidis*. We found that the surface-associated autolysin AtlE from *S. epidermidis* not only is involved in initial attachment of the cells to an unmodified polystyrene surface (see earlier), but also binds to the extracellular matrix protein Vn (Heilmann et al., 1997). Hence, AtlE may play a role in the initial colonisation and persistent growth on host-factor-coated foreign material and possibly host tissue. Thus, AtlE is a multifunctional, surface-associated protein having both autolytic and adherence functions (Heilmann et al., 1997). The importance of AtlE in *S. epidermidis* pathogenicity has been demonstrated: the *atlE* mutant strain was significantly less virulent than the wild type O-47 in an intravascular catheter-associated infection model in rats (Rupp et al., 2001). Recently, the gene encoding an autolysin (Aas) from *S. saprophyticus* has been cloned and sequenced (Hell, Meyer, and Gatermann, 1998); it exhibits significant homology to AtlE and Atl (Figure 6.3). Aas binds to Fn and agglutinates sheep erythrocytes (Hell et al., 1998). The discovery of multifunctional roles of AtlE and Aas has led to the proposal of a new class of staphylococcal adhesins – the autolysin/adhesins. Generally, bacterial autolysins are peptidoglycan hydrolases that play important roles in cell-wall turnover, cell division, cell separation, and antibiotic-induced lysis of bacterial cells. Autolysins are also considered to be virulence factors; some autolysins seem to be involved in adherence. The invasion-associated protein P60 of *Listeria monocytogenes* has autolytic activity and is involved in the adherence to, and intracellular invasion of, mouse fibroblasts (Bubert et al., 1992; Kuhn and Goebel, 1989). The autolysin Ami of *L. monocytogenes* is involved in the adherence to eukaryotic cells (Milohanic et al., 2000). Recently, the eukaryotic cell-binding activity of Ami has been localised to the cell wall-anchoring domain, which contains multiple repeats with GW-motifs (Milohanic et al., 2001). Such GW-modules are also found in the three repetitive domains of AtlE and Aas and potentially participate in the ligand-binding activity of these staphylococcal autolysins (W. Hell, personal communication; Milohanic et al., 2001). We have recently found another member of the family of staphylococcal autolysin/adhesins. AaE from *S. epidermidis* has bacteriolytic activity and binds to Fg, Fn, and Vn (Heilmann, Chhatwal, and Peters, 2000).

Proteins other than the staphylococcal autolysin/adhesins also seem to play roles in the binding of *S. epidermidis* to host proteins. Using the phage display technique, the 119-kDa Fg-binding protein Fbe was identified in the *S. epidermidis* strain HB (Nilsson et al., 1998). Fbe is a member of another recently identified protein family, the SD (serine-aspartate)-repeat-containing (Sdr) family of cell wall-anchored surface proteins. The first member of the Sdr protein family to be described was the Fg-receptor ClfA (clumping factor A) from *S. aureus* (McDevitt et al., 1994). Also from *S. aureus*, other members of the Sdr protein family include a second clumping factor (ClfB; Ni Eidhin et al., 1998) and SdrC, SdrD, and SdrE, whose functions are unknown (Josefsson et al., 1998). Common features of these proteins are an N-terminal signal peptide, a known or putative ligand-binding region A, the SD-repeat region that is predicted to span the cell wall, and a C-terminal cell wall anchor domain. The gene encoding Fbe has been cloned and sequenced (Nilsson et al., 1998). The overall organisation of Fbe corresponds to that of the other Sdr proteins. Sequence comparison revealed that Fbe shows similarity to the clumping factors ClfA and ClfB. This similarity is not only found within the SD-repeats, but also within the Fg-binding region A. Despite the similarity of Fbe to ClfA,

S. epidermidis strains harbouring the *fbe* gene do not exhibit the clumping reaction, another typical feature mediated by ClfA. In contrast to ClfA, which binds to the γ chain of Fg, and like ClfB, Fbe binds to the β chain of Fg (Pei et al., 1999). ClfA and ClfB have been shown to contribute to *S. aureus* virulence in a rat model of catheter-induced endocarditis; whether Fbe plays a similar role in *S. epidermidis* infections has yet to be elucidated. However, experimental results suggest a role in adherence. Construction of an *fbe*-deficient mutant and adherence analysis delineated the importance of Fbe in the interaction of *S. epidermidis* with Fg. Compared with the wild-type strain, the *fbe* mutant showed reduced adherence to Fg immobilised on a polyethylene surface and to peripheral venous catheters that were removed from patients. In contrast, the adherence to immobilised Fn was unchanged (Pei and Flock, 2001b). Furthermore, heterologous expression of *fbe* in *Lactococcus lactis*, which normally does not bind to Fg, conferred strong binding to immobilised Fg (Hartford et al., 2001). Addition of the purified protein Fbe, expressed as a fusion protein with glutathione S-transferase, completely abolished the adherence of *S. epidermidis* to immobilised Fg. Antibodies against Fbe also efficiently inhibited that adherence reaction, suggesting a possible use as an immunoprophylactic therapy against foreign-body-associated infections (Pei et al., 1999; Pei and Flock, 2001a). This immunoprotection might be useful since, as shown by polymerase chain reaction (PCR) analysis, the *fbe* gene is present in forty out of forty-three clinical isolates of *S. epidermidis* (Nilsson et al., 1998).

Recently, other cell-surface-associated proteins containing SD-repeats, SdrG (approximately 95 per cent identical to Fbe), SdrF, and SdrH, were identified in *S. epidermidis* (McCrea et al., 2000) (Figure 6.4). The *sdrG* gene was

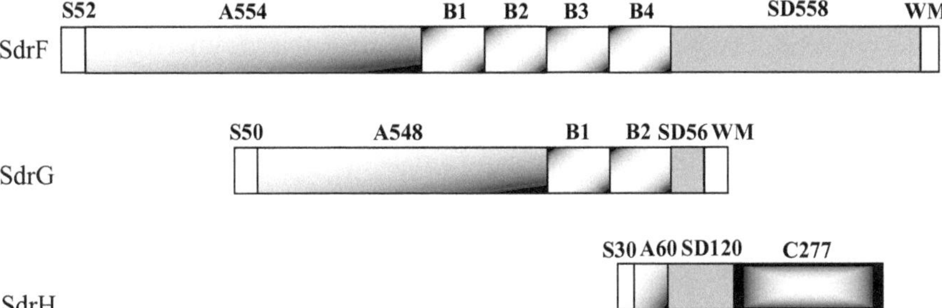

Figure 6.4: Schematic representation of the *S. epidermidis* SdrG, SdrF, and SdrH proteins (modified from McCrea et al., 2000). The relative positions and the sizes (number of amino acids) of their signal sequences (S), A regions (A), B-repeat regions (B_n, 110–113 amino acids each), SD-repeat regions (SD), C region (C) (SdrH only), and cell wall/membrane spanning regions (WM) are indicated.

cloned from strain K28, and the *sdrF* and *sdrH* genes were cloned from strain 9491. Like SdrC, SdrD, and SdrE from *S. aureus* and unlike ClfA and ClfB, the 179-kDa SdrF and the 97.6-kDa SdrG contain an additional repeat region, the B-repeats, which is located between the A region and the SD-repeats. Within each B-repeat, a putative Ca^{2+}-binding EF-hand motif is present. On average, the B-repeats in all Sdr proteins share 55 per cent identical amino acids with each other. The 50.5-kDa SdrH does not contain a B region, and its overall organisation differs from that of other Sdr proteins. It is composed of an N-terminal signal sequence, a short 60 amino acid N-terminal A region followed by an SD-repeat region, a unique 277 amino acid C region, and a C-terminal hydrophobic stretch which lacks an LPXTG motif. The *sdrG* and *sdrH* genes were present in all sixteen strains tested, whereas *sdrF* was present in only twelve strains. Antibodies against SdrG and SdrH proteins were found in sixteen convalescent patient sera, implying that their genes are expressed during infection (McCrea et al., 2000). Like Fbe, SdrG binds to the β chain in Fg. More specifically, SdrG was found to bind to the N-terminal segment (peptide β6-20) of this polypeptide, which is located proximal to the thrombin cleavage site. It was also demonstrated that SdrG inhibits thrombin-induced Fg clotting by interfering with the release of fibrinopeptide B (Davis et al., 2001). It was speculated that the reason for the binding activity of SdrG might be to prevent the release of chemotactic elements, such as fibrinopeptide B. This may reduce the influx of phagocytic neutrophils, aiding the survival of the bacteria in the host (Davis et al., 2001). The function of the other members of this new protein family in *S. epidermidis* remains unknown.

Most recently, it has been shown that aside from proteins, cell wall teichoic acid is involved in the adherence of *S. epidermidis* to Fn (Hussain et al., 2001). Adherence of *S. epidermidis* to immobilised Fn is significantly promoted by teichoic acid in a dose-dependent fashion. Preincubation of either the bacteria or Fn-coated surfaces with teichoic acid promoted *S. epidermidis* adherence, suggesting a possible function of teichoic acid as a bridging molecule between the bacteria and Fn-coated polymer material.

2.3 Accumulation Phase

Once the bacteria have adhered to the foreign body surface, they proliferate and accumulate as multilayered cell clusters. This accumulation step requires intercellular adhesion. Tn*917* mutants that were unable to accumulate on a polymer surface were found to lack a specific polysaccharide antigen referred to as polysaccharide intercellular adhesin (PIA) (Heilmann et al., 1996a). In those mutants, PIA production depends on the expression of the *icaADBC* gene cluster and correlates with autoaggregation, which is indicative of intercellular adhesion, and the formation of black colonies on Congo red agar (Heilmann and Götz, 1998). Accumulation-negative Tn*917* mutants are characterised by the formation of red colonies on Congo red agar. Purification and structural analysis of PIA revealed that it consists of a major polysaccharide I (>80 per cent) and a minor polysaccharide II (<20 per cent) (Mack et al., 1996). Polysaccharide I is a linear β-1,6-linked glucosaminoglycan mainly composed of at least 130 2-deoxy-2-amino-D-glucopyranosyl residues of which 80–85 per cent are *N*-acetylated. Polysaccharide II is structurally related to polysaccharide I, but has a lower content of non-*N*-acetylated glucosaminyl residues and contains a small amount of phosphate and ester-linked succinyl residues. Thus, PIA represents a unique structure so far.

The *icaABC* genes that mediate cell clustering and PIA synthesis in *S. epidermidis* were cloned and sequenced (Heilmann et al., 1996b). Later, an additional open reading frame (ORF) (*icaD*) was identified; it is located between *icaA* and *icaB* and overlaps both genes (Gerke et al., 1998). The functions of the respective proteins in PIA synthesis were analysed. There is evidence that the proposed *N*-acetylglucosaminyltransferase activity is carried out by IcaA. However, IcaA alone exhibits only low transferase activity. Coexpression of *icaA* encoding the catalytic enzyme together with *icaD* leads to a significant increase in activity and to synthesis of *N*-acetylglucosamine oligomers with a maximal length of twenty residues. Only in the presence of *icaC* does IcaAD catalyse the synthesis of long-chain oligomers that react with PIA-specific antisera. The importance of PIA as a virulence factor has recently been demonstrated: a PIA-negative mutant was significantly less virulent than the isogenic wild-type strain in a mouse model of subcutaneous foreign body infection as well as in a rat model of central venous catheter-associated infection (Rupp et al., 1999a,b). Furthermore, heterologous expression of the *icaADCB* operon in *S. carnosus* conferred haemagglutinating ability on the haemagglutination-negative *S. carnosus* wild type. Thus, the *icaADCB* operon mediated biofilm accumulation, PIA production, and haemagglutination (Fey et al., 1999). In a study designed to investigate the pathogenic properties of strains obtained from patients with polymer-associated septicaemia, there was a strong correlation between pathogenesis and both biofilm formation and presence of the *ica* gene cluster. Meanwhile, those factors were rarely observed in saprophytic skin and mucosal isolates (Ziebuhr et al., 1997).

It has been reported that PS/A production is also determined by the *ica* gene cluster and that PS/A is chemically related to PIA (McKenney et al.,

1998). Both antigens are characterised by the common β-1,6-linked polyglucosamine backbone, but PS/A was distinguished from PIA by molecular size (>250,000 kDa PS/A versus ~28 kDa PIA) and the presence of succinate groups on the majority of the amino groups of the glucosamine residues. Recently, it has been shown that the synthesis of a similar, if not identical, polysaccharide from *S. aureus* is mediated by a homologous *ica* gene cluster (Cramton et al., 1999). However, analysis of that polysaccharide suggests that the majority of the amino groups of the glucosamine residues are acetylated as described for PIA, rather than succinylated (G. Pier, personal communication).

Formerly, another antigenic marker of slime production and accumulation was identified and designated as slime-associated antigen (SAA). Changes in the purification procedure have shown that the composition of SAA differs from that originally described and that SAA mainly consists of *N*-acetylglucosamine. Hence, it has been concluded that SAA and PIA may have the same antigenic structure (Baldassarri et al., 1996).

Proteins also seem to be essential for accumulation and biofilm formation in *S. epidermidis.* The Bap-homologous protein Bhp may be involved in biofilm accumulation (see earlier). The accumulation-associated protein (AAP), a 140-kDa extracellular protein that is missing in the accumulation-negative mutant M7, was shown to be essential for accumulative growth in certain *S. epidermidis* strains on polymer surfaces (Hussain et al., 1997; Schumacher-Perdreau et al., 1994). An antiserum specific for AAP inhibited accumulation by up to 98 per cent of the wild-type strain RP62A. Cloning and DNA sequence analysis of *aap* revealed that AAP has a predicted molecular mass of 132 kDa and consists of 1,245 amino acids. It has features typical of Gram-positive surface proteins, for example, an N-terminal signal peptide, multiple repeat domains, and a C-terminal cell wall anchor (Hussain et al., 2000). Biochemical and functional properties clearly differentiate AAP from other factors that have been implicated in biofilm formation. It is proposed that AAP plays a role in the anchoring of PIA to the cell surface, as the mutant M7 produces PIA that is only loosely attached to the cell surface in contrast to the wild type (D. Mack, personal communication).

3 OTHER POTENTIAL VIRULENCE FACTORS INVOLVED IN SKIN COLONISATION, BIOFILM FORMATION, AND POLYMER-ASSOCIATED INFECTION

Although the role of biofilm formation in prosthetic device-related *S. epidermidis* disease is obvious, other mechanisms are involved in the infectious process. A predisposing factor is the ubiquitous presence of the bacterial species as dominant members of the microbiota of human skin and mucous membranes. Ultimately, the virulence associated with a *S. epidermidis* biofilm is due to its persistence on the medical device in spite of the host's immune response.

3.1 Lipases and Lantibiotics

The contamination of the polymer surface of a medical device with *S. epidermidis* cells from either the patient or the hospital staff most often occurs during implantation. Thus, the ability of *S. epidermidis* to colonise the human skin seems to be an important factor in the development of polymer-associated infections. Extracellular products of *S. epidermidis* that may not be directly involved in adherence to the polymer material may play a role in successful and persistent skin colonisation.

Lipases have been proposed to be involved in skin colonisation. Their substrates are lipids that are largely composed of sebum-derived triacylglycerides and are found ubiquitously on the surface of human skin. It is, therefore, not surprising that many organisms that colonise human skin express lipolytic activity. Lipases are believed to catalyse the hydrolysis of sebaceous lipids, resulting in the release of free fatty acids onto the cutaneous surface. Those free fatty acids are believed to sustain the acid mantle of the human skin, thereby maintaining a niche that is tolerated by *S. epidermidis* while being unfavourable for less acid-resistant species. Other possible roles for the liberated fatty acids could be as nutrients for colonising *S. epidermidis* or the promotion of its adherence. The genes of the two lipases *gehC* and *gehSE1* from *S. epidermidis* strains 9 and RP62A, respectively, have been cloned and sequenced (Farrell, Foster, and Holland, 1993; Simons et al., 1998). They exhibit a high degree of similarity (97.8 per cent identical amino acids) and encode preproenzymes that are proteolytically processed into a 43-kDa mature lipase upon secretion into the surrounding medium. The lipase GehSE1 has a pH optimum around 6, a high stability at low pH, and a strong preference for tributyrylglycerol as a substrate and does not hydrolyze phospholipids. Recently, a second lipase gene, *gehD*, from *S. epidermidis* strain 9 has been cloned and sequenced (Longshaw et al., 2000). The molecular mass of the mature form of GehD is approximately 45 kDa, yet GehD and GehC share only 51 per cent identical amino acids. Southern blot hybridisations indicated that usually both genes are present in the same strain and that they are widely distributed within the *S. epidermidis* population. In databases of *S. aureus* strains Col and 8325, homologous DNA regions to *gehC* and *gehD* were found, indicating that a number of *S. aureus* strains also produce two lipases (Longshaw et al., 2000).

The production of lantibiotics is another feature of *S. epidermidis* that may explain its successful colonisation of skin. Lantibiotics are bacteriocins which are produced by *S. epidermidis* and other Gram-positive bacteria, for example, *Bacillus subtilis* and lactobacilli, but not by *S. aureus*. Lantibiotics, such as the well-characterised epidermin (Kupke and Götz, 1996) and Pep5 (Meyer et al., 1995) and the newly identified epilancin K7 (van de Kamp et al., 1995) and epicidin 280 (Heidrich et al., 1998), are antibiotic peptides that contain the rare thioether amino acids lanthionine and/or methyllanthionine and are active against Gram-positive bacteria. Lantibiotic production may play a substantial role in bacterial interference on skin and mucous membranes by excluding competing organisms that are sensitive to their bactericidal activities. In general, these peptides are gene encoded and posttranslationally modified. The genes involved are organised in biosynthetic gene clusters located on plasmids (Heidrich et al., 1998; Kupke and Götz, 1996; Kupke et al., 2000; Meyer et al., 1995; van de Kamp et al., 1995).

3.2 Factors Involved in the Destruction of Host Tissue and Interference with Host Defences

Following the predisposing steps of commensal colonisation and prosthetic device contamination, the subsequent establishment of an infection and the survival of the bacteria in the host depend on the ability of pathogens to invade host tissues and to evade host defence systems. For this, staphylococci, in particular *S. aureus*, have evolved multiple mechanisms, including production of: (a) a variety of extracellular proteins and enzymes, such as protein A, lipases, proteases, esterases, phospholipases, fatty-acid modifying enzymes (FAME); (b) haemolysins; and (c) toxins with superantigenic properties, such as enterotoxins, exfoliative toxins, and TSST-1. The effects of these staphylococcal products are destruction of the host tissue, thereby facilitating invasiveness. Proteases may also play a role in proteolytic inactivation of host defence mechanisms such as antibodies and platelet microbicidal proteins (PMPs).

In *S. epidermidis*, an extracellular metalloprotease with elastase activity has been detected, and its gene has been cloned and sequenced (Teufel and Götz, 1993). Previously, an elastase from *S. epidermidis* that degrades human sIgA, IgM, serum albumin, Fg, and Fn had been identified as a cysteine protease and thus was assumed to be a virulence factor (Sloot et al., 1992). However, the corresponding gene has not yet been cloned. An extracellular serine protease is involved in the processing of the lantibiotic epidermin (see earlier) (Geissler, Götz, and Kupke, 1996). The characterisation and expression of FAME in *S. epidermidis* has also been described (Chamberlain and Brueggemann, 1997).

In contrast to *S. aureus*, which produces numerous toxins (see earlier) in a strain-dependent manner, *S. epidermidis* is much less toxigenic. *S. epidermidis* can produce δ-toxin, which differs from the *S. aureus* δ-toxin in only three amino acids (McKevitt et al., 1990). The δ-toxin is encoded by *hld*, which is a component of the regulatory *agr* system (RNAIII, see Section 4.3) and acts by formation of pores in the membrane, leading to the lysis of erythrocytes and other mammalian cells. Reports on unusual *S. epidermidis* strains producing enterotoxin C or TSST-1 (Becker et al., 2001; Marin, de la Rosa, and Cornejo, 1992) are controversial.

A serious consequence of *S. epidermidis* polymer-associated infection is septicaemia. In the pathophysiology of inflammatory events in septicaemia, the production of cytokines is thought to play a major role. Human monocytes release tumor necrosis factor-α (TNF-α), interleukin-1β (IL-1), and interleukin-6 (IL-6) in a concentration-dependent manner when stimulated by cell wall components such as peptidoglycan and teichoic acid. Further studies revealed that the addition of human serum to the assay strongly increased peptidoglycan-induced TNF-α release by human monocytes (Mattsson et al., 1994).

In contrast to the possibility that cell surface components of *S. epidermidis* elicit non-specific activation of immunity, extracellular products of *S. epidermidis* appear to interfere with several immune functions. Extracellular material produced by *S. epidermidis* has been shown to reduce the blastogenic response of human peripheral mononuclear cells to T cell mitogens in a dose-dependent manner (Gray et al., 1984). The underlying mechanisms are still unclear, as is the biological relevance. Another effect of extracellular products of *S. epidermidis* is interference with neutrophils. Although extracellular slime induced a significant chemotactic response in human polymorphonuclear neutrophils (PMNs) (Johnson et al., 1986), when PMNs were preincubated with increasing amounts of crude slime, their responsiveness to known chemotactic stimuli was inhibited. In addition, the preincubation of PMNs with slime stimulated PMN degranulation. This may lead to depletion of the PMNs' antibacterial cellular products after contact with slime, which together with reduced chemotactic responsiveness results in a decreased ability for intracellular killing. Indeed, in a biomaterial-associated opsonophagocytosis assay using human PMNs, the survival of accumulation-positive strain *S. epidermidis* RP62A was significantly greater than for its accumulation-negative mutant M7 (G. Johnson, personal communication). SdrG may similarly inhibit host immunity by prevention of the release of chemotactic elements such as fibrinopeptide B (see earlier).

4 REGULATORY MECHANISMS INVOLVED IN BIOFILM FORMATION

In contrast to *S. epidermidis*, the regulation of *S. aureus* virulence factors has been well studied in recent years. At least two global regulators, the *agr* (accessory gene regulator) locus and the *sar* (staphylococcal accessory regulator) locus, are involved in regulation of *S. aureus* virulence factors (for a review, see Novick, 2000). The expression of both *agr* and *sar* is influenced by the alternative transcription factor σ^B, which is the global regulator of stress responses in many bacterial species. Recently, homologous regulatory mechanisms have been identified in *S. epidermidis*.

4.1 Environmental Factors and Stress

It has been previously shown that *S. epidermidis* biofilm formation is modulated by environmental conditions and stress, such as high osmolarity, detergents, urea, ethanol, oxidative stress, and subinhibitory concentrations of antibiotics. Recent studies demonstrated that the expression of the *icaADBC* gene cluster, and thus PIA production and biofilm accumulation, is influenced by such factors. Use of a reporter gene fusion between the *ica* promoter and *lacZ* from *Escherichia coli* revealed that external stress, such as high temperature and high osmolarity, induced *icaADBC* gene expression and biofilm formation (Rachid et al., 2000b). Subinhibitory concentrations of some antibiotics had a similar effect: tetracycline and a semisynthetic streptogramin (quinopristin-dalfopristin) led to a 9- to 11-fold increase of *ica* expression, whereas erythromycin increased *ica* expression only 2.5-fold (Rachid et al., 2000b). The presence of ethanol or high concentrations of sodium chloride in the growth medium also created stress conditions, which led to increased biofilm formation and PIA production in *S. epidermidis* (Knobloch et al., 2001). Furthermore, the production of PIA and synthesis of the *ica*-specific mRNA was significantly enhanced under anaerobic *in vitro* growth conditions (Cramton et al., 2001). The effect of oxygen limitation may play an important role in the *in vivo* situation, where anaerobiosis occurs during localised infections and leads to increased biofilm formation. The modulation of gene expression by environmental factors and stress has been shown to be at least partly mediated by the alternative sigma factor σ^B.

4.2 The *sigB* (sigma factor B) Operon

Many bacterial species express various alternative sigma factors to adapt gene expression to altered environmental conditions. Currently, in *S. aureus* only the alternative sigma factor, σ^B, has been described. The gene encoding σ^B is part of an operon (*sigB* operon) that contains four ORFs: *rsbU, rsbV, rsbW,* and *sigB*. RsbU, RsbV, and RsbW are involved in regulation of σ^B, with RsbU being a required activator of σ^B. σ^B-mediated regulation is not functional in *S. aureus* strains that naturally carry a mutated *rsbU* (Bischoff, Entenza, and Giachino, 2001; Kullik, Giachino, and Fuchs, 1998).

Recently, a homologous *sigB* operon has been identified in *S. epidermidis* (Kies et al., 2001; Knobloch et al., 2001). Tn*917* insertion into *rsbU* resulted in a biofilm-negative phenotype due to dramatically reduced PIA synthesis, strongly suggesting that these factors are σ^B dependent (Knobloch et al., 2001). Ethanol and salt stress are both known activators of σ^B. However, the presence of ethanol in the growth medium of the *rsbU* mutant completely restored biofilm formation, whereas salt stress did not. Thus, different regulatory mechanisms are involved in *S. epidermidis* PIA production in response to ethanol and salt stress. Similar findings were obtained for *S. aureus* (Rachid et al., 2000a). However, in contrast, with a σ^B deletion mutant in strain *S. epidermidis* Tü3298, no significant effect of σ^B on either biofilm formation or *agr* activity was detected (Kies et al., 2001). This may be due to a nonfunctional RsbU-mediated signal transduction pathway in that particular strain.

4.3 The *agr* (accessory gene regulator) Locus

The expression of genes for many extracellular and surface proteins in *S. aureus* is regulated in a growth-phase-dependent manner by the global regulator *agr*. At the onset of the stationary growth phase, the production of surface-associated proteins is downregulated, whereas the production of extracellular proteins and toxins is upregulated. The extracellular signalling molecule of this quorum-sensing system is a thiolactone-containing peptide. The *agr* locus is transcribed into two divergent transcripts, RNAII and RNAIII. RNAIII is transcribed from the promoter P3; it mediates the regulation of its target genes (i.e., genes for many cell-surface and secreted proteins) by an unknown mechanism. Besides having a regulatory function, RNAIII also specifies the *hld* gene encoding the δ-toxin (see earlier). The four genes, *agrB, agrD, agrC,* and *agrA*, are arranged in an operon and transcribed into RNAII from the promotor P2. *agrD* codes for the autoinducing signalling peptide, which is excised from the AgrD protein and then modified and secreted into the surrounding environment. The maturation and secretion process seems to be carried out by AgrB. AgrA and AgrC form a classical two-component regulatory system with AgrC being a membrane-associated signal-dependent histidine kinase, which is autophosphorylated upon binding of the signalling peptide. Next, the response regulator protein, AgrA, is phosphorylated and activates the transcription of RNAII and RNAIII (together with SarA, see Section 4.4) in a manner that is not completely understood.

An *agr* homolog in *S. epidermidis* has been identified and sequenced (Otto et al., 1998; Van Wamel et al., 1998). DNA sequence analysis revealed a pronounced similarity between the *S. epidermidis* and *S. aureus agr* systems. The extracellular signalling molecule produced by a typical *S. epidermidis* strain is a cyclic octapeptide (DSVCASYF), which is encoded by *agrD* and contains a thiolester linkage between the central cysteine and the C-terminal carboxyl group (Otto et al., 1998). This octapeptide exhibits activity at nanomolar concentrations. Sequence comparison revealed no striking similarity between the signalling peptides of *S. epidermidis*, *S. aureus*, or *S. lugdunensis* (hepta-, octa-, or nonapeptides) except for the central cysteine and its distance to the C-terminus. Therefore, these conserved structural features are thought to be necessary for thiolactone formation. The AgrD proteins of *S. epidermidis* and *S. aureus* show evident similarity in the region located C-terminal of the signalling peptides, suggesting that this region represents a structural element important for the modifying reaction probably mediated by AgrB. AgrB shows an overall identity of 51.3 per cent between both species. Its location within the cytoplasmic membrane has recently been demonstrated (Saenz et al., 2000). The *S. epidermidis* histidine kinase AgrC shares 50.5 per cent identical amino acids with the *S. aureus* AgrC, with pronounced similarity in the C-terminal portion and low similarity in the N-terminal portion. These sequence data are also in agreement with the *S. lugdunensis* data, leading to the hypothesis that the N-terminal part of AgrC represents the region binding the signalling peptides which differ in sequence, whereas the C-terminal part interacts with the highly conserved response regulator AgrA (87.3 per cent identity between *S. epidermidis* and *S. aureus*). In addition, an RNAIII homolog in *S. epidermidis* (560 nucleotides) was shown to regulate virulence gene expression in *S. aureus* (Tegmark, Morfeldt, and Arvidson, 1998). The *S. epidermidis* RNAIII had the ability to completely repress transcription of protein A and to activate transcription of the α-toxin (*hla*) and serine protease (*ssp*) genes in an RNAIII-deficient *S. aureus* mutant. However, the stimulatory effect was reduced compared to that of the homologous *S. aureus* RNAIII. Especially the first 50 and last 150 nucleotides of RNAIII were found to be highly similar in *S. epidermidis* and *S. aureus*. Analysis of *S. epidermidis*-*S. aureus* RNAIII hybrid molecules showed that both the 5′ and 3′ halves of the RNA molecule are important for regulation.

Recently, an *agr* mutant of *S. epidermidis* was constructed and characterised (Vuong, Götz, and Otto, 2000a). In comparison to the wild type, the *agr* mutant showed significantly altered protein expression: the expression of surface-associated proteins was increased, whereas the expression of extracellular proteins, as shown for the exoenzymes lipase and protease, was decreased. Thus, the function of the *S. epidermidis agr* system in growth-phase-dependent regulation of protein synthesis may correspond to that of *S. aureus* and may be summarised as follows: in an early stage of infection, the cell density is low and surface-associated proteins with adhesive functions are expressed, allowing colonisation of the polymer surface and host protein-coated material. Upon proliferation of the cells on the surface, the autoinducing signalling peptides accumulate and eventually reach the critical concentration necessary for the activation of the synthesis of RNAIII. RNAIII then downregulates the expression of surface protein genes and upregulates the production of tissue-degrading enzymes and other secreted proteins, which facilitates the maintenance of the infection. This potential role of *agr* in *S. epidermidis* disease suggests the use of quorum-sensing blockers as alternative antistaphylococcal drugs. However, *agr* was also found to influence biofilm formation. In *S. aureus*, defined *agr* mutants had an increased ability to form a biofilm (Vuong et al., 2000b). Inhibition of *agr* by quorum-sensing blockers had a similar effect. This makes it doubtful that the proposed use of quorum-sensing blockers as antistaphylococcal drugs would be of any great benefit (Otto, 2001). However, after a biofilm has formed and an *S. epidermidis* infection has been diagnosed, they might be useful in the downregulation of exoproteins that inhibit host defences.

4.4 The *sar* (staphylococcal accessory regulator) Locus

In *S. aureus*, another global regulator, *sar*, also controls exoprotein synthesis by modulating the expression of *agr*. The *sar* locus in *S. aureus* contains a major ORF, *sarA*, preceded by two smaller ORFs. DNA mobility shift assays demonstrated that the *sar* gene products bind to an *agr* P2 promotor fragment, probably leading to activation of transcription of RNAII and subsequently RNAIII. A *sar* homolog of *S. epidermidis* has been cloned and sequenced, which revealed that the SarA protein of *S. epidermidis* is nearly identical (84 per cent) to SarA of *S. aureus* (Fluckiger, Wolz, and Cheung, 1998). In contrast, the *sarA* flanking DNA sequence shows only 50 per cent identity between both strains, and the two smaller ORFs are absent in *S. epidermidis*.

Remarkably, an *S. epidermidis sar* fragment, including *sarA* and the upstream flanking region, interacts with an *agr* promoter fragment of *S. aureus*. Moreover, functional analysis confirmed that the *S. epidermidis sar* homolog was able to restore α-toxin production in an *S. aureus sar* mutant (Fluckiger et al., 1998). Because most of the typical virulence determinants of *S. aureus* are missing in *S. epidermidis*, it needs to be clarified which genes are under the control of *agr* and *sar* in *S. epidermidis*. Possible candidates include genes encoding the Fg-binding protein Fbe, SdrG, SdrF, SdrH, Bhp, AAP, autolysins, lipases, PIA and PS/A production, proteases, and δ-toxin.

4.5 Phase Variation

It has been observed for many years that biofilm-forming *S. epidermidis* strains can undergo phase variation, resulting in biofilm-negative phenotypes. A biofilm-negative variant of the clinical reference strain RP62A was deficient for the production of an extracellular polysaccharide material and was less virulent in animal models of foreign body infections (Christensen, Baddour, and Simpson, 1987). More recently, biofilm- and autoaggregation-negative phase variants were isolated from biofilm-producing blood culture strains using Congo red agar (Ziebuhr et al., 1997). These variants occurred at a frequency of 10^{-5}, and the phenotype of the wild-type strain could be restored after repeated passages. Genetic analysis revealed that in approximately 30 per cent of those variants, the occurrence of the biofilm-negative phenotype was due to the inactivation of either the *icaA* or the *icaC* gene by the insertion sequence element IS256 (Ziebuhr et al., 1999). This insertion is a reversible process and involves the exact excision of the IS256 element, including the 8 base pairs of the target sites, which resulted from duplication during the transposition of the IS element. The complete excision of the IS256 element led to the restoration of the intact *icaA* and *icaC* genes and the biofilm-positive phenotype, thereby explaining the phase variation and modulation of expression of an *S. epidermidis* virulence factor at a molecular level. It may be speculated that a switch from a biofilm-forming phenotype to a biofilm-negative variant may serve in the dissemination from an infected medical device, leading to circulation in the host organism and thereby allowing the colonisation of another potential site of infection.

5 FUTURE ASPECTS

Our knowledge of the molecular basis of biofilm formation by *S. epidermidis* has increased significantly during recent years. Nevertheless, we are still far away from a sufficient understanding of this 'versatile' microorganism. We still know very little about the biology of *S. epidermidis* in its commensal habitat. Increased research in this area is obviously necessary to gain more insight into the complex balance mechanisms between the bacterium and its human host. This will result in an understanding of when and how *S. epidermidis* changes from a saprophyte to a pathogen. The continual improvement of molecular methods will enable the analysis of not only the genome, but also the proteome of *S. epidermidis*. However, this has to be complemented by further research at the functional level, including the development of better animal models.

REFERENCES

Baldassarri, L., Donnelli, G., Gelosia, A., Voglino, M. C., Simpson, A. W. and Christensen, G. D. (1996). Purification and characterization of the staphylococcal slime-associated antigen and its occurrence among *Staphylococcus epidermidis* clinical isolates. *Infection and Immunity*, 64, 3410–3415.

Becker, K., Haverkamper, G., von Eiff, C., Roth, R. and Peters, G. (2001). Survey of staphylococcal enterotoxin genes, exfoliative toxin genes, and toxic shock syndrome toxin 1 gene in non-*Staphylococcus aureus* species. *European Journal of Clinical Microbiology and Infectious Diseases*, 20, 407–409.

Bischoff, M., Entenza, J. M. and Giachino, P. (2001). Influence of a functional *sigB* operon on the global regulators *sar* and *agr* in *Staphylococcus aureus*. *Journal of Bacteriology*, 183, 5171–5179.

Bubert, A., Kuhn, M., Goebel, W. and Kohler, S. (1992). Structural and functional properties of the p60 proteins from different *Listeria* species. *Journal of Bacteriology*, 174, 8166–8171.

Chamberlain, N. R. and Brueggemann, S. A. (1997). Characterisation and expression of fatty acid modifying enzyme produced by *Staphylococcus epidermidis*. *Journal of Medical Microbiology*, 46, 693–697.

Chhatwal, G. S., Preissner, K. T., Muller-Berghaus, G. and Blobel, H. (1987). Specific binding of the human S protein (vitronectin) to streptococci, *Staphylococcus aureus*, and *Escherichia coli*. *Infection and Immunity*, 55, 1878–1883.

Christensen, G. D., Baddour, L. M. and Simpson, W. A. (1987). Phenotypic variation of *Staphylococcus epidermidis* slime production in vitro and in vivo. *Infection and Immunity*, 55, 2870–2877.

Christensen, G. D., Simpson, W. A., Bisno, A. L. and Beachey, E. H. (1982). Adherence of slime-producing strains of *Staphylococcus epidermidis* to smooth surfaces. *Infection and Immunity*, 37, 318–326.

Cottonaro, C. N., Roohk, H. V., Shimizu, G. and Sperling, D. R. (1981). Quantitation and characterization of competitive protein binding to polymers. *Transactions of the American Society for Artificial Internal Organs*, 27, 391–395.

Cramton, S. E., Gerke, C., Schnell, N. F., Nichols, W. W. and Götz, F. (1999). The intercellular adhesion (*ica*) locus is present in *Staphylococcus aureus* and is required for biofilm formation. *Infection and Immunity*, 67, 5427–5433.

Cramton, S. E., Ulrich, M., Götz, F. and Döring, G. (2001). Anaerobic conditions induce expression of polysaccharide intercellular adhesin in *Staphylococcus aureus* and *Staphylococcus epidermidis*. *Infection and Immunity*, 69, 4079–4085.

Cucarella, C., Solano, C., Valle, J., Amorena, B., Lasa, I. and Penades, J. R. (2001). Bap, a *Staphylococcus aureus* surface protein involved in biofilm formation. *Journal of Bacteriology*, 183, 2888–2896.

Davis, S. L., Gurusiddappa, S., McCrea, K. W., Perkins, S. and Höök, M. (2001). SdrG, a fibrinogen-binding bacterial adhesin of the microbial surface components recognizing adhesive matrix molecules subfamily from *Staphylococcus epidermidis*, targets the thrombin cleavage site in the beta chain. *Journal of Biological Chemistry*, 276, 27799–27805.

Dickinson, G. M. and Bisno, A. L. (1989). Infections associated with indwelling devices: concepts of pathogenesis; infections associated with intravascular devices. *Antimicrobial Agents and Chemotherapy*, 33, 597–601.

Farrell, A. M., Foster, T. J. and Holland, K. T. (1993). Molecular analysis and expression of the lipase of *Staphylococcus epidermidis*. *Journal of General Microbiology*, 139, 267–277.

Fey, P. D., Ulphani, J. S., Götz, F., Heilmann, C., Mack, D. and Rupp, M. E. (1999). Characterization of the relationship between polysaccharide intercellular adhesin and hemagglutination in *Staphylococcus epidermidis*. *Journal of Infectious Diseases*, 179, 1561–1564.

Fluckiger, U., Wolz, C. and Cheung, A. L. (1998). Characterization of a *sar* homolog of *Staphylococcus epidermidis*. *Infection and Immunity*, 66, 2871–2878.

Franson, T. R., Sheth, N. K., Rose, H. D. and Sohnle, P. G. (1984). Scanning electron microscopy of bacteria adherent to intravascular catheters. *Journal of Clinical Microbiology*, 20, 500–505.

Geissler, S., Götz, F. and Kupke, T. (1996). Serine protease EpiP from *Staphylococcus epidermidis* catalyzes the processing of the epidermin precursor peptide. *Journal of Bacteriology*, 178, 284–288.

George, R., Leibrock, L. and Epstein, M. (1979). Long-term analysis of cerebrospinal fluid shunt infections. A 25-year experience. *Journal of Neurosurgery*, 51, 804–811.

Gerke, C., Kraft, A., Sussmuth, R., Schweitzer, O. and Götz, F. (1998). Characterization of the N-acetylglucosaminyltransferase activity involved in the biosynthesis of the *Staphylococcus epidermidis* polysaccharide intercellular adhesin. *Journal of Biological Chemistry*, 273, 18586–18593.

Gray, E. D., Peters, G., Verstegen, M. and Regelmann, W. E. (1984). Effect of extracellular slime substance from *Staphylococcus epidermidis* on the human cellular immune response. *Lancet*, 1, 365–367.

Gross, M., Cramton, S. E., Götz, F. and Peschel, A. (2001). Key role of teichoic acid net charge in *Staphylococcus aureus* colonization of artificial surfaces. *Infection and Immunity*, 69, 3423–3426.

Hartford, O., O'Brien, L., Schofield, K., Wells, J. and Foster, T. J. (2001). The Fbe (SdrG) protein of *Staphylococcus epidermidis* HB promotes bacterial adherence to fibrinogen. *Microbiology*, 147, 2545–2552.

Hartleib, J., Kohler, N., Dickinson, R. B., Chhatwal, G. S., Sixma, J. J., Hartford, O. M.,

Foster, T. J., Peters, G., Kehrel, B. E. and Herrmann, M. (2000). Protein A is the von Willebrand factor binding protein on *Staphylococcus aureus*. *Blood*, 96, 2149–2156.

Heidrich, C., Pag, U., Josten, M., Metzger, J., Jack, R. W., Bierbaum, G., Jung, G. and Sahl, H. G. (1998). Isolation, characterization, and heterologous expression of the novel lantibiotic epicidin 280 and analysis of its biosynthetic gene cluster. *Applied and Environmental Microbiology*, 64, 3140–3146.

Heilmann, C., Chhatwal, S. and Peters, G. (2000). In *Microbiology 2000*, Abstr. 15.P.18.14, Munich, Germany.

Heilmann, C., Gerke, C., Perdreau-Remington, F. and Götz, F. (1996a). Characterization of Tn*917* insertion mutants of *Staphylococcus epidermidis* affected in biofilm formation. *Infection and Immunity*, 64, 277–282.

Heilmann, C. and Götz, F. (1998). Further characterization of *Staphylococcus epidermidis* transposon mutants deficient in primary attachment or intercellular adhesion. *Zentralblatt fur Bakteriologie*, 287, 69–83.

Heilmann, C., Hussain, M., Peters, G. and Götz, F. (1997). Evidence for autolysin-mediated primary attachment of *Staphylococcus epidermidis* to a polystyrene surface. *Molecular Microbiology*, 24, 1013–1024.

Heilmann, C. and Peters, G. (2000). In *Gram Positive Pathogens* (Eds. Fischetti, V. A., Novick, R. P., Ferretti, J. J., Portnoy, D. A., and Rood, J. I.), American Society for Microbiology, Washington, DC, pp. 442–449.

Heilmann, C., Schweitzer, O., Gerke, C., Vanittanakom, N., Mack, D. and Götz, F. (1996b). Molecular basis of intercellular adhesion in the biofilm-forming *Staphylococcus epidermidis*. *Molecular Microbiology*, 20, 1083–1091.

Hell, W., Meyer, H. G. and Gatermann, S. G. (1998). Cloning of *aas*, a gene encoding a *Staphylococcus saprophyticus* surface protein with adhesive and autolytic properties. *Molecular Microbiology*, 29, 871–881.

Herrmann, M., Hartleib, J., Kehrel, B., Montgomery, R. R., Sixma, J. J. and Peters, G. (1997). Interaction of von Willebrand factor with *Staphylococcus aureus*. *Journal of Infectious Diseases*, 176, 984–991.

Herrmann, M., Vaudaux, P. E., Pittet, D., Auckenthaler, R., Lew, P. D., Schumacher-Perdreau, F., Peters, G. and Waldvogel, F. A. (1988). Fibronectin, fibrinogen, and laminin act as mediators of adherence of clinical staphylococcal isolates to foreign material. *Journal of Infectious Diseases*, 158, 693–701.

Hussain, M., Heilmann, C., Herrmann, M. and Peters, G. (2000). In *100th General Meeting of the American Society for Microbiology*, Abstr. B-229, Los Angeles, CA.

Hussain, M., Heilmann, C., Peters, G. and Herrmann, M. (2001). Teichoic acid enhances adhesion of *Staphylococcus epidermidis* to immobilized fibronectin. *Microbial Pathogenesis*, 31, 261–270.

Hussain, M., Herrmann, M., von Eiff, C., Perdreau-Remington, F. and Peters, G. (1997). A 140-kilodalton extracellular protein is essential for the accumulation of *Staphylococcus epidermidis* strains on surfaces. *Infection and Immunity*, 65, 519–524.

Hussain, M., Wilcox, M. H. and White, P. J. (1993). The slime of coagulase-negative staphylococci: biochemistry and relation to adherence. *FEMS Microbiology Review*, 10, 191–207.

Johnson, G. M., Lee, D. A., Regelmann, W. E., Gray, E. D., Peters, G. and Quie, P. G. (1986). Interference with granulocyte function by *Staphylococcus epidermidis* slime. *Infection and Immunity*, 54, 13–20.

Josefsson, E., McCrea, K. W., Ni Eidhin, D., O'Connell, D., Cox, J., Höök, M. and Foster, T. J. (1998). Three new members of the serine-aspartate repeat protein multigene family of *Staphylococcus aureus. Microbiology*, 144, 3387–3395.

Karchmer, A. W., Archer, G. L. and Dismukes, W. E. (1983). *Staphylococcus epidermidis* causing prosthetic valve endocarditis: microbiologic and clinical observations as guides to therapy. *Annals of Internal Medicine*, 98, 447–455.

Kies, S., Otto, M., Vuong, C. and Götz, F. (2001). Identification of the *sigB* operon in *Staphylococcus epidermidis*: construction and characterization of a *sigB* deletion mutant. *Infection and Immunity*, 69, 7933–7936.

Knobloch, J. K., Bartscht, K., Sabottke, A., Rohde, H., Feucht, H. H. and Mack, D. (2001). Biofilm formation by *Staphylococcus epidermidis* depends on functional RsbU, an activator of the *sigB* operon: differential activation mechanisms due to ethanol and salt stress. *Journal of Bacteriology*, 183, 2624–2633.

Kochwa, S., Litwak, R. S., Rosenfield, R. E. and Leonard, E. F. (1977). Blood elements at foreign surfaces: a biochemical approach to the study of the adsorption of plasma proteins. *Annals of the New York Academy of Science*, 283, 37–49.

Kuhn, M. and Goebel, W. (1989). Identification of an extracellular protein of *Listeria monocytogenes* possibly involved in intracellular uptake by mammalian cells. *Infection and Immunity*, 57, 55–61.

Kullik, I., Giachino, P. and Fuchs, T. (1998). Deletion of the alternative sigma factor sigmaB in *Staphylococcus aureus* reveals its function as a global regulator of virulence genes. *Journal of Bacteriology*, 180, 4814–4820.

Kupke, T. and Götz, F. (1996). Expression, purification, and characterization of EpiC, an enzyme involved in the biosynthesis of the lantibiotic epidermin, and sequence analysis of *Staphylococcus epidermidis epiC* mutants. *Journal of Bacteriology*, 178, 1335–1340.

Kupke, T., Uebele, M., Schmid, D., Jung, G., Blaesse, M. and Steinbacher, S. (2000). Molecular characterization of lantibiotic-synthesizing enzyme EpiD reveals a function for bacterial Dfp proteins in coenzyme A biosynthesis. *Journal of Biological Chemistry*, 275, 31838–31846.

Longshaw, C. M., Farrell, A. M., Wright, J. D. and Holland, K. T. (2000). Identification of a second lipase gene, *gehD*, in *Staphylococcus epidermidis*: comparison of sequence with those of other staphylococcal lipases. *Microbiology*, 146, 1419–1427.

Lopes, J. D., dos Reis, M. and Brentani, R. R. (1985). Presence of laminin receptors in *Staphylococcus aureus. Science*, 229, 275–277.

Mack, D., Fischer, W., Krokotsch, A., Leopold, K., Hartmann, R., Egge, H. and Laufs, R. (1996). The intercellular adhesin involved in biofilm accumulation of *Staphylococcus epidermidis* is a linear beta-1,6-linked glucosaminoglycan: purification and structural analysis. *Journal of Bacteriology*, 178, 175–183.

Marin, M. E., de la Rosa, M. C. and Cornejo, I. (1992). Enterotoxigenicity of *Staphylococcus* strains isolated from Spanish dry-cured hams. *Applied and Environmental Microbiology*, 58, 1067–1069.

Mattsson, E., Rollof, J., Verhoef, J., Van Dijk, H. and Fleer, A. (1994). Serum-induced potentiation of tumor necrosis factor alpha production by human monocytes in

response to staphylococcal peptidoglycan: involvement of different serum factors. *Infection and Immunity*, 62, 3837–3843.
McCrea, K. W., Hartford, O., Davis, S., Eidhin, D. N., Lina, G., Speziale, P., Foster, T. J. and Höök, M. (2000). The serine-aspartate repeat (Sdr) protein family in *Staphylococcus epidermidis*. *Microbiology*, 146, 1535–1546.
McDevitt, D., Francois, P., Vaudaux, P. and Foster, T. J. (1994). Molecular characterization of the clumping factor (fibrinogen receptor) of *Staphylococcus aureus*. *Molecular Microbiology*, 11, 237–248.
McKenney, D., Hubner, J., Muller, E., Wang, Y., Goldmann, D. A. and Pier, G. B. (1998). The *ica* locus of *Staphylococcus epidermidis* encodes production of the capsular polysaccharide/adhesin. *Infection and Immunity*, 66, 4711–4720.
McKevitt, A. I., Bjornson, G. L., Mauracher, C. A. and Scheifele, D. W. (1990). Amino acid sequence of a deltalike toxin from *Staphylococcus epidermidis*. *Infection and Immunity*, 58, 1473–1475.
Meyer, C., Bierbaum, G., Heidrich, C., Reis, M., Suling, J., Iglesias-Wind, M. I., Kempter, C., Molitor, E. and Sahl, H. G. (1995). Nucleotide sequence of the lantibiotic Pep5 biosynthetic gene cluster and functional analysis of PepP and PepC. Evidence for a role of PepC in thioether formation. *European Journal of Biochemistry*, 232, 478–489.
Milohanic, E., Jonquieres, R., Cossart, P., Berche, P. and Gaillard, J. L. (2001). The autolysin Ami contributes to the adhesion of *Listeria monocytogenes* to eukaryotic cells via its cell wall anchor. *Molecular Microbiology*, 39, 1212–1224.
Milohanic, E., Pron, B., Berche, P. and Gaillard, J. L. (2000). Identification of new loci involved in adhesion of *Listeria monocytogenes* to eukaryotic cells. European *Listeria* Genome Consortium. *Microbiology*, 146, 731–739.
Muller, E., Hubner, J., Gutierrez, N., Takeda, S., Goldmann, D. A. and Pier, G. B. (1993). Isolation and characterization of transposon mutants of *Staphylococcus epidermidis* deficient in capsular polysaccharide/adhesin and slime. *Infection and Immunity*, 61, 551–558.
Ni Eidhin, D., Perkins, S., Francois, P., Vaudaux, P., Höök, M. and Foster, T. J. (1998). Clumping factor B (ClfB), a new surface-located fibrinogen-binding adhesin of *Staphylococcus aureus*. *Molecular Microbiology*, 30, 245–257.
Nilsson, M., Frykberg, L., Flock, J. I., Pei, L., Lindberg, M. and Guss, B. (1998). A fibrinogen-binding protein of *Staphylococcus epidermidis*. *Infection and Immunity*, 66, 2666–2673.
Novick, R. P. (2000). In *Gram-Positive Pathogens* (Eds. Fischetti, V. A., Novick, R. P., Ferretti, J. J., Portnoy, D. A. and Rood, J. I.), American Society for Microbiology, Washington, D.C., pp. 392–407.
Otto, M. (2001). *Staphylococcus aureus* and *Staphylococcus epidermidis* peptide pheromones produced by the accessory gene regulator *agr* system. *Peptides*, 22, 1603–1608.
Otto, M., Sussmuth, R., Jung, G. and Götz, F. (1998). Structure of the pheromone peptide of the *Staphylococcus epidermidis agr* system. *FEBS Letters*, 424, 89–94.
Pei, L. and Flock, J. I. (2001a). Functional study of antibodies against a fibrogenin-binding protein in *Staphylococcus epidermidis* adherence to polyethylene catheters. *Journal of Infectious Diseases*, 184, 1.

Pei, L. and Flock, J. I. (2001b). Lack of *fbe*, the gene for a fibrinogen-binding protein from *Staphylococcus epidermidis*, reduces its adherence to fibrinogen coated surfaces. *Microbial Pathogenesis*, 31, 185–193.
Pei, L., Palma, M., Nilsson, M., Guss, B. and Flock, J. I. (1999). Functional studies of a fibrinogen binding protein from *Staphylococcus epidermidis*. *Infection and Immunity*, 67, 4525–4530.
Peters, G., Locci, R. and Pulverer, G. (1981). Microbial colonization of prosthetic devices. II. Scanning electron microscopy of naturally infected intravenous catheters. *Zentralblatt fur Bakteriologie*, 173, 293–299.
Peters, G., Locci, R. and Pulverer, G. (1982). Adherence and growth of coagulase-negative staphylococci on surfaces of intravenous catheters. *Journal of Infectious Diseases*, 146, 479–482.
Peters, G., Saborowski, F., Locci, R. and Pulverer, G. (1984). Investigations on staphylococcal infection of transvenous endocardial pacemaker electrodes. *American Heart Journal*, 108, 359–365.
Rachid, S., Ohlsen, K., Wallner, U., Hacker, J., Hecker, M., and Ziebuhr, W. (2000a). Alternative transcription factor sigma(B) is involved in regulation of biofilm expression in a *Staphylococcus aureus* mucosal isolate. *Journal of Bacteriology*, 182, 6824–6826.
Rachid, S., Ohlsen, K., Witte, W., Hacker, J. and Ziebuhr, W. (2000b). Effect of subinhibitory antibiotic concentrations on polysaccharide intercellular adhesin expression in biofilm-forming *Staphylococcus epidermidis*. *Antimicrobial Agents and Chemotherapy*, 44, 3357–3363.
Rupp, M. E., Fey, P. D., Heilmann, C. and Götz, F. (2001). Characterization of the importance of *Staphylococcus epidermidis* autolysin and polysaccharide intercellular adhesin in the pathogenesis of intravascular catheter-associated infection in a rat model. *Journal of Infectious Diseases*, 183, 1038–1042.
Rupp, M. E., Ulphani, J. S., Fey, P. D., Bartscht, K. and Mack, D. (1999a). Characterization of the importance of polysaccharide intercellular adhesin/hemagglutinin of *Staphylococcus epidermidis* in the pathogenesis of biomaterial-based infection in a mouse foreign body infection model. *Infection and Immunity*, 67, 2627–2632.
Rupp, M. E., Ulphani, J. S., Fey, P. D. and Mack, D. (1999b). Characterization of *Staphylococcus epidermidis* polysaccharide intercellular adhesin/hemagglutinin in the pathogenesis of intravascular catheter-associated infection in a rat model. *Infection and Immunity*, 67, 2656–2659.
Saenz, H. L., Augsburger, V., Vuong, C., Jack, R. W., Götz, F. and Otto, M. (2000). Inducible expression and cellular location of AgrB, a protein involved in the maturation of the staphylococcal quorum-sensing pheromone. *Archives of Microbiology*, 174, 452–455.
Schneewind, O., Mihaylova-Petkov, D. and Model, P. (1993). Cell wall sorting signals in surface proteins of gram-positive bacteria. *EMBO Journal*, 12, 4803–4811.
Schumacher-Perdreau, F., Heilmann, C., Peters, G., Götz, F. and Pulverer, G. (1994). Comparative analysis of a biofilm-forming *Staphylococcus epidermidis* strain and its adhesion-positive, accumulation-negative mutant M7. *FEMS Microbiology Letters*, 117, 71–78.
Shiro, H., Muller, E., Gutierrez, N., Boisot, S., Grout, M., Tosteson, T. D., Goldmann, D.

and Pier, G. B. (1994). Transposon mutants of *Staphylococcus epidermidis* deficient in elaboration of capsular polysaccharide/adhesin and slime are avirulent in a rabbit model of endocarditis. *Journal of Infectious Diseases*, 169, 1042–1049.

Simons, J. W., van Kampen, M. D., Riel, S., Götz, F., Egmond, M. R. and Verheij, H. M. (1998). Cloning, purification and characterisation of the lipase from *Staphylococcus epidermidis* – comparison of the substrate selectivity with those of other microbial lipases. *European Journal of Biochemistry*, 253, 675–683.

Sloot, N., Thomas, M., Marre, R. and Gatermann, S. (1992). Purification and characterisation of elastase from *Staphylococcus epidermidis*. *Journal of Medical Microbiology*, 37, 201–205.

Tegmark, K., Morfeldt, E. and Arvidson, S. (1998). Regulation of *agr*-dependent virulence genes in *Staphylococcus aureus* by RNAIII from coagulase-negative staphylococci. *Journal of Bacteriology*, 180, 3181–3186.

Teufel, P. and Götz, F. (1993). Characterization of an extracellular metalloprotease with elastase activity from *Staphylococcus epidermidis*. *Journal of Bacteriology*, 175, 4218–4224.

van de Kamp, M., van den Hooven, H. W., Konings, R. N., Bierbaum, G., Sahl, H. G., Kuipers, O. P., Siezen, R. J., de Vos, W. M., Hilbers, C. W. and van de Ven, F. J. (1995). Elucidation of the primary structure of the lantibiotic epilancin K7 from *Staphylococcus epidermidis* K7. Cloning and characterisation of the epilancin-K7-encoding gene and NMR analysis of mature epilancin K7. *European Journal of Biochemistry*, 230, 587–600.

Van Wamel, W. J., van Rossum, G., Verhoef, J., Vandenbroucke-Grauls, C. M. and Fluit, A. C. (1998). Cloning and characterization of an accessory gene regulator (*agr*)-like locus from *Staphylococcus epidermidis*. *FEMS Microbiology Letters*, 163, 1–9.

Veenstra, G. J., Cremers, F. F., van Dijk, H. and Fleer, A. (1996). Ultrastructural organization and regulation of a biomaterial adhesin of *Staphylococcus epidermidis*. *Journal of Bacteriology*, 178, 537–541.

Vuong, C., Götz, F. and Otto, M. (2000a). Construction and characterization of an *agr* deletion mutant of *Staphylococcus epidermidis*. *Infection and Immunity*, 68, 1048–1053.

Vuong, C., Saenz, H. L., Götz, F. and Otto, M. (2000b). Impact of the *agr* quorum-sensing system on adherence to polystyrene in *Staphylococcus aureus*. *Journal of Infectious Diseases*, 182, 1688–1693.

West, T. E., Walshe, J. J., Krol, C. P. and Amsterdam, D. (1986). Staphylococcal peritonitis in patients on continuous peritoneal dialysis. *Journal of Clinical Microbiology*, 23, 809–812.

Ziebuhr, W., Heilmann, C., Götz, F., Meyer, P., Wilms, K., Straube, E. and Hacker, J. (1997). Detection of the intercellular adhesion gene cluster (*ica*) and phase variation in *Staphylococcus epidermidis* blood culture strains and mucosal isolates. *Infection and Immunity*, 65, 890–896.

Ziebuhr, W., Krimmer, V., Rachid, S., Lössner, I., Götz, F. and Hacker, J. (1999). A novel mechanism of phase variation of virulence in *Staphylococcus epidermidis*: evidence for control of the polysaccharide intercellular adhesin synthesis by alternating insertion and excision of the insertion sequence element IS256. *Molecular Microbiology*, 32, 345–356.

CHAPTER SEVEN

Biofilm Complications of Urinary Tract Devices

Sean P. Gorman and David S. Jones

1 INTRODUCTION

The anatomy of the urinary tract, unfortunately, allows ready access by pathogens to the urethra and beyond when normal defences are breached. The potential for urinary tract infection (UTI) is considerably enhanced by the presence of an indwelling device such as a urethral catheter that provides a conduit to the bladder (Tunney, Jones, and Gorman, 1999). This allows not only the voiding of urine, but also the ingress of microorganisms that colonise the device material and adopt a biofilm growth mode. Such 'device-related' infection is a frequent occurrence in the urinary tract, requiring considerable time and resource in its management. It is estimated that over 40 per cent of nosocomial infections are related to the urinary tract (Nickel, Downey, and Costerton, 1989). Despite careful aseptic management, bacteriuria arises in approximately 50 per cent of patients within 10–14 days and in all those undergoing long-term catheterisation for management of urinary retention and incontinence by 28 days. Additional complications such as blocking encrustations, stone formation, pyelonephritis, and bladder cancer may also arise in patients with asymptomatic infection (Gorman and Tunney, 1997). The elderly are particularly prone to urinary device-related infection. Residents of nursing homes undergoing long-term catheterisation are three times more likely to receive antibiotics, be hospitalised, and die within a year than matched non-catheterised residents (Kunin, Chin, and Chambers, 1987).

The obstruction of urine flow in urinary devices by crystalline encrustation is an additional clinical complication. Although insertion and withdrawal of urinary devices can cause pain and tissue trauma (Jones, Garvin, and Gorman, 2001a), this may be exacerbated by the presence of encrustation. Withdrawal of encrusted catheters causes trauma to the bladder mucosa and urethra, whereas obstruction of the eyelet and lumen blocks the drainage of urine from the bladder with attendant acute pain and distress in the patient. Encrustation in the urinary tract may arise by several mechanisms, one of which involves urease-producing bacteria present on urinary devices (Cox, Hukins, and Sutton, 1989; Gorman, Woolfson, and McCafferty, 1991a).

Similar problems of infection and encrustation are encountered in the ureter wherein ureteral stents are widely used in urological practice to provide internal urine drainage for patients with obstructive uropathy. Most patients with stones require stenting for a short period to relieve acute obstruction; however, those with malignant disease, retroperitoneal fibrosis, or chronic renal failure require stenting for prolonged periods. We have shown that contact between the device material and urine can produce material degradation, encrustation, and stone formation, with encrustation and blockage of the lumen occurring in approximately 75 per cent of stents by 24 weeks (Bonner, Keane, and Gorman, 1993; Keane et al., 1994). Patients requiring long-term ureteral stenting need careful medical supervision and monitoring of stent patency and frequent admissions to the hospital for changes of stents. Stent fracture or stone formation, when it occurs, may require major surgery to remedy the situation.

A priority for fundamental and applied researchers in this field must be to develop biomaterials and systems for use as medical devices exhibiting significantly improved resistance to microbial attachment, biofilm formation, and blocking encrustations. This would constitute a major advance in decreasing the morbidity of patients with, especially, long-term indwelling catheters and stents and would contribute greatly to cost savings in medical care.

2 BACTERIAL ADHERENCE

Adherence to surfaces leading to biofilm formation has evolved as a natural growth and survival strategy for bacteria and is the preferred mode of existence in many areas over planktonic (suspension) populations. Bacteria readily colonise the uroepithelium (Gorman et al., 1987). Although it is readily appreciated that the presence of infecting bacteria is responsible for UTI and much of the observed encrustation of devices and stone formation, it is less apparent that *adherence* of bacteria to a surface plays a major role in the pathogenesis of these complications. Infecting bacteria produce extensive exopolysaccharide 'glycocalyces', often in combination with derivatives from the host environment, to form a confluent microbial biofilm on tissues and

Figure 7.1: Adherence of *Escherichia coli* (rods approximately 2 μm in length) to a latex rubber urethral catheter.

medical devices (Gorman et al., 1993b; Tunney, Gorman, and Patrick, 1996a). The initiation of infection via adherence of bacteria to a material can depend on a number of factors, including the surface charge and hydrophobic character of the pathogen and surface, specific adhesion genes, and surface roughness (Gorman, 1991; Liedl, 2001). From our own observations, Figure 7.1 illustrates bacterial adherence to a urethral catheter surface.

Of considerable concern for clinicians is the magnitude of the resistance, observed in our laboratories amongst others, of biofilm bacteria to high and prolonged levels of antibiotics, which normal means of culture and sensitivity testing have indicated should be therapeutic (Adair et al., 1993; Gorman et al., 2001). In this respect, a clear distinction must be made between bacteria with a planktonic phenotype that are readily eliminated with antibiotic therapy and those having a biofilm phenotype that present particular difficulties for patients with indwelling urinary devices (Morris, Stickler, and McClean, 1999).

3 CATHETER-ASSOCIATED UTI

Urethral catheters may be retained for weeks to years in patients with neurogenic bladders and those with chronic urinary incontinence or urethral obstruction. Bacteriuria usually accompanies this practice. Patients with chronic indwelling catheters may have $>10^5$ viable bacteria per milliliter of urine (Mobley and Warren, 1987). In catheter-induced bacteriuria, the urethra becomes colonised with Gram-negative rods and enterococci derived from the faecal flora. These bacteria enter the bladder through the mucosal sheath around the catheter. It is likely that mechanical irritation of the urethral and bladder mucosa by the catheter enhances the susceptibility of these structures to invasion by organisms. Bacteria may be introduced into the bladder of patients with indwelling urinary catheters by three major routes:

1. Introduction at the time of catheter insertion
2. Migration of bacteria present in the urethra around the catheter
3. Ascent of bacteria through the lumen of the catheter from a contaminated drainage system

The development and increased use of closed methods of catheter drainage have contributed to reduction of the latter problem and markedly improved the management of UTI. However, a closed catheter system merely postpones the inevitable; a recent study by Liedl and Hofstetter (2000) showed that bacteriuria began developing in a patient group after the second day, with 100 per cent of the patients developing bacteriuria by 1 month.

Although a small number of urethral bacteria may be introduced into the bladder when the catheter is inserted, they appear to be washed out by urine flow or are unable to adhere and multiply in the bladder. This is thought to account for the relatively low incidence of bacteriuria after single or intermittent catheterisation. An illuminating study by Nickel, Grant, and Costerton (1985) into the mechanism of bacterial invasion of the urinary tract via devices showed that the intraluminal ascent of bacteria is faster (32–48 hours) than extraluminal ascent (72–168 hours). Colonisation of the luminal surface, particularly of urinary devices, has been shown in a number of studies (Keane et al., 1994; Nickel et al., 1989; Ramsay et al., 1999), but is not unique to the urinary tract. We have also demonstrated this problem in endotracheal tubes and peritoneal catheters retrieved from patients (Gorman et al., 1993a; Gorman, Adair, and Mawhinney, 1994). This, naturally, presents difficulties for treatment as antibiotics administered by normal routes will not have access to the infection. Further work by Nickel and his colleagues (1992) established that *Pseudomonas aeruginosa* makes substantial progress against the urine flow in the catheter lumen. The speed of the bacterial ascent in what was described as the 'creep phase' was measured at 1–2 $cm.h^{-1}$. When the aminoglycoside tobramycin was introduced, the biofilm still developed and ascended the catheter surface, albeit at a reduced rate of 0.2–0.3 $cm.h^{-1}$.

4 BIOFILM FORMATION ON URINARY DEVICES

Microbial biofilm formation is not unique to devices used within the urinary tract or, indeed, to medicine. However, a considerable body of information has been compiled pertinent to studies within the urinary tract. Particularly thorough reviews have been published recently in relation to the fundamental research in this area (Costerton 1999; Morris et al., 1999). A mature biofilm on a retrieved device from our own studies is shown in Figure 7.2.

Adherence of bacteria to the device surface is the first step in infection, but our understanding of the cause of this interaction is important for prevention of device-related infection. Stickler et al. (1998) have shown that planktonic bacteria can produce signalling molecules such as acylated homoserine lactones with the ability to adsorb to surfaces, thereby serving to attract the bacteria. Gene expression can also be triggered (Fuqua and Greenberg, 1998).

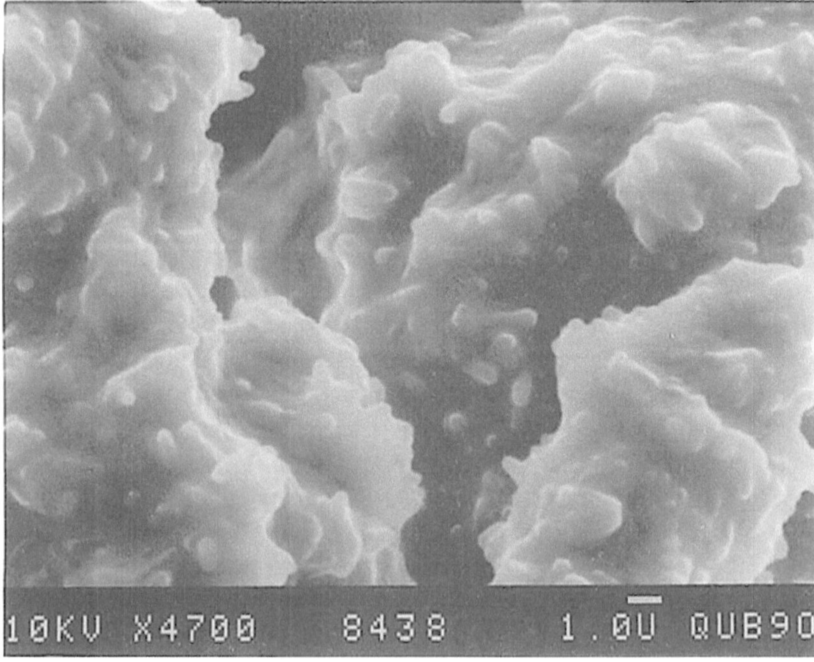

Figure 7.2: A mature biofilm on a retrieved device showing Gram-negative bacteria (approximately 1–2 μm in length) embedded within the confluent glycocalyx.

During the critical stage in attachment of reversible adherence, the bacterial cells can upregulate specific adhesion genes. Then, in order to ensure their successful colonisation of the surface, extrapolymeric substances, chiefly exopolysaccharide, are synthesised to provide a protective glycocalyx as the fledgling community of a new biofilm develops. As discussed previously, rapidly dividing bacteria can spread along the surface of a device within the glycocalyx of the biofilm. Some of the surface bacteria can be shed and become free to attach to a new, non-colonised surface (Nickel et al., 1994). Techniques such as confocal scanning laser microscopy and atomic force microscopy have allowed detailed examination of the biofilm structure (Adair et al., 2000). The biofilm is a complex unit. It can be up to 500 cells deep, but may only be composed of a small percentage of bacterial cells with the remainder being comprised of extrapolymeric substances. Within the urinary tract, the matrix may also contain large quantities of encrusting minerals (Gorman and Tunney, 1997). The biofilm may be single or multispecies and often resembles a mushroom in form. Microcolonies of bacterial cells are found within the biofilm matrix, and water-filled channels observed within the system may allow ingress of nutrients and removal of waste products and toxins (Reid, 1999).

5 ANTIMICROBIAL RESISTANCE OF DEVICE-RELATED BIOFILMS

Comparison of planktonic and biofilm phenotypes shows that radical differences exist. For example, the proteins of the cell envelope fractions can show distinctions of 30–40 per cent between these cell types (Morris et al., 1999). The implication of such differences for resistance to antimicrobial agents is obvious, and the reality of this can be seen where antibiotic therapy reduced the incidence of bacteriuria only during the first 4 days of catheterisation (Hustinx and Verbrugh, 1994). Beyond this time, antibiotics were of no benefit and, in fact, predisposed the development of antibiotic-resistant species. Consequently, most urologists do not recommend prophylactic use of antibiotics, but only if UTI is confirmed (Adams, 1994). The nature of a biofilm indicates how increased resistance is achieved in comparison to the planktonic counterpart. Attachment of bacteria to surface ligands is secured with a mechanical barrier formed against the host's immune defences. An ion-exchange matrix is formed with greater access to organic nutrients and with the ability to counter cationic antimicrobials within a microenvironment with optimal pH and high enzyme concentration. The biofilm structure also provides close proximity of other bacterial cells, facilitating high bacterial cell density cooperative activities such as cross-feeding and promoting genetic exchange and resistance transfer. Consequently, Elves and Feneley (1997) regarded biofilms as highly evolved, functional consortiums in which a protective microenvironment is maintained.

Another very important characteristic of the bacteria within the biofilm is the ability to grow and metabolise at very different rates at different locations in the same biofilm. It has been shown that bacterial growth can be slow or almost absent (Elves and Feneley, 1997). Consequently, the potency of antibiotics is reduced significantly if their mode of action is dependant on bacterial growth. Bacteria within a biofilm are 50–500 times more resistant to antibiotics than their planktonic counterparts (Costerton et al., 1993). In the manner that stationary growth phase bacteria are more resistant to many antibiotics than are logarithmic phase cells, bacteria become significantly more resistant immediately after they adhere to a surface and subsequently increase this resistance over the ensuing days of biofilm formation (Gorman et al., 2001).

6 CONDITIONING FILM AND URINARY DEVICES

Gristina (1987) coined the term 'race for the surface' to describe the competition between uropathogens and host cells for position on the surface of the urinary device. If host tissue cells win this race, the device surface will be defended by a layer of living, integrated cells against bacterial colonisation. However, if the race is won by uropathogens, biofilms are formed on the device surface, which can lead to obstruction of urine flow and possibly sepsis (Wollin et al., 1998).

In this context, it is important to consider the role of a 'conditioning film' formed on the surface of the device not by the bacteria, but by the host itself (Jones et al., 2001). A conditioning film may be composed of various extracellular matrix proteins such as fibronectin, fibrin, collagen, immunoglobulins, electrolyte materials, and other, still unidentified, molecules (Bryers and Hendricks, 1997; McGovern et al., 1997). The device biomaterial properties can influence the sequence of protein deposition on the surface. The main function of the conditioning film is to provide adhesion receptor sites for bacteria or host tissue cells (Gristina, 1987). Each constituent of the conditioning film has a distinct role for different uropathogens or the host cell. Reid, Davidson and Denstedt (1994a) observed that *in vitro*, a conditioning film had been adsorbed on the biomaterial surface within 24 hours, making it quite different from the surface of the original device. Some components of human urine may have antiadherence properties, as bacterial adhesion studies demonstrate the ability of *Escherichia coli, Staphylococcus epidermidis,* and *Proteus mirabilis* to adhere to a lesser degree to a biomaterial surface exposed to human urine (Reid et al.,1992a). The presence of urinary inhibitors such as Tamm-Horsfall Protein, which has the ability to bind *E. coli,* may reduce its adherence to a device surface and facilitate the elimination of bacteria during urination. Recently, Santin et al. (1999) showed albumin to be present on all ureteral stents examined from patients, but Tamm-Horsfall Protein and α1-microglobulin adsorption was limited to non-encrusted devices. In an *in vitro* study of encrustation, they observed increased crystal precipitation on biomaterials with an organic conditioning film composed of these proteins.

Due to the influence of the conditioning fluid on device infection and encrustation, the main factors influencing these events on the device surface require further consideration.

6.1 Electrostatic and Hydrophobic Interactions

Device biomaterials and the outer surface of the bacterial cell wall are generally negatively charged. Repulsive electrostatic forces will, consequently, position the bacterial cell at a distance of approximately 10 nm from the device surface (Reid et al., 1992b). However, positively charged ions within the surrounding urine will provide a counterbalance to the repulsive forces. Hydrophobic forces also play a vital role in enabling the bacterium to colonise the device surface. These hydrophobic forces are attractive and may be 10–100 times greater than the repulsive electrostatic forces, thereby bringing the bacterial cell much closer (<3 nm) to the surface of the hydrophobic device (Gristina, 1987). A number of other binding forces – ionic, hydrogen, or covalent – become operative (Gilbert, Evans, and Brown, 1993). This form of attachment is referred to as 'reversible', as bacteria in the urine and bacteria adhered to the device surface can exchange and the bacteria will still exhibit typical Brownian motion (Van Loosdrecht et al., 1990).

Bonner et al. (1997) showed that bacteria vary in their relative cell surface hydrophobicity. Of two bacteria obtained from biofilms on ureteral stents retrieved from patients, the *E. coli* isolate was significantly less hydrophobic than the *Enterococcus faecalis* isolate. It may logically be concluded that a hydrophobic bacterium may be more readily attracted to a non-aqueous, hydrophobic surface, pointing the way for development of novel device materials and coatings to combat this. However, despite numerous studies in this area, there is a lack of consensus as to the role of hydrophilic coatings in preventing the adherence of bacteria to devices. Similarly, more hydrophilic uropathogens might have a greater affinity for hydrophilic surfaces. A catheter surface may become significantly, and rapidly, altered when inserted in the urinary tract as the conditioning film is formed (Reid et al., 1992a). Consequently, for device manufacture, bacterial cell surface properties may be of greater importance for evaluation of the potential for bacterial adherence than the properties of the biomaterials (Cormio et al., 1995).

6.2 Surface Characteristics of Biomaterial

The key physicochemical characteristics of device biomaterials governing the adherence of bacteria to their surface are roughness (microrugosity), texture, hydrophobicity, surface charge, atomic structure, and composition. Scanning electron microscopy, confocal laser scanning microscopy, and atomic force microscopy studies provide an excellent insight into the surface roughness of various biomaterials used for the manufacture of medical devices (Gorman et al., 1993b, 2001). Surface irregularities may serve as a starting point for biofilm formation as the surface area and charge density are increased. Glycocalyx may adhere better to rougher areas, and bacteria are protected from the shear forces arising from fluid flow as in a urine environment (Baldassarri et al., 1994). Devices such as ureteral stents and peritoneal catheters, when retrieved from patients after providing fluid flow *in situ* for months or years, show significant increases in microrugosity with associated increased levels of biofilm formation (Bonner et al., 1997). The availability of hydrogel-coated catheters offers a smoother surface and, therefore, the possibility of reduced damage to the uroepithelium. This may decrease the likelihood of bacterial adherence, as a traumatised uroepithelium provides an environment rich in nutrient materials and ligands that promote adherence to surfaces (Gristina, 1987; Jones et al., 2001a).

7 ENCRUSTATION AND STONE FORMATION IN THE URINARY TRACT

The formation of crystalline deposits, and associated stones, on the surfaces of urinary catheters and ureteral stents is a major problem in the long-term management of these devices (Choong et al., 2001). These deposits can obstruct the lumen of devices, leading to retention of urine; painful distension of the bladder; bacteriuria; or even more severe complications such as pyelonephritis, septicaemia, and shock. Additionally, the hard nature of these deposits may result in permanent damage to the uroepithelium. Bacteria play an important role in certain aspects of encrustation and stone formation.

Five types of urinary encrustation and stones are commonly encountered: calcium oxalate, calcium phosphate, cystine, uric acid, and magnesium ammonium phosphate ($MgNH_4PO_46H_2O$). Of these, calcium phosphate encrustation may present as brushite ($CaHPO_4$), hydroxyapatite [$Ca_{10}(PO_4)_6(OH)_2$], or carbonate-apatite ($Ca_{10}[PO_4]_6CO_3$) complexes. The first four types listed are often referred to as metabolic encrustation, as they normally result from metabolic dysfunction, whereas magnesium ammonium phosphate encrustation (struvite) has an infectious origin and progresses to form infection stones (Tunney et al., 1999). Metabolic stones often pass spontaneously from the urinary tract; however, infection stones require more effective means of removal, such as shockwave lithotripsy or surgical intervention. Infection stones account for 15–20 per cent of all urinary stones and may be manifested as 'mushy' stones consisting mainly of an amorphous organic matrix and isolated struvite crystals or 'staghorn' stones, which are formed in the renal pelvis and contain within the amorphous matrix significantly high levels of struvite crystals (Figure 7.3).

Bladder stones occur frequently in association with long-term indwelling urinary catheters or stents. As a result of similar aetiology, catheter encrustation resembles bladder and infection-induced renal stones (Cox et al., 1989).

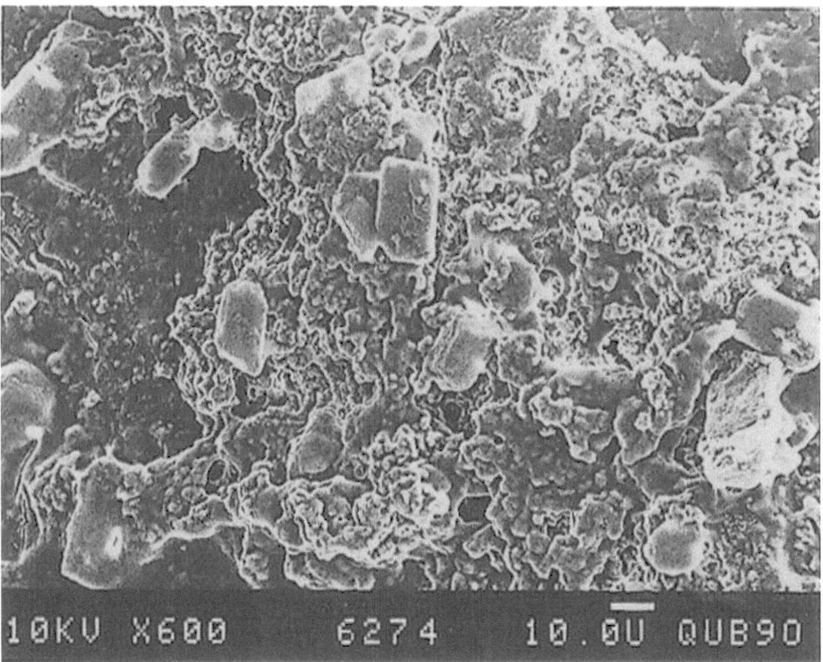

Figure 7.3: Scanning electron micrograph of a heavily encrusted urethral catheter. Large 'coffin'-shaped struvite crystals are observed within a mass of smaller hydroxyapatite and brushite crystals.

8 MECHANISM OF ENCRUSTATION

The process of encrustation has been outlined recently by Morris and Stickler (1998) as involving the following:

8.1 Infection of the urinary tract by urease-producing bacteria

Approximately 200 bacterial species have the ability to produce the enzyme urease – the main cause of mineral deposition on the surface of urinary devices. Some of the pathologically significant urease producers are listed in Table 7.1.

Microbiological analyses of urine specimens from 32 long-term catheterised patients showed 86 per cent of these had a urease-producing species at more than 10^5 cfu/ml (Mobley and Warren, 1987). The most frequently isolated microorganisms were *P. mirabilis* and *Morganella morganii*. The importance of these findings is reflected in a study by Stickler et al. (1993a), who also isolated *P. mirabilis* and *M. morganii*, in addition to *Pseudomonas aeruginosa* and *Klebsiella pneumoniae*, but this time from the surfaces of encrusted catheters. *Proteus mirabilis* has the ability to hydrolyse urea 6–10 times faster than ureases from other species (Mobley and Warren, 1987).

8.2 Adherence of Urease-Producing Bacteria and Development of Biofilm

As encrustation arising from bacterial infection is often essentially a mineralised bacterial biofilm, the process of biofilm formation is pertinent to the development of these crystalline deposits. Stickler et al. (1993b) observed development of *P. mirabilis* biofilm on the surface of long-term indwelling

Table 7.1: Urease-producing bacteria*

Organism	% Positive
Proteus vulgarus	99.6
P. mirabilis	98.7
P. morganii	91.8
P. rettgeri	99.0
Providencia stuarti	97.1
Klebsiella pneumoniae	63.6
Pseudomonas aeruginosa	32.6
Serratia marcescens	29.0
S. liquefaciens	5.0
Enterobacter aerogenes	2.6
Escherichia coli	0

*Modified from Giffiths, 1978.

catheters in eighteen of sixty-nine patients. Although present in the biofilm, bacteria may not be detected in the urine. In 1994, Dumanski et al. confirmed the relationship between the *P. mirabilis* capsule and the enhanced struvite crystallisation *in vitro*. Further work by Stickler and Hughes (1999) showed that differentiation of *P. mirabilis* in swarming cells plays an important role in adherence and biofilm formation. Swarmer cells that were elongated and highly flagellated were identified with the ability to secrete polysaccharide slime and to arrange themselves into rafts of parallel cells allowing rapid movement over surfaces. *P. mirabilis* and *P. vulgaris* bacilli when differentiated into swarmers exhibit a thirty-fold increase in the production of urease.

8.3 Elevation of Urine pH by the Action of Urease

Early work by Griffith, Musher, and Itin in 1976 described a key role for urease-producing bacteria in the process of encrustation. Their *in vitro* study showed that non-urease-producing bacteria could not initiate encrustation, nor could *Proteus* spp. in urine deprived of urea. Kunin et al. (1987) observed that 'blockers', patients presenting high level encrustation, excreted more alkaline urine. Analysis of such encrustation on retrieved ureteral stents by us showed mainly struvite and hydroxyapatite to be present (Keane et al., 1994).

If produced in the urinary tract, the urease catalyses the hydrolyses of urea, the main source of nitrogen for bacteria, to ammonia and carbon dioxide (Griffith, 1978):

$$NH_2CONH_2 + H_2O \rightarrow 2NH_3 + CO_2$$

Under neutral or slightly acidic conditions, as in normal urine, the ammonia becomes protonated, thereby inducing alkalisation of the urine:

$$NH_3 + H_2O \rightarrow NH_4^+ + OH^-$$

Carbon dioxide further reacts with water, producing a weak carbonic acid:

$$CO_2 + H_2O \rightarrow H_2CO_3$$

Depending on the pH of the urine, the carbonic acid may dissociate:

$$H_2CO_3 \leftrightarrow HCO_3^- + H^+ \leftrightarrow CO_3^{2-} + H^+$$

Ammonia production in the urine with the consequent dramatic increase in urine pH induces precipitation of the poorly soluble Mg^{2+} and Ca^{2+} salts in the form of struvite, hydroxyapatite, or carbonate apatite, initiating the process of surface deposition and crystal formation (Gorman and Tunney, 1997).

8.4 Attraction of Calcium and Magnesium Ions into the Matrix

The role of the bacterial capsule is important in the formation of biofilm in the presence of mineral elements present in the urine. Clapham et al. (1990) showed that the *P. mirabilis* capsule creates a gel capable of stabilising growing crystalline deposits and the capsule exopolysaccharide has a unique ability to bind magnesium, the key element in struvite (Dumanski et al., 1994). Biofilm is not only a bacterial protective barrier, but it also serves as a nutrient delivery system. As it is anionic in nature, it has the ability to attract and bind positively charged metal ions such as magnesium and calcium, causing supersaturation in this microenvironment and their eventual precipitation (Clapham et al., 1990).

8.5 Crystallisation of Calcium and Magnesium Phosphates

Precipitation and crystallisation of compounds from the urine is a complex, multistep process. Precipitation of amorphous salts arises if the process of nucleation is initiated. The nucleus serves as a pivot point for the growth of the crystal that is characterised by the deposition and the binding of new molecules onto the crystal surfaces (McLean et al., 1991a). The onset of *P. mirabilis*-induced struvite production is characterised by the appearance of an amorphous precipitate when the urine pH rises above 7.19 (Mc Lean et al., 1990). These amorphous precipitates form quite readily, as there is no ordered structure. Rearrangement into a more ordered array may arise to form X-shaped crystals with a bacterial cell as a nucleation site. Hence, these rapidly growing crystals are typically X-shaped due to the preferential growth along one crystal axis (Dumanski et al., 1994). In contrast, some slow growing crystals are more 'coffin' shaped due to a more balanced growth along all crystal axes. In contrast, crystallisation of hydroxyapatite occurs in two phases. At neutral pH, unstable, soluble amorphous calcium phosphate (ACP1) is formed, followed by its transformation to a less soluble amorphous phase, ACP2. If nucleation is initiated, ACP2 converts into crystalline hydroxyapatite. The latter precipitates in the presence of albumin as tiny, poorly crystalline flakes, whereas in the absence of albumin, hydroxyapatite is more crystalline and coexists with brushite (Cox and Hukins, 1989). A 'biogel growth' theory has been proposed in which the role of bacteria is to produce a microenvironment in the urine where struvite crystals grow more rapidly than in other parts of the urine, resulting in crystal growth within the biofilm matrix (Clapham et al., 1990). Interestingly, McLean et al. (1991b) demonstrated, *in vitro*, that X-shaped struvite crystals can grow more rapidly inside a biofilm. They also established that struvite crystals within a biofilm are not affected by an increase in the urine flow rate.

9 STRUCTURE AND COMPOSITION OF THE ENCRUSTED DEPOSITS

Closer observation of encrustation shows its structural complexity. Soon after the urinary device is inserted in the body, proteins, inorganic salts, inflammatory debris, electrolytes, and other molecules race for the biomaterial surface. Scanning electron microscope (SEM) examination of encrusted deposits reveals large struvite crystals surrounded by much smaller crystals of hydroxyapatite and bacteria (Cox et al., 1989). An interesting observation, *in vivo*, is the presence of bacterial cells in the hydroxyapatite crystals themselves. Cox et al. (1989) proposed that bacteria may engulf the crystals once they are formed and further promote binding of crystals to each other and to the surface of the biomaterial. Struvite crystals have been found to lie in the body of mineralised biofilm matrix and in direct contact with the biomaterial surface with amorphous calcium phosphate salts being distributed throughout the biofilm (Stickler, 1996). It is interesting to note that, even in infection stones removed from patients, bacteria are also present throughout the stone (Griffith, 1978). Our own studies of retrieved ureteral stents show gross encrustation in the presence of biofilm (Keane et al., 1994). Figure 7.4 typically illustrates this presentation.

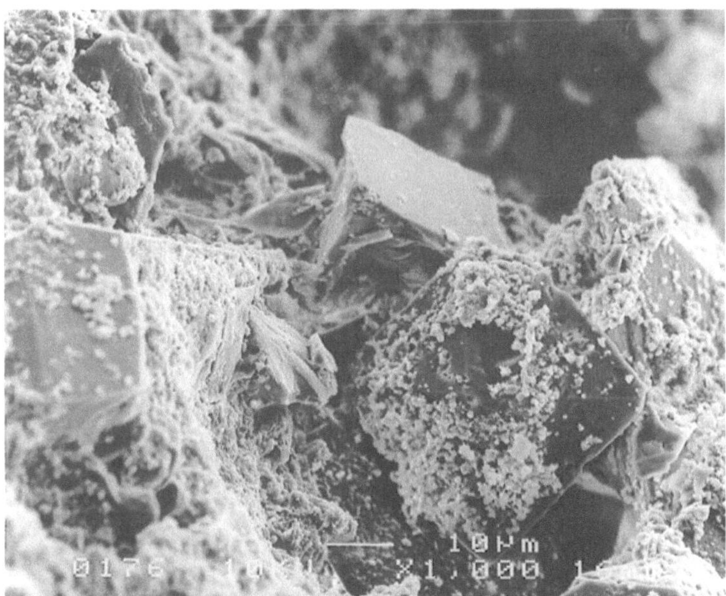

Figure 7.4: Scanning electron micrograph of the deposit on a ureteral stent retrieved from a patient showing large struvite crystals in the presence of biofilm.

A number of *in vitro* models have been developed to simulate encrustation on biomaterials (Gorman and Tunney, 1997). These simulate either the relatively static urine condition of the bladder for portions of devices therein or the dynamic flow as experienced by devices in the urethra and ureter (Tunney et al., 1996b,c). Such models can be applied to the assessment of device biomaterials for resistance to encrustation (Tunney, Keane, and Gorman, 1997a; Tunney, Jones, and Gorman, 1997b).

10 DEVELOPMENTAL APPROACHES

In recent years, many attempts have been made to modify biomaterial surfaces to decrease bacterial adherence and/or encrustation. Most Foley urethral catheters are constructed from silicone or rubber latex. Various materials, for example, Teflon, have been layered onto the surface of these to reduce, in the main, friction and aid insertion. Attempts to lower the incidence of infection have largely been unsuccessful in practice, though silver-releasing polymer coatings can inhibit bacterial growth for up to 96 hours (Stickler, 2000). However, the leaching of polymer molecules can irritate the urethra and cause strictures (Talja, Virtanen, and Andersson, 1986). Similar problems are encountered with the generally employed ureteral stent materials of silicone or polyurethane. Rigidity of the stent material can be a problem, and this is particularly the case with polyurethane causing back pain and bladder irritation. This material is also prone to fracture and stone formation with prolonged usage. Device biomaterials can have a marked influence on the degree of bacterial adherence. Our studies have demonstrated a significant increase in bacterial adherence to polyurethane compared to silicone over a 5-day period (Gorman, Mawhinney, and Adair, 1993c). To negate these material influences, hydrogel coatings have been developed comprising polymers that adsorb water onto the surface. In one such coating developed by us (Gorman, Woolfson, and McCafferty, 1991b), the hydrophobic character of the latex base material is enveloped by a highly hydrophilic surface coating that allows much easier insertion of the catheter and reduces encrustation but becomes attractive to infecting hydrophilic bacteria. Similar findings have been reported for a manufactured hydrogel-coated catheter (Reid et al., 1993), although significant reductions in adherence, specifically staphylococci, were observed for polyvinylpyrrolidone (PVP) and poly(2-hydroxyethylmethacrylate) (PHEMA) hydrogel-coated materials (Denyer et al., 1993). Substantial reductions in *in vitro* encrustation have been observed with a poly(ethylene oxide)/polyurethane composite ureteral stent biomaterial (Gorman et al., 1998). A major complicating factor in developing hydrogel coatings against bacterial adherence is the bacterial cell surface hydrophobicity (as described earlier). This factor differs substantially between genera and species, making selection of an anti-adherent coating for other than a very limited clinical use very difficult (Gorman, 1991; Jones et al., 1991). Other biomaterial surface modifications, such as glow discharge treatment and ion implantation, have been described by us, but are associated with certain limitations, including the cost of the process (Denyer et al., 1993).

11 PHYSICOCHEMICAL METHODS FOR OVERCOMING BIOFILM FORMATION ON MEDICAL DEVICES

In recent years, there has been a marked interest in the design and development of biomaterials as medical devices that offer resistance to the formation of biofilm. Indeed, whereas in previous years the main concern of the medical device industry was to develop high performance materials that could be readily engineered, the design brief of new materials now involves consideration of the propensity for microbial biofilm formation on such surfaces. Interestingly, toxicity and the possibility of medical device-related infection have been cited as the two main criteria that limit the clinical use of medical device biomaterials (Bilbruck, Hanlon, and Martin, 1993). In spite of these problems, there have been significant advances in the design and development of novel medical device biomaterials. The following approaches will be examined in more detail:

1. Direct incorporation into, or coating of, medical device biomaterials with antimicrobial agents
2. Use of novel antiadherent coatings of medical devices
3. Chemical bonding of the antimicrobial agent to the polymer structure

11.1 Direct Incorporation into or Coating of Medical Device Biomaterials with Antimicrobial Agents

It is accepted that the resolution of microbial biofilms on medical devices using conventional antimicrobial chemotherapy, whether oral or parenteral, is rarely achieved (Tunney et al., 1996a). This lack of efficacy may be attributed to several factors, including the poor penetration of the antimicrobial agent through the biofilm matrix, the metabolic state of the microorganisms within the biofilm, and the inability of the antimicrobial agent to accumulate in sufficient concentration at the interface between the biological bathing fluid and the microbial biofilm. Systemic or oral delivery of antimicrobial agents to resolve medical device-related infection is generally poor due to both limited vascularisation at the implant site and, also on some occasions, the requirement of the antimicrobial agent to pass through the device wall to reach the microbial biofilm adherent to the lumen of the device. Therefore, it has been suggested that the success of antimicrobial chemotherapy for inhibition/reduction of microbial biofilm formation on medical devices may be markedly improved by maintaining sufficiently high concentrations of the antimicrobial agent in the vicinity of the device. To facilitate this requirement, there have been several studies in which antimicrobial agents have been directly incorporated into, or onto, medical device biomaterials. The antimicrobial agent, therefore, will be subsequently released directly to the microbial biofilm, resulting in dramatically higher concentrations than may be achieved by conventional methods of administration (Jansen and Peters, 1991; Jones et al., 2002).

11.1.1 Loading of Medical Devices with Antimicrobial Agents by Immersion Techniques

One of the most straightforward methods by which antimicrobial agents may be loaded into medical devices involves the immersion of the device in a solution of an antimicrobial agent. If the polymeric biomaterial has limited affinity for the soaking solution, the degree of polymeric swelling is limited, and, accordingly, the majority of the antimicrobial agent will be adsorbed on the surface and not dispersed throughout the polymer matrix. The clinical success of this approach has been varied. For example, grafts that had been previously immersed in a saline solution containing rifampicin prior to placement in animal models exhibited a greater resistance to a microbial challenge of *S. aureus*, *S. epidermidis*, and *Escherichia coli* than was achieved using systemic antibiotics (GoeauBrissonniere et al., 1999). Similarly, the immersion of segments of silicone, silver-coated, and hydrogel-coated stents in ceftriaxone and ciprofloxacin was reported to significantly reduce bacterial adherence. This was suggested to be due to adsorption/absorption of the antibiotics to the surface of the biomaterial. However, following treatment with a tobramycin solution, reduced bacterial adherence to these biomaterials was not observed (Cormio et al., 1997). Pretreatment of urinary silicone latex catheters *in vitro* with 0.1 and 0.5 μg/mL of ciprofloxacin for 1, 24, and 48 hours significantly decreased adherence of the clinical isolate *Pseudomonas aeruginosa*. Ultraviolet (UV) spectroscopy and high performance liquid chromatography confirmed the presence of ciprofloxacin adsorbed onto the catheter surfaces and that up to 56 per cent of the drug leached into the surrounding fluid within 24 hours (Reid et al., 1994b). Silicone-coated latex catheter sections soaked in ciprofloxacin, although inhibiting bacterial colonisation, have been shown to allow colonisation by *E. coli* and *Proteus mirabilis* (Stickler, Howe, and Winters, 1994). Catheters immersed in teicoplanin solution revealed that although this treatment did not prevent initial adherence, any bacteria adhering to the surface were subsequently eliminated (Jansen et al., 1992a). The authors concluded that such loaded catheters could be suitable for early onset, catheter-related infections.

In general, these studies have illustrated that the immersion of polymers in antimicrobial solutions, although reducing early-onset colonisation, would be unlikely to prevent biofilm formation on long-term indwelling catheters. Unfortunately, many studies in this area have failed to appreciate the mechanism by which drugs will interact with the biomaterial substrate. Hence, drug loading of medical devices following immersion in aqueous antibiotic solutions should only be performed whenever the medical device is hydrophilic in nature, that is, possesses a hydrophilic coating. Under these circumstances, the aqueous solution (containing antibiotic) will be absorbed into the aqueous (hydrogel) coating. Following insertion into the patient, the antimicrobial agent will be released into the adjacent biological fluids in a relatively controlled fashion. Immersion of hydrophobic medical devices into aqueous antibiotic solutions will result in a weak and limited surface attachment of antibiotic to the device. Following insertion into the patient, the drug will be rapidly desorbed and may result in a limited clinical effect. Interestingly, although this approach offers a straightforward method by which antimicrobial agents may be loaded into medical devices, little attention has been paid to experimental factors that influence the process of drug absorption, for example, choice of soaking solution and time of immersion. However, in one study Beckett et al. (1997) examined the effects of time of immersion and solvent blend (aqueous/hydroalcoholic) on the sorption and resultant desorption of chlorhexidine into/from poly(methylmethacrylate), poly(hydroxyethylmethacrylate) and their co-polymers. The authors illustrated the importance of polymer type, time of immersion, concentration of antimicrobial agent in solution, and the chemical nature of the solvent on these properties. Similarly, Kohnen et al. (1998) described a process by which silicone was immersed in a solution of antimicrobial agent that was a non-solvent for the polymer matrix. Consequently, the polymer exhibited swelling, thus facilitating entry of the antimicrobial agent into the polymer matrix. Controlled evaporation of the solvent left the agent trapped in the matrix. Upon exposure to body fluids, the antimicrobial agent is released by diffusion.

Therefore, it may be concluded that if proper consideration is given to these physicochemical properties, immersion of medical devices in antimicrobial solutions may be a useful method to produce antimicrobial-loaded medical devices. Interestingly, in the clinical setting, medical devices can be treated immediately prior to placement (Lachapelle, Graham, and Symes, 1994), with this process being controlled by the medical staff. This flexibility, therefore, will be of interest to many practitioners.

11.1.2 Coating Medical Devices with Antimicrobial Agents

Several studies have examined methods by which antimicrobial agents may be effectively coated onto the surface of medical devices and, in addition, the *in vitro*/*in vivo* efficacy of these systems. As the antimicrobial agent is required at the surface of the device, adjacent to the proposed site of microbial biofilm formation, the incorporation of the agent into coatings for medical devices offers a novel approach to reduce the problem of medical device-related infection.

The loading of antibiotics onto the surface of a device can be enhanced by precoating the polymeric medical device with a substance which can act as a 'tie' layer for the antimicrobial agent. In these systems the interaction between the antimicrobial agent and the tie layer component is facilitated by electrostatic interactions. Anionic antibiotics can be electrostatically bonded to the surface of the medical device via a tie layer composed of cationic surfactants such as tri-dodecylmethylammonium chloride (TDMAC). The long alkyl chains of TDMAC are thought to interact with the polymer chains of the medical device by hydrophobic interactions, exposing the ammonium group at the surface where it can electrostatically attract anions. Interestingly, polytetrafluoroethylene (PTFE) surfaces that had been treated with TDMAC exhibited a 100-fold increase in loading of benzylpenicillin and cefazolin when compared to the untreated polymer (Harvey, Tesoriero, and Greco, 1984). TDMAC was also found to bind the antibiotics if they were added, after device placement, either by local irrigation or by systemic administration. Furthermore, benzalkonium chloride was also found to have a similar binding property, but to a lesser extent than TDMAC. There have been several studies that have examined the clinical efficacy of catheters that have been coated with antimicrobial agents using this coating technology. For example, polyurethane catheters have been coated with minocycline/ethylenediaminetetraacetate in an attempt to reduce recurrent vascular catheter-related bacteraemia; *in vitro* and *in vivo* results tentatively suggested that these coatings may prevent catheter colonisation (Raad et al., 1997a). Minocycline and rifampicin coatings were also shown to exhibit significant antimicrobial activity *in vitro*, and in an animal model catheter infection was prevented (Raad et al., 1996). The same authors reported that catheters that had been coated with minocycline and rifampicin significantly reduced the risk of catheter-related colonisation and infections of the bloodstream (Raad et al., 1997b) and maintained effective antimicrobial activity against *S. epidermidis* adherence to indwelling central catheters for at least 2 weeks (Raad et al., 1998). Minocycline, coated onto bladder catheters, has also been shown to provide some protection against colonisation (Darouiche et al., 1997). More recently, Johnson, Delavari, and Azar (1999) described the *in vitro* antimicrobial activity of a commercially available nitrofurazone-coated silicone catheter against multidrug-resistant clinical bacterial isolates. The activity of this system against these problematic clinical isolates may offer considerable clinical promise.

The possible antibacterial activity of silver-containing compounds as antimicrobial coatings for medical devices has been widely investigated. In an *in vitro* assay, no biofilm of *Pseudomonas aeruginosa* had developed on silver-coated latex discs after 10 hours (Liedberg and Lundeberg, 1989). Adherence of four strains of *E. coli* to a hydrogel/silver-coated catheter was found to be decreased by 50–99 per cent compared to latex, silicone, and hydrogel-coated latex catheters (Gabriel et al., 1996). Ion-beam-assisted deposition of silver onto indwelling intravenous catheter materials has also been shown to have inhibitory effects on bacterial growth (Bambauer et al., 1997; Woodyard et al., 1996). A novel, silver-releasing device has been investigated by Stickler, Morris, and Williams (1996); the incorporation of this device into urinary catheter drainage systems protected the catheterised bladder from contamination for 10 days. Silver coating of devices has also been investigated in other areas to attempt to decrease bacterial adherence; a silver coating on orthopaedic external fixation pins was also found to reduce adhesion of *E. coli* to their surfaces compared to uncoated stainless steel (Wassall et al., 1997). These results were contradicted by a large *in vivo* study in which a significant increase in the incidence of bacteriuria was detected on silver-oxide-coated urinary catheters (Riley et al., 1995). Cormio et al. (1997) have also shown that neither silver- nor hydrogel-coated silicone reduced uropathogenic adherence to ureteral stents.

The feasibility of other antimicrobial agents included in coatings of medical devices has been investigated, for example, teicoplanin coatings on hydrophilic intravenous catheters. This antibiotic is retained on the surface of the coatings for a short time (36 hours) and so may limit clinical efficacy (Bach et al., 1996). Chlorhexidine has received particular interest as a catheter coating, mainly due to its broad spectrum of activity against organisms that cause catheter infections. Indeed, Maki et al. (1997) reported that the use of a central venous catheter coated with silver sulphadiazine and chlorhexidine was associated with a lower incidence of catheter-related infection when compared to the control catheter. However, chlorhexidine-coated polyurethane vascular catheters did not show any decrease in infection rates compared to uncoated polyurethane catheters in intensive care unit patients (Sherertz, 1997). Similarly, the incidence of catheter-related sepsis with the use of uncoated central venous catheters was statistically similar to that associated with catheters coated with silver sulphadiazine and chlorhexidine gluconate (Pemberton et al., 1996). A clinical evaluation of the incidence of catheter-related bacteraemia associated with triple lumen central venous catheters coated with chlorhexidine and silver sulphadiazine showed that the antimicrobial coating did not offer any significant improvement (Heard et al., 1998). In a further interesting study, Darouiche et al. (1999) reported on the rates of catheter-related bloodstream infection associated with the use of two coated catheters, one with minocycline and rifampicin and the other with chlorhexidine and silver sulphadiazine. The incidence of infection associated with minocycline/rifampicin catheters was significantly lower than for their silver sulphadiazine/chlorhexidine-coated comparators. However, the authors did recognise that this difference may be due to the nature of the coating on these catheters. Accordingly, in minocycline/rifampicin catheters both the external and the internal surfaces were coated, whereas the silver sulphadiazine/chlorhexidine catheters were coated only on the external surface. Additionally, microbial colonisation of the external surface of the silver sulphadiazine/chlorhexidine-coated catheter was significantly lower than for uncoated catheters.

These examples illustrate the importance of the nature of the catheter coating on the resultant clinical efficacy of such systems. One significant point concerning the use of antimicrobial coatings on medical devices concerns the possible neutralisation of the associated antimicrobial activity by the shielding effects of conditioning films that coat biomaterials *in vivo*. It has been suggested that although antimicrobial surfaces may kill bacteria initially adhering, these dead cells and other debris may provide a foundation for subsequent colonisation, shielded from the effects of the antimicrobial (Stickler et al., 1994). Nevertheless, antibiotic coating of devices is still a major focus of attention in antimicrobial research to combat device-related infection (Reid, 1997).

11.1.3 Antimicrobial Matrix Medical Devices

In addition to the problems outlined previously concerning the loading of antimicrobial agents into medical devices by immersion or coating technologies, two further problems are associated with these techniques. First, in general, the mass of drug that can be incorporated into the medical device is limited and may often be insufficient for a prolonged antimicrobial effect *in vivo*. Second, the subsequent release of the antimicrobial agent following clinical insertion of the device is rapid and relatively uncontrolled. These problems may be obviated by direct incorporation of the antimicrobial agent into the polymeric matrix of the medical device. This may be performed either at the polymer synthesis stage in which the antimicrobial agent is mixed with the monomer and polymerisation of the monomers is conducted in the presence of the antimicrobial agent or at the device manufacture stage in which the antimicrobial agent may be mixed with the polymeric ingredients. In so doing, a greater mass of antimicrobial agent may be incorporated into the polymer matrix and a greater prolongation of drug release may be achieved.

There have been several reports of the *in vitro* and *in vivo* activity of antimicrobial-impregnated medical devices. Chlorhexidine and silver sulphadiazine have been impregnated into polyurethrane (PU) central venous catheters. These impregnated catheters prevented bacterial adherence and biofilm formation when implanted in swine (Greenfeld et al., 1995). Salicylic acid incorporated into catheters has been shown to inhibit the adherence of a range of bacteria (Farber and Wolff, 1993). The authors suggested that this inhibition may be due to the ability of the salicylates to inhibit polysaccharide formation, resulting in less bacterial adherence and biofilm formation. Among other agents that have been used to effectively control microbial adhesion to biomaterials are methyl and propyl parabens (Golomb and Shpigelman, 1991).

A silicone-elastomer material compounded with a chlorhexidine matrix has also been developed. The chlorhexidine has been shown to be released in a sustained fashion over a 4-week period and has demonstrated significant inhibitory activity against *E. coli, Proteus mirabilis*, and *S. epidermidis* (Whalen et al., 1997). In an attempt to prevent infection of cerebrospinal fluid shunts, incorporation of rifampicin into silicone biomaterials significantly reduced adherence of *S. epidermidis* (Schierholz et al., 1994). Drug release from the biomaterial exceeded minimum biocidal concentration levels, and this was deemed a prerequisite in order to kill sessile biofilm organisms and hence prevent device-related infection. Similarly, the inclusion of antimicrobial agents into poly(vinyl chloride) (PVC) was observed to decrease the subsequent adherence of *Pseudomonas fluorescens*, the effect being more marked for materials containing higher loadings of antimicrobial agent (Jones et al., 1996).

Due to its excellent antimicrobial properties, iodine has been incorporated into polymeric systems for use as medical device biomaterials. For example,

Kristinsson et al. (1991) co-polymerised polyvinyl fluoride films with PVP to which iodine was complexed. These films were able to resist bacterial adhesion of *S. aureus*, *E. coli*, and *P. aeruginosa* for a period of up to 6 days. Similarly, Jansen et al. (1992b) coated a Secalon-Hydrocath central venous catheter with PVP which was further complexed with iodine. In this study, the adherence of *S. epidermidis* was completely inhibited for 2 days, whereas the inhibitory effect of iodine on planktonic bacteria lasted for a period of more than 4 days. More recently, Djokic and co-workers (1998, 1999) described the resistance of biodegradable poly(e-caprolactone)-poly(vinylpyrrolidone) blends as urinary device biomaterials. Blends with a greater rate of biodegradation exhibited a greater resistance to urinary encrustation *in vitro*. Accordingly, the process of biomaterial degradation assisted in the resistance of the biomaterial to the development of urinary encrustation. The microbial antiadherent properties of these materials were also demonstrated. This approach may offer new directions for medical device design due to the bifunctionality of the system. Hence, the controlled release of antimicrobial agents to the biological fluids that bathe the medical device or the attached biofilm and, second, the controlled degradation of the surface of the device (and hence removal of adherent microorganisms) will reduce the propensity for medical device-related infection. In a related study, Multanen et al. (2000) described the bacterial antiadherence properties of a biodegradable matrix composed of ofloxacin-blended polylactone that was coated onto self-reinforced poly L-lactide.

One disadvantage of the direct incorporation of antimicrobial agents into the matrix of the medical device concerns the possible deleterious effects of the incorporated agent on the mechanical properties of the host polymer (Jones et al., 2002; Schierholz et al., 1997). The effect of the incorporated agent on the mechanical properties is dependent on the chemical properties of both the incorporated agent and the polymer matrix and, in addition, the method by which the polymeric matrix is fabricated or synthesised. There have been reports of interactions between antimicrobial agents and the polymer matrix. Schierholz et al. (1997) described an interaction between rifampicin and PVC. However, in some studies, incompatibilities between the incorporated agent and the host polymer have been observed that ultimately will affect the performance of the system. For example, Golomb and Shpigelman (1991) described the precipitation of paraben antimicrobials in PU. More recently, the deleterious effects of hexetidine, a non-antibiotic antimicrobial agent, on the curing and mechanical properties of silicone and PVC have been reported (Brown, Jones, and Woolfson, 1998; Jones et al., 2002). Therefore, as the mechanical properties of the medical device are essential to ensure the optimal performance of the medical device, these examples serve to illustrate the potential hazards of incorporation of antimicrobial (and related) agents on the essential mechanical performance of medical devices.

11.2 Use of Novel Antiadherent Coatings of Medical Devices

If long-term resistance to bacterial adherence is to be achieved, the surfaces of device biomaterials need to be made unfavourable to bacterial adherence and biofilm formation. Biomaterials with 'antiadhesive' surfaces would not lose their antiinfective properties during use, and, therefore, late onset infections could be avoided (Kohnen and Jansen, 1995).

As early as 1974 this concept had been recognised, and a hydrophilic-coated catheter (Hydron), composed of poly(hydroxyethylmethacrylate), had been evaluated in the prevention of infection (Monson and Kunin, 1974). However, no significant difference in bacterial colonisation was demonstrated. Similar results were seen in bacterial adherence to hydrophilic-coated PU biliary stents (Jansen, Goodman, and Ruiten, 1993). Conversely, Tebbs, Sawyer, and Elliott (1994) compared the adherence of five strains of *S. epidermidis* to a PU and to a hydrophilic-coated PU catheter and reported that adhesion of three of the strains was considerably reduced by the coating on the catheters. Bridgett, Davies, and Denyer (1992) reported the reduced *in vitro* adherence of three strains of *S. epidermidis* to polystyrene surfaces modified with a co-polymer of poly(ethylene oxide) (PEO) and poly(propylene oxide). Similar results were reported by Desai, Hossainy, and Hubbell (1992), who observed reductions in adherent bacteria on surfaces modified with PEO. More recently, a hydrogel coating has also been shown to decrease the adhesion of *S. epidermidis* and *S. aureus* to PU catheters (John et al., 1995).

The influence of the rugosity of a polymer surface on bacterial adhesion has been studied. McAllister et al. (1993) postulated that irregularities in polymer surfaces promoted bacterial adhesion, and studies have shown that smoother surfaces decrease bacterial adherence to catheters. Accordingly, there have been several reports that have described the development of novel polymers with surface properties that may lead to a decrease or prevention of bacterial adhesion. One approach has been the physicochemical modification of polymer surfaces. Gamma-radiation and glow discharge techniques act by introducing new functional groups onto the polymer surface. This can lead to altered physicochemical surface properties, thus altering interactions with proteins and bacterial cells (Jansen and Peters, 1991). The newly created functional groups may possess intrinsic antimicrobial activity. Alternatively, antimicrobial substances may be covalently linked to the functional surface groups. It has been shown that by radiation modification of PU with 2-hydroxyethylmethacrylate, the PU surface is rendered more hydrophilic and adhesion of *S. epidermidis* is reduced (Jansen et al., 1989). Similar results were observed with glow discharge-modified silicone catheters.

Surface treatment of biomaterials by an ion beam deposition may improve device function and biocompatibility (Sioshansi and Tobin, 1996). The process offers a wide array of beneficial surface property modifications, without adversely affecting bulk properties. Ion-beam-assisted deposition can be used to apply antimicrobial coatings to catheters and other devices. Bambauer et al. (1997) investigated the *in vivo* adherence of bacteria to silicone and PU surfaces treated by ion beam deposition processes. Bacterial colonisation was found on 2.4 per cent of the surface-treated catheter pieces compared to 7.1 per cent in the control.

A photochemical coating of polymers also demonstrated reduced adhesion of a variety of strains (Dunkirk et al., 1991). Rad, Ayhan, and Piskin (1998) found that the plasma glow discharge technique produced surfaces that caused, in general, significant drops in bacterial attachment.

Despite the success of such novel coatings, it is worth remembering that no material or coating has been developed that leads to a complete inhibition of colonisation or 'zero adherence' (Kohnen and Jansen, 1995). Therefore, it is recommended that antiadherent coatings should be combined with the local delivery of antimicrobial agents.

11.3 Chemical Bonding of the Antimicrobial Agent to the Polymer Structure

Although the formation of covalently bound drug-polymer conjugates is relatively novel within the device fabrication industry, conjugates are widely used as polymeric drug carriers. These macromolecules contain at least one drug molecule covalently bound to a polymeric chain. In comparison to the other methods of drug incorporation within medical device biomaterials, covalent attachment of the drug to the polymer has received comparatively little attention. However, an example of such a system may be found in cardiovascular stents. These stents are exposed to constant elution of drug due to the fluid dynamics in the immediate area of the device, hence surface adsorbed materials are rapidly removed from the site of action. Materials for coronary stenting, therefore, have been commonly modified by covalent attachment of the pharmaceutical drug. Heparin has been found to have antithrombogenic activity after percutaneous transluminal coronary angioplasty. Covalent attachment has been achieved by coating the polymer surface with a poly-imine. The exposed amine groups can then be reacted with free aldehyde residues of partly hydrolysed heparin to form amide linkages by reductive amination (Hardhammar et al., 1996). Fallgren et al. (1998) covalently linked heparin to silicone and then entrapped antibiotics in crosslinked collagen bound to the heparinised surface. The modified surfaces reduced the growth and adherence of *S. epidermis* for more than 24 hours *in vitro*.

The covalent linkage of an agent to a monomer prior to polymerisation provides a method of producing perhaps the most resilient drug-polymer material. Co-polymers of methacrylic acid and methacryloyl phosphorylcholine have successfully reduced platelet adhesion and subsequent thrombosis when used as coatings for metallic stents in baboon coronary arteries (Chronos et al., 1995). Analogous results have been shown *in vitro* by Ishihara et al. (1990).

There are several disadvantages to this technique, namely, the expense associated with the synthetic process and, in addition, the selection of therapeutic agents whose chemistry is compatible with the synthetic reaction scheme. However, in light of the utility and success of this approach, it may be predicted that many other examples of this technology will be reported. We have found drug-polymer conjugates to provide significant reductions in bacterial adherence and urinary encrustation (our own unpublished results).

12 CONCLUSION

The key to developing improved devices for use in the urinary tract is the understanding of the problems besetting devices and the important role played by microorganisms in the biofilm growth mode. The comprehensive body of knowledge acquired from fundamental studies of these problems now presents researchers in applied disciplines with an exciting opportunity to deliver the next generation of urinary devices with improved biocompatibility and patient comfort.

REFERENCES

Adair, C. G., Gorman, S. P., O'Neill, F., McClurg, B., Goldsmith, E. C. and Webb, C. H. (1993). Selective decontamination of the digestive tract (SDD) does not attain sufficient antibiotic concentration in tracheal secretions to prevent the formation of microbial biofilm on endotracheal tubes. *Journal of Antimicrobial Chemotherapy*, 31, 689–697.

Adair, C. G., Gorman, S. P., Byers, L. M., Gardner, T. and Jones, D. S. (2000). Confocal laser scanning microscope examination of microbial biofilms. In *Handbook of*

Bacterial Adhesion, eds. Y. H. An and R. J. Friedman, pp. 249–259. New Jersey: Humana Press.

Adams, J. (1994). Renal stents. *Emergency Medical Clinics of North America*, 12, 749–758.

Bach, A., Darby, D., Bottiger, B., Bohrer, H., Motsch, J. and Martin, E. (1996). Retention of the antibiotic teicoplanin on a hydromer-coated central venous catheter to prevent bacterial colonisation in postoperative surgical patients. *Intensive Care Medicine*, 22, 1066–1069.

Baldassarri, L., Gelosia, A., Fiscarelli, E., Donnelli, G., Mignozzi, M. and Rizzoni, G. (1994). Microbial colonisation of implanted silicone and polyurethane catheters. *Journal of Materials Science. Materials in Medicine*, 5, 601–605.

Bambauer, R., Mestres, P., Schiel, R. and Sioshansi, P. (1995). New surface-treatment technologies for catheters used for extracorporeal detoxification methods. *Dialysis and Transplantation*, 24, 228.

Bambauer, R., Mestres, P., Schiel, R., Klinkmann, J. and Sioshansi, P. (1997). Surface-treated catheters with ion-beam process evaluation in rats. *Artificial Organs*, 21, 1039–1041.

Beckett, G., Schep, L. J., Crichton, D. and Jones, D. S. (1997). Chlorhexidine sorption into and release from dental prosthetic biomaterials. *Journal of Pharmacy and Pharmacology*, 49(S4), 135.

Bilbruck, J., Hanlon, G. W. and Martin, G. P. (1993). The effect of polyHEMA coating on the adhesion of bacteria to polymer monofilaments. *International Journal of Pharmaceutics*, 99, 293–301.

Bonner, M., Keane, P. F. and Gorman, S. P. (1993). Characterisation and antibiotic sensitivities of isolates from ureteral stent biofilm. *Journal of Pharmacy and Pharmacology*, 45, 1445.

Bonner, M. C., Tunney, M. M., Jones, D. S. and Gorman, S. P. (1997). Factors affecting *in vitro* adherence of ureteral stent biofilm isolates to polyurethane. *International Journal of Pharmaceutics*, 151, 201–207.

Bridgett, M. J., Davies, M. C. and Denyer, S. P. (1992). Control of *Staphlococcal* adhesion to polystyrene surfaces by polymer surface modification with surfactants. *Biomaterials*, 13, 411–416.

Brown, A. F., Jones, D. S. and Woolfson, A. D. (1998). Effect of hexetidine and triclosan on the curing and mechanical properties of silicone elastomer. *Journal of Pharmacy and Pharmacology*, 50, 105.

Bryers, J. D. and Hendricks, S. (1997). Bacterial infection of biomaterials – experimental protocol for *in vitro* adhesion studies. *Annals of the New York Academy of Sciences*, 831, 127–137.

Choong, S., Wood, S., Fry, C. and Whitfield, H. (2001). Catheter-associated urinary tract infection and encrustation. *International Journal of Antimicrobial Agents*, 17, 305–310.

Chronos, N. A. F., Robinson, K. A., Kelly, A. B., Taylor, A., Yianni, J., King, S. B., Harker, L. A. and Hanson, R. (1995). Thromboresistant phosphorylcholine coating for coronary stents. *Circulation*, 92, I-685

Clapham, L., McLean, R. J. C., Nickel, J. C. and Downey, J. (1990). The influence of bacteria on struvite crystal habit and its importance in urinary stone formation. *Journal of Crystal Growth*, 104, 475–484.

Cormio, L., Vuopio-Varkila, J., Siitonen, A., Talja, M. and Ruutu, M. (1995). Biocompatibility of various indwelling double J stents *in vivo* and *in vivo. Scandinavian Journal of Urology and Nephrology*, 30, 19–24.

Cormio, L., La Forgia, P., La Forgia, D., Siitonen, A. and Ruutu, M. (1997). Is it possible to prevent bacterial adhesion onto ureteric stents? *Urological Research*, 25, 213–216.

Costerton, J. W., Khoury, A. E., Ward, K. H. and Anwar, H. (1993). Practical measures to control device-related bacterial infections. *International Journal of Artificial Organs*, 16, 765–770.

Costerton, J. W. (1999). Introduction to biofilm. *International Journal of Antimicrobial Agents*, 11, 217–221.

Cox, A. J. and Hukins, D. W. L. (1989). Morphology of mineral deposits on encrusted urinary catheters investigating by scanning electron microscopy. *Journal of Urology*, 142, 1347–1350.

Cox, A. J., Hukins, D. W. L. and Sutton, T. M. (1989). Infection of catheterised patients: bacterial colonisation of encrusted foley catheters shown by scanning electron microscopy. *Urological Research*, 17, 349–352.

Darouiche, R. O., Hampel, O. Z., Boone, T. B. and Raad, I. I. (1997). Antimicrobial activity and durability of a novel antimicrobial-impregnated bladder catheter. *International Journal of Antimicrobial Agents*, 8, 243–247.

Darouiche, R. O., Raad, I. M., Heard, S. O., Thornby, J. I., Wenker, O. C., Gabrielli, A., Berg, J., Khardori, N., Hanna, H., Hachem, R., Harris, R. L. and Mayhall, G. (1999). A comparison of two antimicrobial-impregnated central venous catheters. *New England Journal of Medicine*, 340, 1–8.

Denyer, S. P., Hanlon, G. W., Davies, M. C. and Gorman, S. P. (1993). Antimicrobial and other methods for controlling microbial adhesion in infection. In *Microbial Biofilms: Formation and Control*, eds. S. P. Denyer, S. P. Gorman and M. Sussman, pp. 147–165. Oxford: Blackwell Scientific Publications.

Desai, N. P., Hossainy, S. F. and Hubbell, J. A. (1992). Surface-immobilised polyethylene oxide for bacterial repellence. *Biomaterials*, 13, 417–420.

Djokic, J., Jones, D. S. and Gorman, S. P. (1998). Development of a novel polymer coating for urinary medical devices: assessment of biodegradation and resistance to encrustation. *Journal of Pharmacy and Pharmacology*, 50, 172.

Djokic, J., Jones, D. S., Gorman, S. P. and McGrath, S. (1999). Assessment of resistance of poly(-caprolactone) films, impregnated with povidone-iodone, to *Escherichia coli* adherence. *Journal of Pharmacy and Pharmacology*, 51, 33.

Dumanski, A. J., Hedelin, H., Edin-Liljegren, A., Beauchemin, D. and McLean, R. (1994). Unique ability of the *Proteus mirabilis* capsule to enhance mineral growth in infectous urinary calculi. *Infection and Immunity*, 62, 2998–3003.

Dunkirk, S. G., Gregg, S. L., Duran, L. W., Monsfiels, J. D., Haapala, J. E., et al. (1991). Photochemical coatings for the prevention of bacterial colonisation. *Journal of Biomaterials Applications*, 6, 131–155.

Elves, A. W. S. and Feneley, R. C. L. (1997). Long-term uretheral catheterisation and urine biomaterial interferance. *British Journal of Urology*, 80, 1–5.

Fallgren, C., Utt, M., Petersson, A. C., Ljungh, A. and Wadstrom, T. (1998). In vitro anti-staphylococcal activity of heparinized biomaterials bonded with combinations of rifampicin. *Zentralblatt fur Bakteriologie*, 287, 19–31.

Farber, B. F. and Wolff, A. G. (1993). Salicylic acid prevents the adherence of bacteria and yeast to silastic catheters. *Journal of Biomedical Materials Research*, 27, 599–602.

Fuqua, C. and Greenberg, E. P. (1998). Self-perception in bacteria: quorum sensing with acylated homoserine lactones. *Current Opinion in Microbiology*, 1, 183–189.

Gabriel, M. M., Mayo, M. S., May, L. L., Simmons, R. B. and Ahearn, D. G. (1996). *In vitro* evaluation of the efficacy of a silver-coated catheter. *Current Microbiology*, 33, 1–5.

Gilbert, P., Evans, D. J. and Brown, M. R. W. (1993). Formation and dispersal of bacterial biofilms *in vivo* and *in situ*. *Journal of Applied Bacteriology*, 74, S67–S78.

GoeauBrissonniere, O., Leflon, V., Letort, M. and Nicolas, M. H. (1999). Resistance of antibiotic bonded gelatin coated polymer meshes to *Staphylococcus aureus* in a rabbit subcutaneous pouch model. *Biomaterials*, 20, 229–232.

Golomb, G. and Shpigelman, A. (1991). Prevention of bacterial colonisation on polyurethane *in vitro* by incorporated antibacterial agent. *Journal of Biomedical Material Research*, 25, 937–952.

Gorman, S. P. (1991). Microbial adherence and biofilm production. In *Mechanisms of Action of Chemical Biocides*, eds. S. P. Denyer and W. B. Hugo, pp. 271–295. Oxford: Blackwell Scientific Publications.

Gorman, S. P., McCafferty, D. F., Woolfson, A. D. and Jones, D. S. (1987). A comparative study of the microbial anti-adherence capacities of three antimicrobial agents. *Journal of Clinical Pharmacy and Therapeutics*, 12, 393–399.

Gorman, S. P., Woolfson, A. D. and McCafferty, D. F. (1991a). Elemental analysis of latex and polymer-coated urinary catheter encrustation by a novel electron probe and digimap technique. *Proceedings of the 10th Pharmaceutical Technology Conference, Bologna, Italy*, 2, 649–651.

Gorman, S. P., Woolfson, A. D. and McCafferty, D. F. (1991b). Microbial adherence andbiofilm formation on latex and polymer-coated urinary catheters: role of hydrophobicity. *Proceedings of 10th Pharmaceutical Technology Conference, Bologna, Italy*, 2, 661–663.

Gorman, S. P., Adair, C. G., O'Neill, F., Goldsmith, E. C. and Webb, C. H. (1993a). Influence of selective decontamination of the digestive tract on microbial biofilm formation on endotracheal tubes from artificially ventilated patients. *European Journal of Clinical Microbiology and Infectious Diseases*, 12, 9–17.

Gorman, S. P., Mawhinney, W. M., Adair, C. G. and Issouckis, M. (1993b). Confocal scanning laser microscopy of CAPD catheter surface microrugosity in relation to recurrent peritonitis. *Journal of Medical Microbiology*, 38, 411–417.

Gorman, S. P., Mawhinney, W. M. and Adair, C. G. (1993c). The influence of a protein-conditioning film and cell surface hydrophobicity on bacterial adherence to silicone and polyurethane CAPD catheters. *Proceedings of the 12th Pharmaceutical Technology Conference, Copenhagen, Denmark*, 2, 465–468.

Gorman, S. P., Adair, C. G. and Mawhinney, W. M. (1994). Incidence, nature and antibiotic resistance of CAPD catheter biofilm in relation to peritonitis. *Epidemiology and Infection*, 112, 551–559.

Gorman, S. P. and Tunney, M. M. (1997). Assessment of encrustation behaviour on urinary tract biomaterials. *Journal of Biomedical Material Research*, 12, 136–166.

Gorman, S. P., Tunney, M. M., Keane, P. F., van Bladel, K. and Bley, B. (1998). Characterisation and assessment of a novel poly(ethylene oxide)/polyurethane composite hydrogel (Aquavene) as a ureteral stent biomaterial. *Journal of Biomedical Material Research*, 39, 642–650.

Gorman, S. P., McGovern, J. G., Woolfson, A. D., Adair, C. G. and Jones, D. S. (2001). The concomitant development of poly(vinylchloride)-related biofilm and antimicrobial resistance in relation to ventilator-associated pneumonia. *Biomaterials*, 22, 2741–2747.

Greenfeld, J. I., Sampath, L., Popilskis, S. J. and Brunnert, S. R. (1995). Decreased bacterial adherence and biofilm formation on chlorhexidine and silver sulfadiazine-impregnated central venous catheters implanted in swine. *Critical Care Medicine*, 23, 894–900.

Griffith, D. P., Musher, D. M. and Itin, C. (1976). Urease-the primary cause of infection-induced urinary stones. *Investigative Urology*, 13, 346–350.

Griffith, D. P. (1978). Struvite stones. *Kidney International*, 13, 372–382.

Gristina, A. G. (1987). Biomaterial-centered infection: microbial adhesion versus tissue integration. *Science*, 237, 1588–1595.

Hardhammer, P. A., van Beusekom, H. M. M., Emanuelsson, H. U., Hofma, S. H., Albertsson, P. A., Verdouw, P. D., Boersma, E., Serruys, P. W. and van der Giessen, W. J. (1996). Reduction in thrombotic events with heparin-coated Palmaz–Schatz stents in normal porcine coronary arteries. *Circulation*, 93, 423–430.

Harvey, R. A., Tesoriero, J. V. and Greco, R. S. (1984). Non-covalent bonding of penicillin and cefazolin to Dacron. *American Journal of Surgery*, 147, 205–209.

Heard, S. O., Wagle, W., Vijayakumar, E., McLean, S., Brueggemann, A., Napolitanho, L. M., Edwards, P., O'Connell, F. M., Puyana, J. C. and Doern, G. V. (1998). Influence of triple-lumen central venous catheters coated with chlorhexidine and silver sulfadiazine on the incidence of catheter-related bacteremia. *Archives of Internal Medicine*, 158, 81–87.

Hustinx, W. N. M. and Verbrugh, H. A. (1994). Catheter associated urinary tract infections: epidemiological, preventive and therapeuric considerations. *International Journal of Antimicroial Agents*, 4, 117–123.

Ishihara, K., Aragaki, R., Ueda, T., Watenabe, A. and Nakabayashi, N. (1990). Reduced thrombogenicity of polymers having phospholipids polar group. *Journal of Biomedical Materials Research*, 24, 1069–1077.

Jansen, B., Schumacher-Perdreau, F., Peters, G. and Pulverer, G. (1989). New aspects in the pathogenesis and prevention of polymer associated foreign body infections caused by coagulase-negative staphylococci. *Journal of Investigative Surgery*, 2, 361–380.

Jansen, B. and Peters, G. (1991). Modern strategies in the prevention of polymer-associated infections. *Journal of Hospital Infection*, 19, 83–88.

Jansen, B., Jansen, S., Peters, G. and Pulverer, G. (1992a). *In vitro* efficacy of a central venous catheter (Hydrocath) loaded with teicoplanin to prevent bacterial colonization. *Journal of Hospital Infection*, 22, 93–107.

Jansen, B., Kristinsson, K. G., Jansen, J., Peters, G. and Pulverer, G. (1992b). *In vitro* efficacy of a central venous catheter complexed with iodine to prevent bacterial colonisation. *Journal of Antimicrobial Chemotherapy*, 30, 135–139.

Jansen, B., Goodman, L. P. and Ruiten, D. (1993). Bacterial adherence to hydrophilic polymer-coated polyurethane stents. *Gastrointestinal Endoscopy*, 39, 670–673.

John, S. F., Hillier, V. F., Handley, P. S. and Derrick, M. R. (1995). Adhesion of *Staphylococci* to polyurethane and hydrogel-coated polyurethane catheters assayed by an improved radiolabelling technique. *Journal of Medical Microbiology*, 43, 133–140.

Johnson, J. R., Delavari, P. and Azar, M. (1999). Activities of a nitrofurazone-containing urinary catheters and a silver hydrogel catheter against multidrug resistant bacteria characteristic of catheter-associated urinary tract infection. *Antimicrobial Agents and Chemotherapy*, 43, 2990–2995.

Jones, C. R., Handley, P. S., Robson, G. D., Eastwood, I. M. and Greenhalgh, M. (1996). Biocides incorporated into plasticised polyvinyl chloride reduce adhesion of *Pseudomonas fluorescens* B1146 and substratum hydrophobicity. *Journal of Applied Bacteriology*, 81, 553–560.

Jones, D. S., Gorman, S. P., McCafferty, D. F. and Woolfson, A. D. (1991). The effects of three non-antibiotic antimicrobial agents on the surface hydrophobicity of certain micro-organisms evaluated by different methods. *Journal of Applied Bacteriology*, 71, 218–227.

Jones, D. S., Garvin, C. P. and Gorman, S. P. (2001a). Design of a simulated urethra model for the quantitative assessment of urinary catheter lubricity. *Journal of Matererials Science. Materials in Medicine*, 12, 15–21.

Jones, D. S., McGovern, J. G., Adair, C. G., Woolfson, A. D. and Gorman, S. P. (2001b). Conditioning film and environmental effects on the adherence of *Candida* spp. to silicone and poly(vinylchloride) biomaterials. *Journal of Materials Science. Materials in Medicine*, 12, 399–405.

Jones, D. S., McGovern, J. G., Woolfson, A. D., Adair, C. G. and Gorman, S. P. (2002). Physicochemical characterisation of hexetidine impregnated endotracheal tube poly(vinylchloride) and resistance to adherence of respiratory bacterial pathogens. *Pharmaceutical Research*, 19, 818–824.

Keane, P. F., Bonner, M., Johnston, S. R., Zafar, A. and Gorman, S. P. (1994). Characterisation of biofilm and encrustation on ureteric stents *in vivo*. *British Journal of Urology*, 73, 687–691.

Kohnen, W. and Jansen, B. (1995). Polymer material for the prevention of catheter-related infections. *Zentralblatt fur Bakteriologie*, 283, 175–186.

Kohnen, W., Schaper, J., Klien, O., Tieke, B. and Jansen, B. (1998). A silicone ventricular catheter coated with a combination of rifampin and trimethoprim for the prevention of catheter-related infections. *Zentralblatt fur Bakteriologie*, 287, 147–156.

Kristinsson, K. G., Jansen, B., Treitz, U., Schumacher-Perdreau, F., Peters, G. and Pulverer, G. (1991). Antimicrobial activity of polymers coated with iodine-complexed polyvinylpyrrolidone. *Journal of Biomaterials Applications*, 5, 173–184.

Kunin, C. M., Chin, Q. F. and Chambers, S. (1987). Morbidity and mortality associated with indwelling urinary catheters in elderly patients in a nursing home – confounding due to the presence of associated diseases. *Journal of the American Geriatric Society*, 35, 1001–1006.

Lachapelle, K., Graham, A. M. and Symes, J. F. (1994). Antibacterial activity, antibiotic retention and infection resistance of a rifampin-impregnated gelatin sealed Dacron graft. *Journal of Vascular Surgery*, 19, 675–682.

Liedberg, H. and Lundeberg, T. (1989). Silver coating of urinary catheters prevents adherence and growth of *Pseudomonas aeruginosa*. *Urolological Research*, 17, 357–358.

Liedl, B. and Hofstetter, A. (2000). Pathogenese und verhinderung katheterassoziierter harnwegsinfektionen. *Urologie B*, 40, 233–237.

Liedl, B. (2001). Catheter-associated urinary tract infections. *Current Opinion in Urology*, 11, 75–79.

McAllister, E. W., Carey, L. C., Brady, P. G. and Heller, R. (1993). The role of polymeric surface smoothness of biliary stents in bacterial adherence, biofilm deposition, and stent occlusion. *Gastrointestinal Endoscopy*, 39, 422–425.

McLean, R. J. C., Downey, J., Clapham, L. and Nickel, J. C. (1990). A simple technique for studying struvite crystal growth *in vitro*. *Urological Research*, 18, 39–43.

McLean, R. J. C., Lawrence, J. R., Korber, D. R. and Caldwell, D. E. (1991a). *Proteus mirabilis* biofilm protection against struvite crystal dissolution and its implications in struvite urolithiasis. *Journal of Urology*, 146, 1138–1142.

McLean, R. J. C., Downey, J., Clapham, L., Wilson, J. W. L. and Nickel, J. C. (1991b). Pyrophosphate inhibition of *Proteus mirabilis* induced struvite crystallization *in vitro*. *Clinica Chimica Acta*, 200, 107–118.

McGovern, J. G., Garvin, C. P., Jones, D. S., Woolfson, A. D. and Gorman, S. P. (1997). Modification of biomaterial surface characteristics by body fluids *in vitro*. *International Journal of Pharmaceutics*, 149, 251–254.

Maki, D. G., Stolz, S. M., Wheeler, S. and Mermel, L. A. (1997). Prevention of central venous catheter-related bloodstream infection by use of an antiseptic-impregnated catheter. *Annals of Internal Medicine*, 127, 257–266.

Mobley, H. L. T. and Warren, J. W. (1987). Urease-positive bacteriuria and obstruction of long-term urinary catheter. *Journal of Clinical Microbiology*, 25, 2216–2217.

Monson, T. and Kunin, C. M. (1974). Evaluation of a polymer-coated indwelling catheter in prevention of infection. *Journal of Urology*, 111, 220–222.

Morris, N. S. and Stickler, D. J. (1998). Encrustation of indwelling uretheral catheters by *Proteus mirabilis* biofilms growing in human urine. *Journal of Hospital Infection*, 39, 227–234.

Morris, N. S., Stickler, D. J. and McClean, R. J. C. (1999). The development of bacterial biofilms on indwelling urethral catheters. *World Journal of Urology*, 17, 245–350.

Multanen, M., Talja, M., Hallanvuo, S., Siitonen, A., Välimaa, T., Tammela, T. L. J., Seppälä, J. and Törmälä, P. (2000). Bacterial adherence to ofloxacin-blended polylactone-coated self-reinforced L-lactic acid polymer urological stents. *BJU International*, 86, 966–969.

Nickel, J. C., Grant, S. K. and Costerton, J. W. (1985). Catheter-associated bacteriuria. An experimental study. *Urology*, 24, 369–375.

Nickel, J. C., Downey, J. and Costerton, J. W. (1989). Ultrastructural study of microbiologic colonisation of urinary catheters. *Urology*, 34, 284–291.

Nickel, J. C., Downey, J. and Costerton, J. W. (1991). Movement of *Pseudomonas aeruginosa* along catheter surfaces. A mechanism in the pathogenesis of catheter-associated infection. *Urology*, 39, 93–98.

Nickel, J. C., Downey, J. and Costerton, J. W. (1992). Movement of *Pseudomonas aeruginosa* along catheter surfaces. A mechanism in pathogenesis of catheter-associated infection. *Urology*, 39, 93–98.
Nickel, J. C., Costerton, J. W., McLean, R. J. C. and Olson, M. (1994). Bacterial biofilms: influence on the pathogenesis, diagnosis and treatment of urinary tract infections. *Journal of Antimicrobial Chemotherapy*, 33, 31–41.
Pemberton, L. B., Ross, V., Cuddy, P., Kremer, H., Fessler, T. and McGurk, E. (1996). No difference in catheter sepsis between standard and antiseptic central venous catheters. *Archives of Surgery*, 131, 986–989.
Raad, I., Darouiche, R., Hachem, R., Mansouri, N. and Bodey, G. P. (1996). The broad-spectrum activity of catheters coated with minocycline and rifampicin. *Journal of Infectious Diseases*, 173, 418–424.
Raad, I., Buzaid, A., Rhyne, J., Hachem, R., Darouiche, R., et al. (1997a). Minocycline and ethylenediaminetetraacetate for the prevention of recurrent vascular catheter infections. *Clinical Infectious Diseases*, 25, 149–151.
Raad, I., Darouiche, R., Dupuis, J., Abi-Said, D., Gabrielli, A., Hachem, R., Wall, M., Harris, R., Jones, J., Buzaid, A., Robertson, C., Sheaq, S., Curling, P., Burke, T. and Ericsson, C. (1997b). Central venous catheters coated with minocycline and rifampin for the prevention of catheter-related colonisation and bloodstream infections: a randomized, double-blind trial. *Annals of Internal Medicine*, 127, 267–274.
Raad, I. I., Darouiche, R. O., Hachem, R., AbiSaid, D., Safar, H., Darnule, T., Mansouri, M. and Morck, D. (1998). Antimicrobial durability and rare ultrastructural colonization of indwelling central catheters coated with minocycline and rifampin. *Critical Care Medicine*, 26, 219–224.
Rad, A. Y., Ayhan, H. and Piskin, E. (1998). Adhesion of different bacterial strains to low-temperature plasma treated biomedical silicon catheter surfaces. *Journal of Bioactive and Compatible Polymers*, 13, 81–101.
Ramsay, J. W. A., Gernham, A. J., Mulhall, A. B., et al. (1999). Biofilms, bacteria and bladder catheters. *British Journal of Urology*, 64, 395–398.
Reid, G. (1997). Microbial adhesion to biosurfaces. *Current Opinion in Colloid and Interface Science*, 2, 513–516.
Reid G. (1999). Biofilms in infectious disease and on medical devices. *International Journal of Antimicrobial Agents*, 11, 223–226.
Reid, G., Tieszer, C., Foerch, R., Busscher, H. I., Khoury, A. E. and Van Der Mei, H. C. (1992a). The binding of urinary components and uropathogens to a silicone latex urethral catheter. *Cells and Materials*, 2, 253–260.
Reid, G., Khoury, A. E., Neumann, A. W. and Bruce, A. W. (1992b). Components involved in biomaterial-related infections. *American Urology Association. Updates Service, Lesson 18*, 11, 138–143.
Reid, G., Lam, D., Policova, Z. and Neumann, A. W. (1993). Adhesion of two uropathogens to silicone and lubricious catheters: influence of pH, urea and creatinine. *Journal of Materials Science. Materials in Medicine*, 4, 17–22.
Reid, G., Davidson, R. and Denstedt, J. D. (1994a). Analyses of conditioning film deposition onto ureteral stents. *Surface and Interface Analysis*, 21, 581–586.
Reid, G., Sharma, S., Advikolanu, K. and Tieszer, C. (1994b). Effects of ciprofloxacin, norfloxacin and ofloxacin on *in vitro* adhesion and survival of *Pseudomonas*

aeruginosa AK1 on urinary catheters. *Antimicrobial Agents and Chemotherapy*, 38, 1490–1495.

Riley, D. K., Classen, D. C., Stevens, L. E. and Burke, J. P. (1995). A large randomized clinical trial of a silver-impregnated urinary catheter: lack of efficacy and staphylococcal superinfection. *American Journal of Medicine*, 98, 349–356.

Santin, M., Motta, A., Denyer, S. P. and Cannas, M. (1999). Effect of the urine conditioning film on ureteral stent encrustation and characterisation of its protein composition. *Biomaterials*, 20, 1245–1251.

Schierholz, J. M., Jansen, B., Jaenicke, L. and Pulverer, G. (1994). *In vitro* efficacy of an antibiotic-releasing silicone ventricle catheter to prevent shunt infection. *Biomaterials*, 15, 996–1000.

Schierholz, J. M., Steinhouser, H., Rump, A. F. E., Berkels, R. and Pulverer, G. (1997). Controlled release of antibiotics from biomedical polyurethanes. *Biomaterials*, 18, 839–844.

Sherertz, R. J. (1997). Selected thoughts on the development of new medical devices. *Current Opinion in Infectious Diseases*, 10, 330–334.

Sioshansi, P. and Tobin, E. J. (1996). Surface-treatment of biomaterials by ion-beam processes. *Surface and Coatings Technology*, 83, 175–182.

Stickler, D. J. (1996). Bacterial biofilms and the encrustation of urethral catheters. *Biofouling*, 9, 293–305.

Stickler, D. J. (2000). Biomaterials to prevent nosocomial infections: is silver the gold standard? *Current Opinion in Infectious Diseases*, 13, 389–393.

Stickler, D. J., King, J., Nettleton, J. and Winters, C. (1993a). The structure of urinary catheters encrusting bacterial biofilms. *Cells and Materials*, 3, 315–320.

Stickler, D., Ganderton, L., King, J., Nettleton, J. and Winters, C. (1993b). *Proteus mirabilis* biofilms and the encrustation of uretheral stents. *Urological Research*, 21, 407–411.

Stickler, D. J., Howe, N. S. and Winters, C. (1994). Bacterial biofilm growth on ciprofloxacin treated urethral catheters. *Cells and Materials*, 4, 387–398.

Stickler, D. J., Morris, N. S. and Williams, T. J. (1996). An assessment of the ability of a silver-releasing device to prevent bacterial-contamination of urethral catheter drainage systems. *British Journal of Urology*, 78, 579–588.

Stickler, D. J., Morris, N. S., McClean, R. J. C. and Fuqua, C. (1998). Biofilms on indwelling urinary catheters produce quorum-sensing signal molecules *in situ* and *in vitro*. *Applied and Environmental Microbiology*, 64, 3486–3490.

Stickler, D. J. and Hughes, G. (1999). Ability of *Proteus mirabilis* to swarm over urethral catheters. *European Journal of Clinical Microbiology and Infectious Diseases*, 18, 206–208.

Talja, M., Virtanen, J. and Andersson, L. C. (1986). Toxic catheter and diminished uretheral blood circulation in the induction of urethral strictures. *European Urology*, 12, 340–345.

Tebbs, S. E., Sawyer, A. and Elliott, T. S. (1994). Influence of surface morphology on *in vitro* bacterial adherence to central venous catheters. *British Journal of Anaesthesia*, 72, 587–591.

Tunney, M. M., Gorman, S. P. and Patrick, S. (1996a). Infection associated with medical devices. *Reviews in Medical Microbiology*, 74, 195–205.

Tunney, M. M., Bonner, M. C., Keane, P. F. and Gorman, S. P. (1996b). Development of a model for assessment of biomaterial encrustation in the upper urinary tract. *Biomaterials*, 17, 1025–1029.

Tunney, M. M., Keane, P. F., Jones, D. S. and Gorman, S. P. (1996c). Comparative assessment of ureteral stent biomaterial encrustation. *Biomaterials*, 17, 1541–1546.

Tunney, M. M., Keane, P. F. and Gorman, S. P. (1997a). Assessment of urinary tract biomaterial encrustation using a modified Robbins device continuous flow model. *Journal of Biomedical Materials Research*, 38, 87–93.

Tunney, M. M., Jones, D. S. and Gorman, S. P. (1997b). Methacrylate polymers and copolymers as urinary tract biomaterials: resistance to encrustation and microbial adhesion. *International Journal of Pharmaceutics*, 151, 121–126.

Tunney, M. M., Jones, D. S. and Gorman, S. P. (1999). Biofilm and biofilm-related encrustation of urinary tract devices. In *Methods in Enzymology*, ed. R. Doyle, pp. 558–565. San Diego, CA: Academic Press.

Van Loosdrecht, M. C. M., Lyklema, J., Norde, W. and Zehnder, A. J. B. (1990). Influence of interfaces on microbal activity. *Microbiological Reviews*, 54, 75–87.

Wassall, M. A., Santin, M., Isalberti, C., Cannas, M. and Denyer, S. P. (1997). Adhesion of bacteria to stainless steel and silver-coated orthopaedic external fixation pins. *Journal of Biomedical Materials Research*, 36, 325–330.

Whalen, R. L., Cai, C., Thompson, L. M., Sarrasin, M. J., et al. (1997). An infection inhibiting urinary catheter material. *ASAIO Journal*, 43, M843–M847.

Wollin, T. A., Tieszer, C., Riddell, J. V., Denstedt, J. D. and Reid, G. (1998). Bacterial biofilm formation, encrustation, and antibiotic adsorption to ureteral stents indwelling in humans. *Journal of Endourology*, 12, 101–111.

Woodyard, L. L., Bowersock, T. L., Turek, J. J., McCabe, G. P. and Deford, J. A. (1996). Comparison of the effects of several silver-treated intravenous catheters on the survival of *staphylococci* in suspension and their adhesion to the catheter surface. *Journal of Controlled Release*, 40, 23–30.

PART THREE

ORAL BIOFILMS

CHAPTER EIGHT

Novel Microscopic Methods to Study the Structure and Metabolism of Oral Biofilms

David J. Bradshaw and Philip D. Marsh

1 INTRODUCTION

Uniquely in the human body, the tooth surface provides a non-shedding surface for microbial attachment. As a result, large numbers of microorganisms accumulate, together with their metabolic products, to form dental plaque, especially at occluded sites. Dental plaque was the first biofilm described by van Leeuwenhoek (1683), who noted the unusual resistance of the 'animalcules' he observed to treatment with vinegar (McHugh, 1999). Since this time, dental plaque has probably been the most studied of biofilms, with some justification: dental plaque is the most accessible of human-associated biofilms and is responsible for the most common diseases affecting man in the developed world – dental caries and periodontal diseases. This chapter provides a brief historical view of light and electron microscopic studies of dental plaque. The major focus of the chapter is then to describe how novel microscopic approaches are providing new insights into the structure and metabolic activities of dental plaque biofilms.

2 LIGHT AND ELECTRON MICROSCOPIC STUDIES OF DENTAL PLAQUE DEVELOPMENT

Since van Leewenhoek's time, vast numbers of light microscopic studies (especially of dental plaque) have examined the structure of biofilms. Furthermore, the development of electron microscopy (EM) began to unearth the diversity within, and structures of, developing dental plaque (Saxton, 1973; Listgarten, 1976, 1999).

The development of dental plaque can be divided into several arbitrary stages. *Pellicle formation* occurs almost instantaneously on a cleaned tooth surface (Figure 8.1a). After a cleaned tooth surface has been exposed for 4 hours to the oral environment, relatively few bacteria are found. The pellicle is unevenly distributed over the surface and is mainly found in depressions in the enamel. The major constituents of the pellicle are salivary glycoproteins, phosphoproteins, and lipids. Components from gingival crevicular fluid may also be present, and remnants of cell walls of bacteria and microbial products have also been identified.

Monobacterial attachment (0–4 hours: Figure 8.1b) initially occurs via non-specific, long-range, physicochemical forces between bacteria and the pellicle-coated tooth surface, including van der Waals attractive forces and repulsive negative electrostatic forces (DLVO theory). In addition, a high degree of surface hydrophobicity may facilitate attachment. Firm attachment results from specific, stereochemical interactions between molecules (adhesins) on the microbial cell surface and receptors in the pellicle. The only bacteria detected at this early stage are coccoid or coccobacillary types, and these are always located in depressions on the enamel surface.

The growth of attached bacteria (4–24 hours: Figure 8.1c) leads to the formation of distinct microcolonies. Even after 8 hours, only a few small groups of microorganisms are found on the surface. The numbers of bacteria begin to increase rapidly after 8–12 hours and gradually form a monolayer across the surface. Layers of microorganisms several cells thick can form in some areas, with organisms embedded in a glycocalyx matrix. Even at this stage of colonisation, Gram-positive and Gram-negative bacteria are relatively distinct, forming single microcolonies.

Microbial succession (Figure 8.1d) leads to increasing species diversity (1 + days) and the eventual formation of a 'pseudo-stable' climax community (Figure 8.1e). Pioneer bacteria create an environment which is either more attractive for secondary colonisers or increasingly unfavourable to themselves because of, for example, lack of nutrients or accumulation of inhibitory metabolic products. In this way, the resident microbial community is gradually replaced by other species more suited to the modified habitat. The pioneer species themselves become colonised by a wider variety of filamentous organisms, often arranged in a perpendicular orientation to the surface. This stage of development represents the transition to a mature dental plaque, which occurs predominantly as a result of cell division of the filamentous organisms in columnar microcolonies. In addition, adsorption of microorganisms from saliva continues to contribute to the development of plaque.

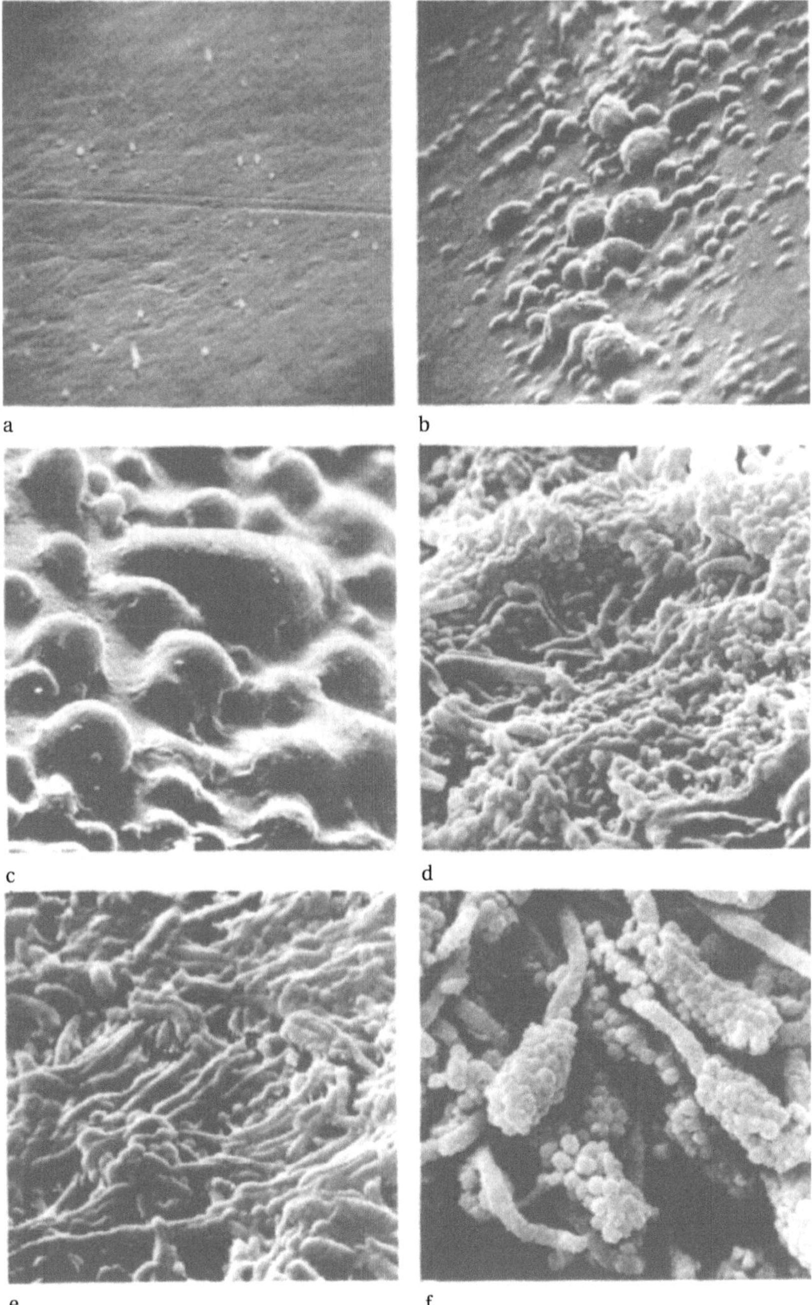

Figure 8.1: Development of dental plaque over time: (a) Recently cleaned tooth surface, with pellicle (<1 hour); (b) mono-bacterial attachment (1–4 hours); (c) bacterial growth and formation of microcolonies (4–24 hours); (d) microbial succession and increasing species diversity (24 hours +); (e) 'climax community' of mature plaque (>24 hours); (f) 'corn-cob' formation in mature plaque.

Co-aggregation (or co-adhesion) with unrelated species can help to develop plaque in an ordered fashion allows the formation of 'corn cob' and other structures (Figure 8.1f), and further contributes to developing plaque complexity (Kolenbrander, 1988). It also appears that some bacteria can use specific strategies to detach from these biofilms.

As plaque becomes older (14 days +), further changes in the depths of plaque are visible. In particular, a layer of densely packed Gram-positive bacteria is often next to the tooth surface. The outer part of mature plaque is more loosely structured and varies in composition (Nyvad and Fejerskov, 1989). The outer layers may represent single types of organisms; alternatively, layers of different bacterial species are sometimes observed parallel to the tooth surface. In still other cases, the outer layers do not show any characteristic structure. A variable matrix of 'glycocalyx' is observed in all layers of plaque. Such a heterogeneous composition is, in part, assumed to be due to the development of concentration gradients in key environmental parameters. Fissure plaque at the entrance to the fissure is composed of cocci and rods arranged in palisades perpendicular to the enamel surface, with some filamentous forms. In the depth of the fissure, filaments are rarer, with cocci and rods usually predominant.

A feature of the studies of the architecture of dental plaque determined from EM studies is the observation of biofilms that are spatially organised with a complex structure and with cells that are densely packed together (Figure 8.2). This view has gradually begun to shift, following the application of novel microscopic techniques, some of which are discussed in the following sections.

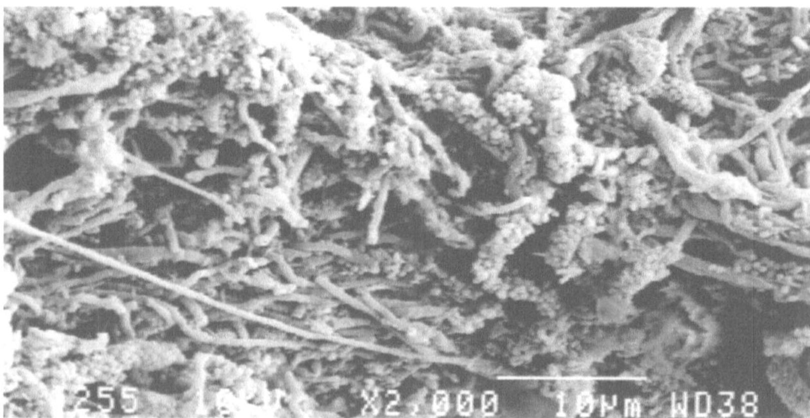

Figure 8.2: Spatial organisation of human dental plaque revealed by EM.

3 NOVEL MICROSCOPIC APPROACHES FOR STUDYING THE STRUCTURE AND PHYSIOLOGY OF DENTAL PLAQUE

3.1 Confocal Microscopy

The application of the confocal laser scanning microscope (CLSM) to the study of biofilms in the early 1990s suggested that some of the structures visualised previously by EM might have incorporated artefacts, particularly because of the dehydration processes often used in EM sample preparation (Lawrence et al., 1991). The first study of dental plaque biofilms using the confocal method was carried out by Cummins et al. (1992). Later, a more comprehensive study was undertaken by Wood et al. (2000), who described the architecture of undisturbed plaque formed in an *in situ* enamel device containing a nylon ring to aid retention. They were able to make measurements of plaque dimensions and noted the heterogeneous structures in plaque. Furthermore, it was also possible to show that the voids, or water channels, often described in the literature on biofilms from other sources (Lawrence et al., 1991; Costerton et al., 1995), were also a feature of natural dental plaque. Although Wood et al. (2000) were able to show that structures within the plaque biofilm layer were amenable to study by this method, it is notable that resolution of the ultrastructure at deeper layers within the biofilm was limited. This is, in part, because conventional CLSM is hampered by the wavelength of laser light used in order to excite fluorescence. In the standard CLSM set-up, blue or ultraviolet (uv) light is usually used to excite fluorescence. This light, of relatively short wavelength, is readily scattered by turbid samples. In two-photon microscopy, red or infrared light is used instead. As illustrated in Section 3.2, this provides significant benefits.

As the biofilm becomes thicker, it is implied that gradients develop in key factors that affect the growth of resident microorganisms. For example, the oxygen concentration may be lowered by metabolism of early colonisers, making the environment more conducive to the growth of obligate anaerobes. Also, it is assumed that gradients in pH develop following the catabolism of dietary carbohydrates, which might result in demineralisation of the tooth surface and can ultimately lead to dental caries formation.

At the gum margin, the later stages of plaque development are therefore characterised by a shift from mainly aerobic and facultatively anaerobic species dominant in the early stages of developing plaque to a situation in which facultatively and obligately anaerobic organisms predominate in the 'climax community' of mature plaque. In contrast, in individuals with a high frequency of sugar intake, and thus with plaque, which regularly undergoes cycles of low pH, there may be selection of acidogenic and aciduric species in dental plaque, which may ultimately lead to dental caries. Although the preceding sections describe plaque formation in stages, it is clear that these processes are highly dynamic and that attachment, growth, removal, and reattachment of bacteria occur simultaneously.

3.2 Two-Photon Excitation Microscopy

The limited depth penetration of CLSM can be overcome by employing two-photon excitation microscopy (TPEM) (Denk, Strickler, and Webb, 1990; Centonze and White, 1998), in which the fluorescent molecule is excited by the simultaneous absorption of two red or (near) infrared photons. Because red light has a longer wavelength than the blue/UV light used in conventional fluorescence excitation, TPEM gives a far greater penetration into the sample than is obtained in CLSM. Such a penetration benefit will vary in proportion to the scattering properties of the specimen studied. In a study of *in vitro* oral biofilms, we demonstrated that the benefit was at least fourfold in terms of penetration depth (Figure 8.3). This fourfold estimate is conservative, since it only compares average fluorescence intensities, and the image quality is not taken into account. In the CLSM images, attempts to visualise deeper layers suffer a significant loss of contrast compared with TPEM (illustrated

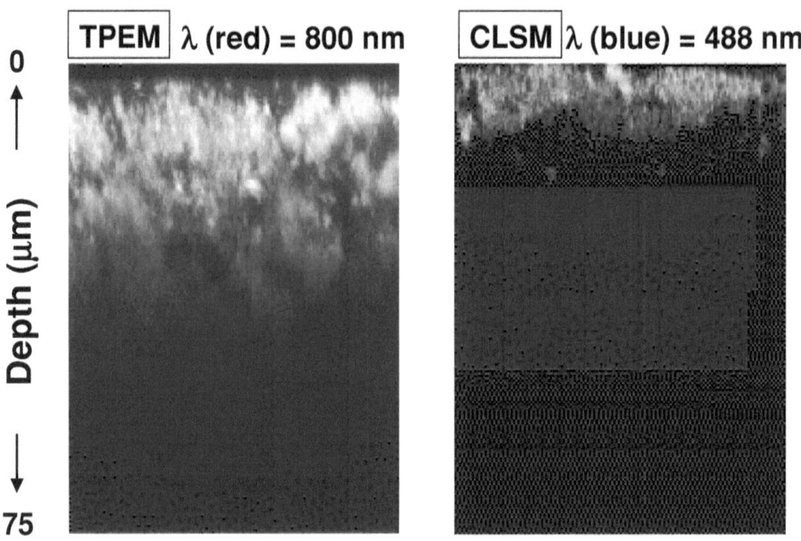

Figure 8.3: Fluorescence intensity *versus* depth in *x*-*z* section through mixed culture *in vitro* oral biofilms.

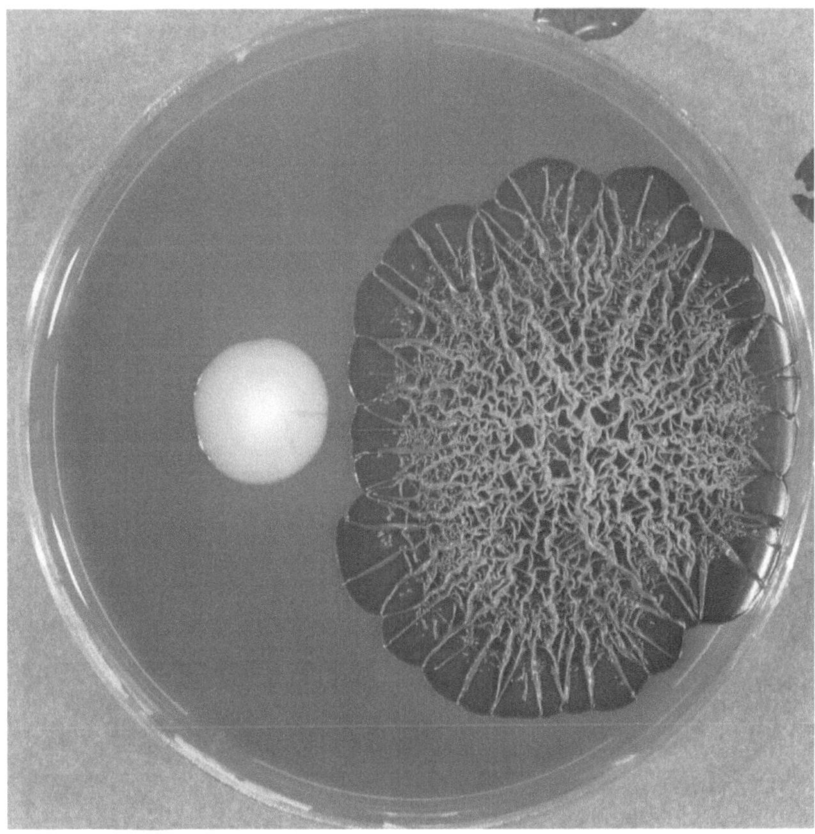

Figure 11.1: The rdar morphotype of *S. enterica* serotype Typhimurium. Left: *S. enterica* serotype Typhimurium ATCC14028 colony with regulated rdar morphotype expression. Right: *S. enterica* serotype Typhimurium ATCC 14028 colony with semi-constitutive rdar morphotype expression. The two strains are isogenic mutants and differ by the insertion of one nucleotide in the promoter region of *agfD*. Strains were grown on Luria Bertani (LB) agar medium without salt at 37°C for 4.5 days. To enhance the appearance of the rdar morphotype, Congo red was added to the agar.

The following figures are available in color for download from
www.cambridge.org/9781107403451

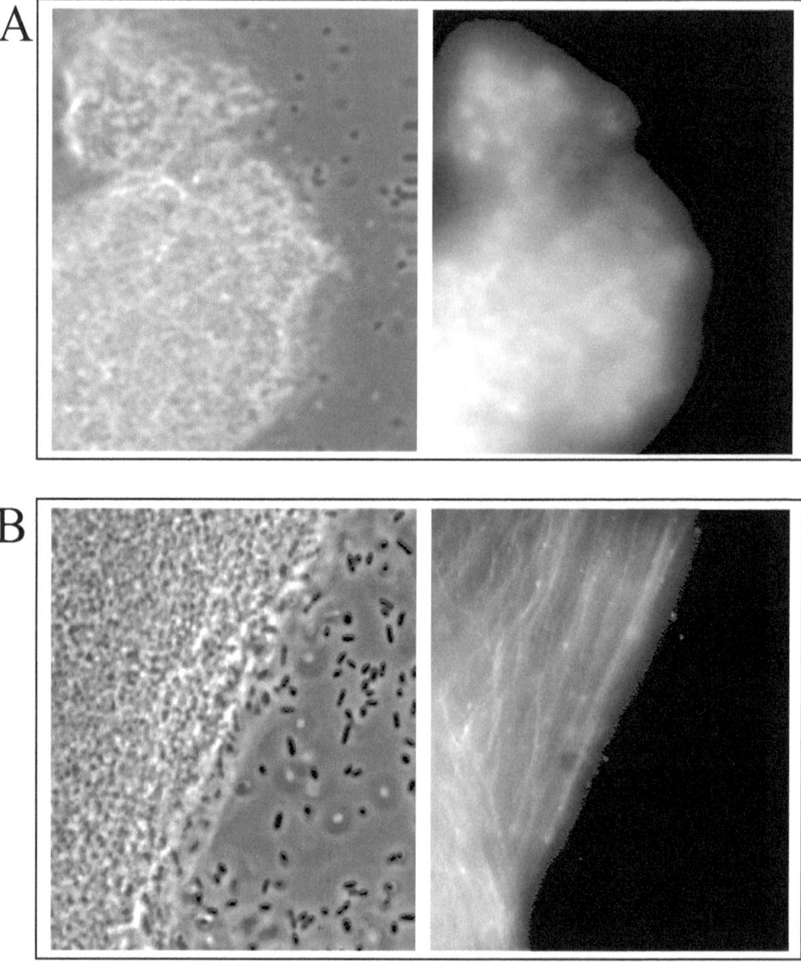

Figure 11.5: Interaction of thin aggregative fimbriae with cellulose as visualised by fluorescence microscopy. (A) Arrangement of cellulose fibres when co-expressed with thin aggregative fimbriae. Cellulose fibres are tightly wrapped around the cells. (B) Cellulose fibres, but not thin aggregative fimbriae, are expressed. The cellulose fibres run in long bands and are only loosely connected with the cell. Magnification × 400.

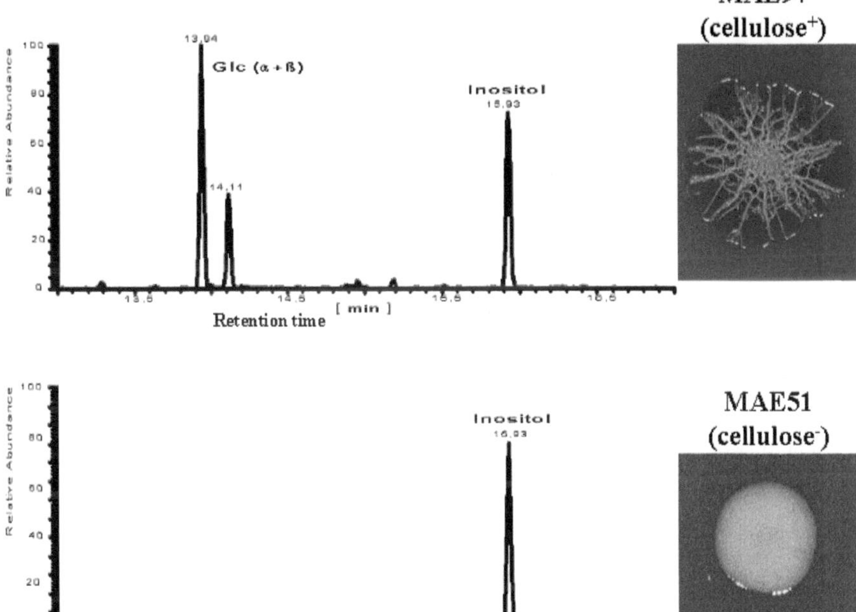

Figure 11.8: Detection of glucose monomers after isolation of crystalline cellulose. *S. enterica* serotype Typhimurium strain MAE97 expresses cellulose (upper right). When grown on LB medium without salt and containing Congo red, strain MAE97 absorbs the dye and expresses the pdar (pink, dry and rough) morphotype indicative of cellulose expression. After isolation of crystalline cellulose, the macromolecule was totally hydrolysed with 4 *N* Tri-fluoro-acetic acid at 100°C for 72 hours and pertrimethylsilylated. The glucose derivative was detected by GC/MS. MAE51, the *agfD* mutant, is cellulose negative. MAE51 remained white on CR medium (lower right); no sugar was detected by GC/MS.

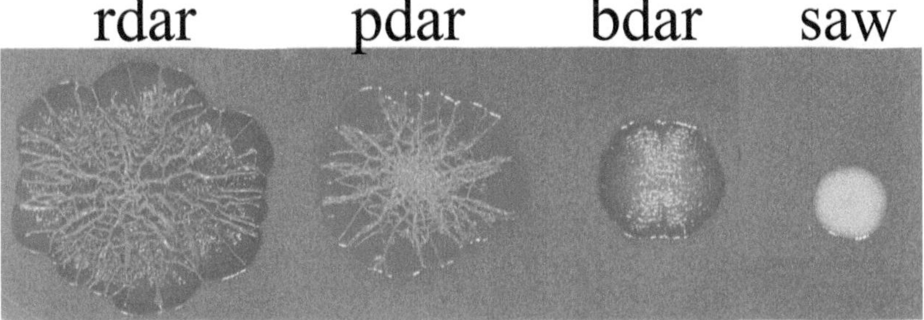

Figure 11.10: Morphotypes of *S. enterica* serotype Typhimurium expressing different matrix components. Rdar: thin aggregative fimbriae and cellulose are expressed; pdar: expression of cellulose; bdar: expression of thin aggregative fimbriae; saw: no expression of matrix components. For a pronounced visualisation, an ATCC14028 derivative with semi-constitutive expression of the rdar morphotype and respective mutants were used. Cells were incubated on LB plates without salt supplemented with Congo red dye at 37°C.

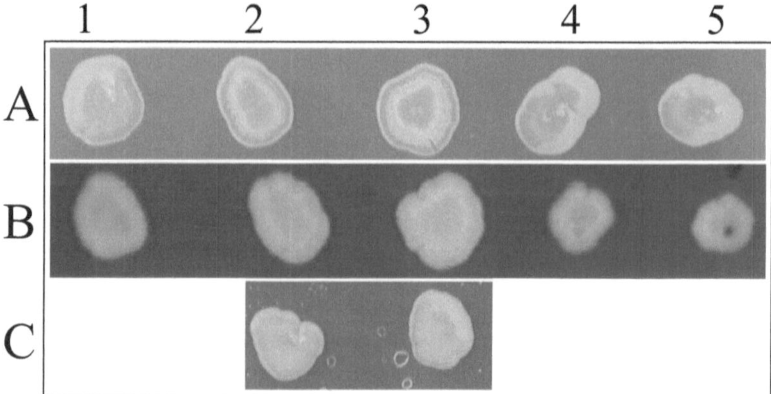

Figure 11.11: The most common regulatory pattern of *S. enterica* serotype Typhimurium/Enteritidis strains expressing the rdar morphotype. (A) Shown are the rdar morphotype of *S. enterica* serotype Typhimurium strains incubated on LB agar plates without salt supplemented with Congo red dye at 28°C for 48 hours. The rdar morphotype is expressed weakly. The strains are 1, *S. enterica* serotype Enteritidis PT21/1; 2, *S. enterica* serotype Enteritidis PT4/6; 3, *S. enterica* serotype Enteritidis PT4/6; 4, *S. enterica* serotype Typhimurium DT 104; and 5, *S. enterica* serotype Typhimurium DT 104. (B) Shown are the same strains incubated on LB plates without salt substituted with Calcofluor at 28°C for 48 hours. Expression of cellulose is indicated by the binding of Calcofluor to the colonies, resulting in white fluorescence. (C) The strains do not show expression of the rdar morphotype when incubated on CR plates at 37°C for 24 hours.

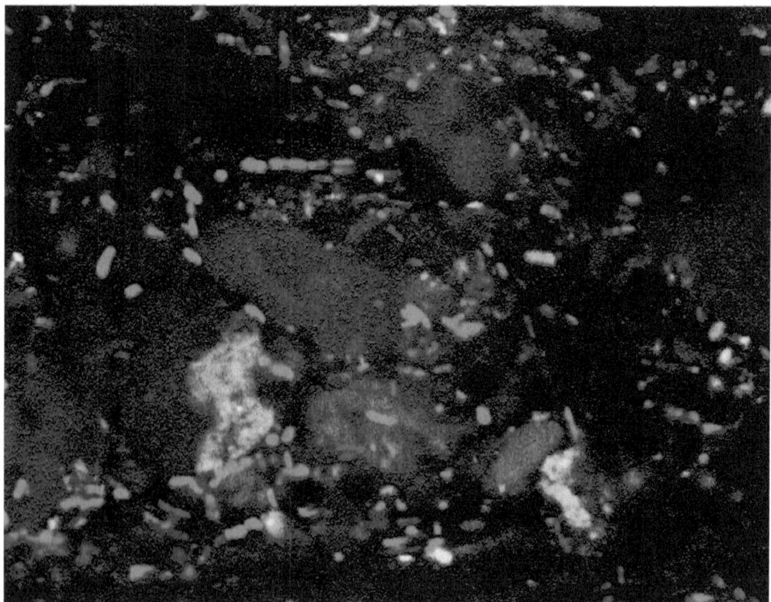

Figure 12.1: Light micrograph of the colonic mucosal surface stained with 16S rRNA oligonucleotide probes targeted against bacteroides (red, cy3), bifidobacteria (blue, cy5), and escherichia (green, fluorescein isothiocyanate).

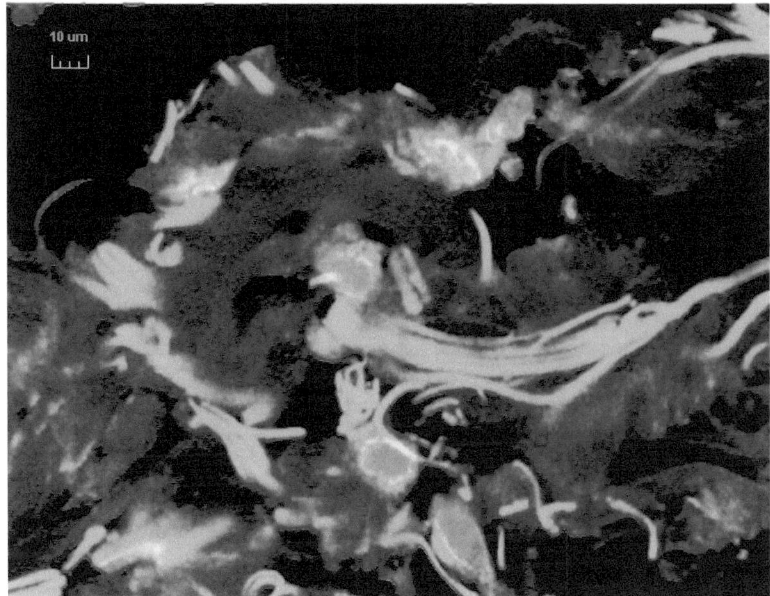

Figure 12.2: Light micrograph of the mucosal surface in tissue taken from the colon and stained with a eubacterial 16S rRNA oligonucleotide probe labelled with FITC.

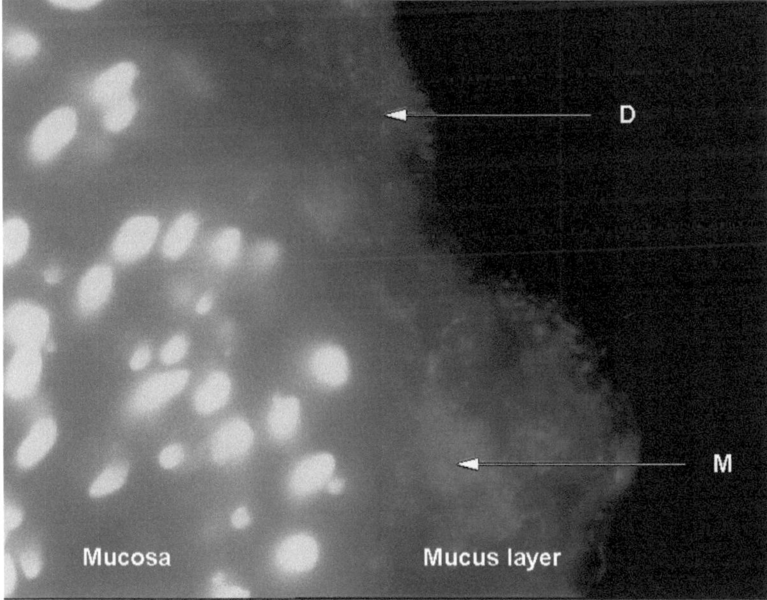

Figure 12.4: Light micrograph of a tranverse section of mucosal tissue taken from the colon and stained with 4,6-diamidino-2-phenylindole. The bacteria can be seen occurring as microcolonies (M) and diffusely (D) in the mucus layer.

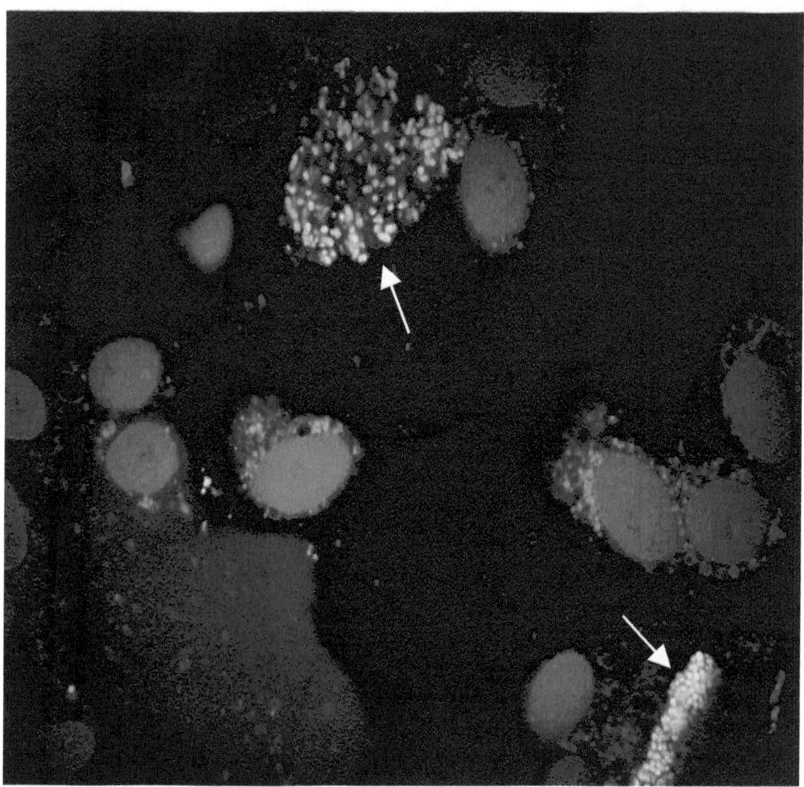

Figure 12.5: Light micrograph showing a live/dead stain of two bacterial microcolonies (arrows) on the surface of the rectal mucosa. Yellow cells are live, and red bacteria are dead.

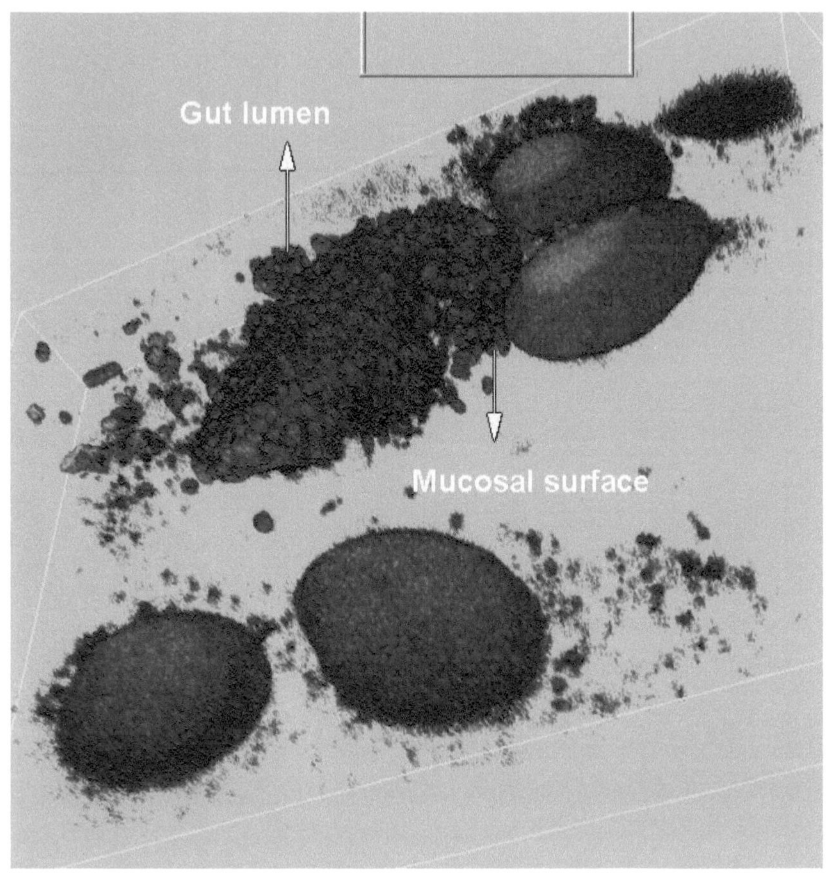

Figure 12.6: Confocal image of the small bacterial microcolony in Figure 12.5. Green bacteria are living cells, and red bacteria are dead.

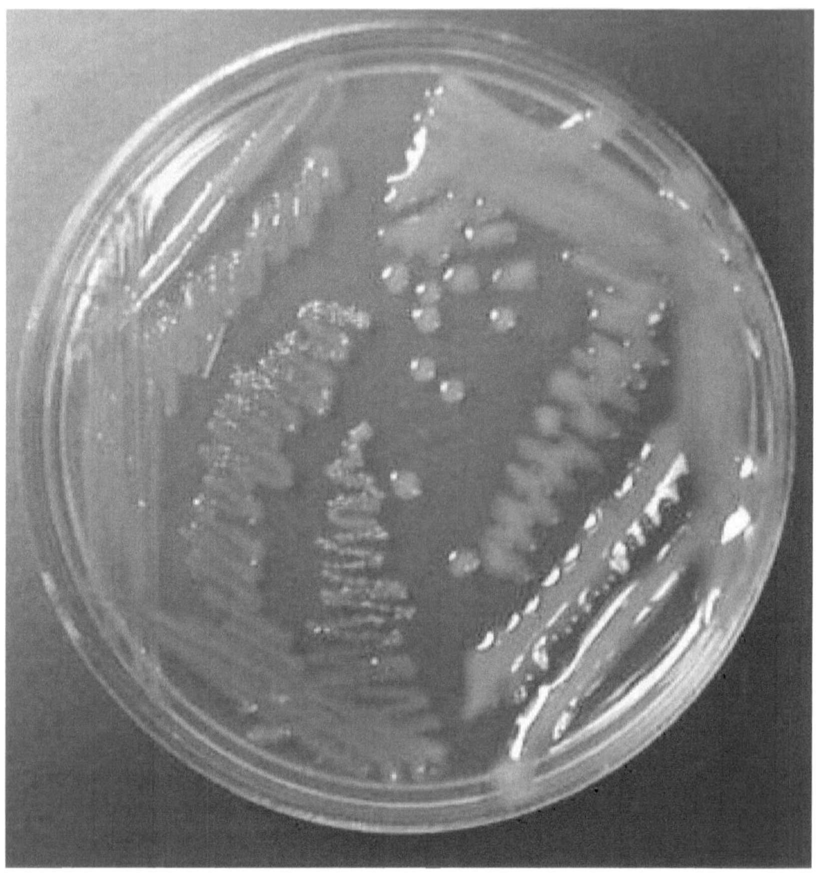

Figure 13.2: CF patients are initially colonised by non-mucoid *P. aeruginosa* (left half of plate). Eventually, the mucoid phenotype emerges (right half of plate). Once established in the CF lung, mucoid *P. aeruginosa* is very difficult, if not impossible, to eradicate.

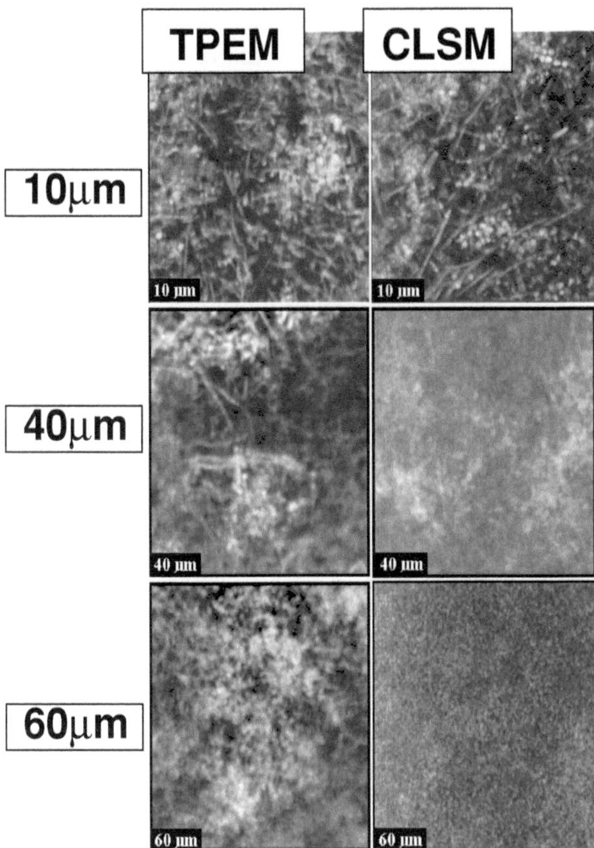

Figure 8.4: Depth resolution and image quality at different depths in mixed culture *in vitro* oral biofilms.

in Figure 8.4). This effect is well known in confocal imaging and is caused by scattering effects (Vroom et al., 1999). In addition to these image benefits, the nature of the excitation process in TPEM means that only fluorescent molecules in the focal plane are excited, which results in markedly reduced photobleaching of the fluorescent probes. These two features of TPEM make the technique well suited for in-depth imaging studies of strongly scattering samples such as biofilms.

3.3 Fluorescence Lifetime Imaging and TPEM

One property of biofilms which seems especially important is the development of gradients of key biochemical parameters (Wimpenny, 1982). As a result, microenvironments within biofilms may differ from the planktonic phase. In dental plaque biofilms, it seems clear that localised low pH (pH < 5.5) must occur in dental plaque following sugar consumption for demineralisation (and ultimately caries) to take place, while the pH of saliva overlying plaque remains near neutrality (6.75 < pH < 7.25). The presence of such microenvironments will be important in determining where species may be able to grow and be located within plaque, thereby influencing the role of plaque in health and disease. The TPEM technique described previously can be extended to include fluorescence lifetime imaging (FLIM) to determine localised variation in ion concentrations. In conventional methods, ion-sensitive probes have been used, where the fluorescence intensity of the probe is used as a putative measure of the ion concentration. Compartmentalisation or concentration of the probe can compromise this method. In the early 1990s, FLIM was introduced as an alternative contrast microscopic method. The FLIM method uses differences in fluorescence decay time, over extremely short time periods (< 10 ns), to produce contrast images corresponding with ion concentration. Unlike intensity methods, FLIM yields results that are unaffected by probe concentration. In previous studies, we were able to show that using carboxyfluorescein as a probe, TPEM-FLIM was well suited to demonstrate the development of pH gradients in-depth sections through *in vitro* oral biofilms over relatively short distances (Vroom et al., 1999) (Figure 8.5).

3.4 pH Mapping in *in Vitro* Oral Biofilms

The TPEM-FLIM method can be extended to allow pH mapping of multiple areas of a biofilm in real time. In more recent studies, we have been able to demonstrate that data extracted from TPEM-FLIM pH images give comparable pH-fall data to a conventional 'micro'-electrode measurement (Figure 8.6). Although previously we had suggested that gradient development in the x-z (i.e., depth) plane was greater than in the x-y (i.e., cross-sectional) plane, more recent work has shown that there are not notable differences in these planes. Instead, studies continue to suggest highly localised development of low pH microzones, which may develop over a few minutes within *in vitro* oral biofilms after glucose supplementation (Figure 8.7). Indeed, the images presented in Figure 8.7 show the presence of a mosaic of microenvironments in these oral biofilms. These microenvironments may explain the coexistence of species that would be incompatible with one another in homogeneous environments.

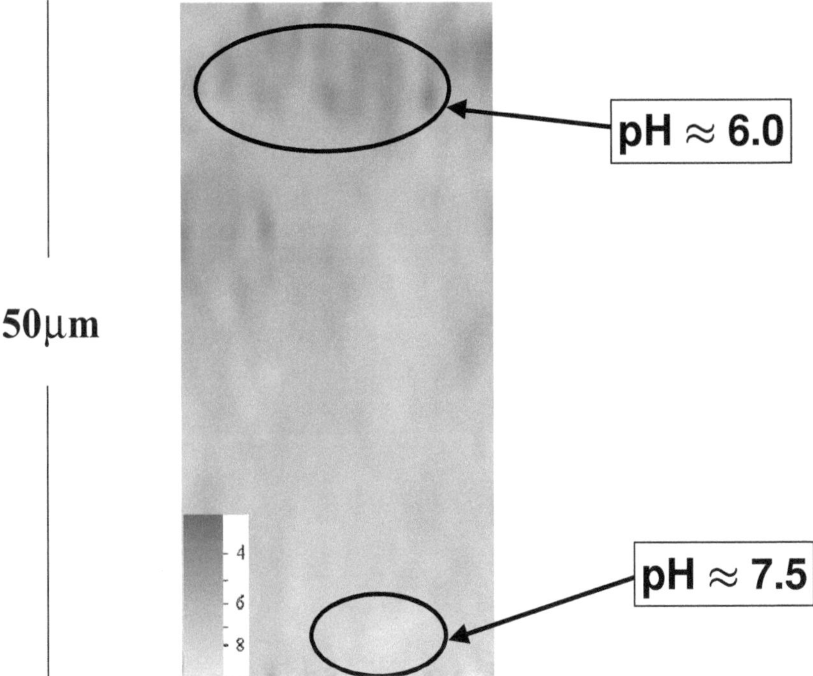

Figure 8.5: pH gradients resolved by TPEM-FLIM in-depth (x-z) section through mixed culture *in vitro* oral biofilm.

The images presented pose a tantalising question regarding the species distribution within these mixed culture biofilms. We therefore assessed the potential for TPEM-FLIM to co-localise simultaneously particular species within zones of acid production in a proof-of-principle experiment. Instead of our more complex biofilms, a simulated 'biofilm' was created. Cultures of *Lactobacillus rhamnosus* and *Fusobacterium nucleatum* were grown. The *L. rhamnosus* culture was then incubated with a specific mouse monoclonal antibody solution (supplied by Dr. Rudolf Gmur, Zurich, Switzerland). This was subsequently labelled with a Pacific Blue (Molecular Probes, Eugene, OR) fluorescently tagged goat-anti-mouse IgG secondary antibody. The labelled *L. rhamnosus* and unlabelled *F. nucleatum* cultures were mixed and then centrifuged to yield a mixed culture pellet. This simple simulation of a dense, two-species biofilm was then examined for acid production after glucose addition by TPEM-FLIM.

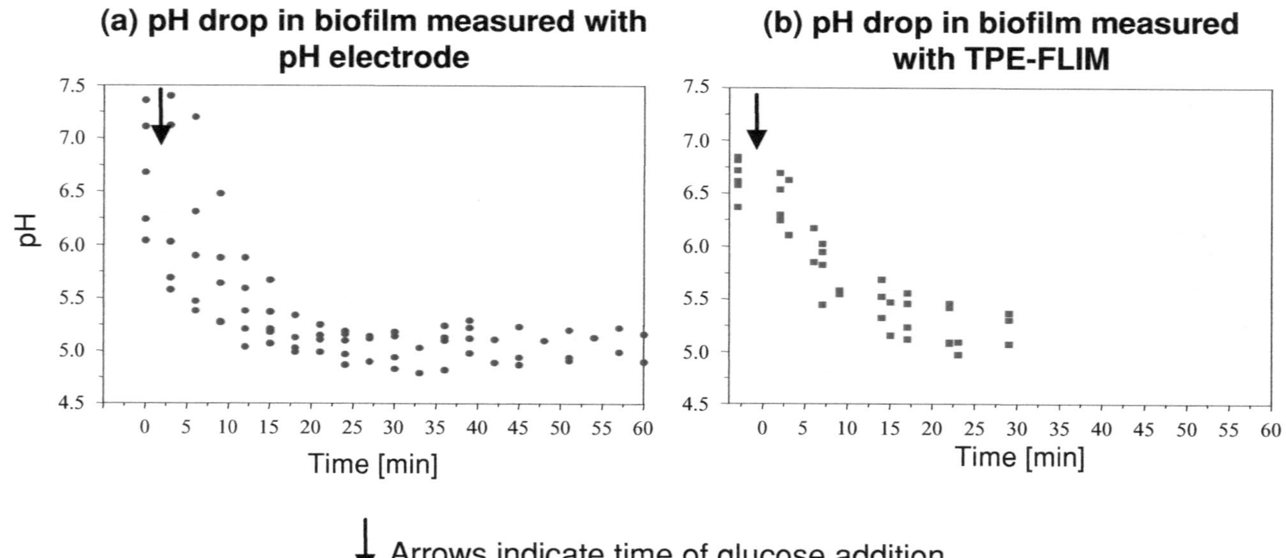

Figure 8.6: Measurement of biofilm pH using microelectrode *versus* TPEM-FLIM.

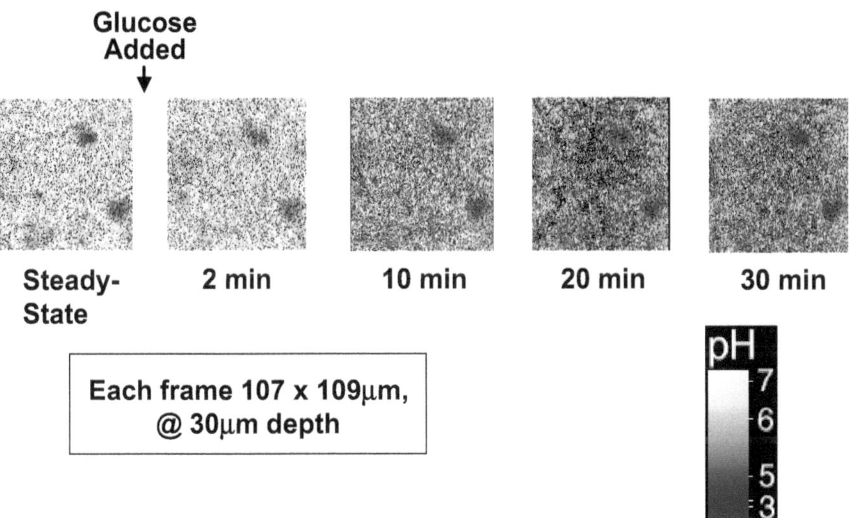

Figure 8.7: pH fall over time in cross-sectional (x-y) images through *in vitro* oral biofilms.

An image of a binary pelleted culture, detecting Pacific Blue fluorescence, following glucose addition is shown in Figure 8.8a. A dense cluster of *L. rhamnosus* cells is visible near the centre of the image. A clear co-localisation of acid production with the *L. rhamnosus* cluster is seen in the equivalent carboxyfluorescein lifetime (pH) image in Figure 8.8b.

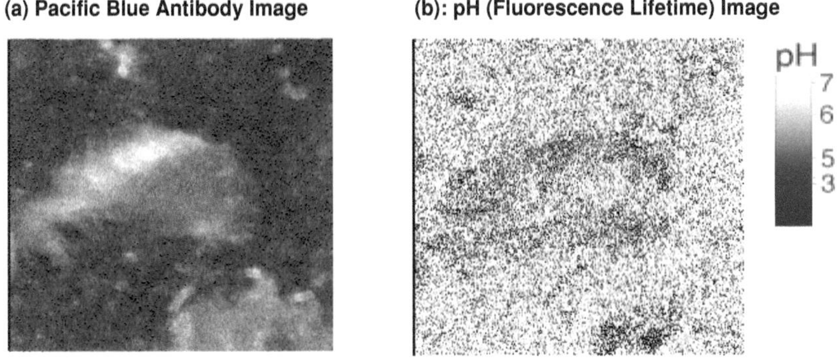

Figure 8.8: Co-localisation of Pacific Blue-labelled *L. rhamnosus* and pH in a simulated two-species biofilm with *F. nucleatum*.

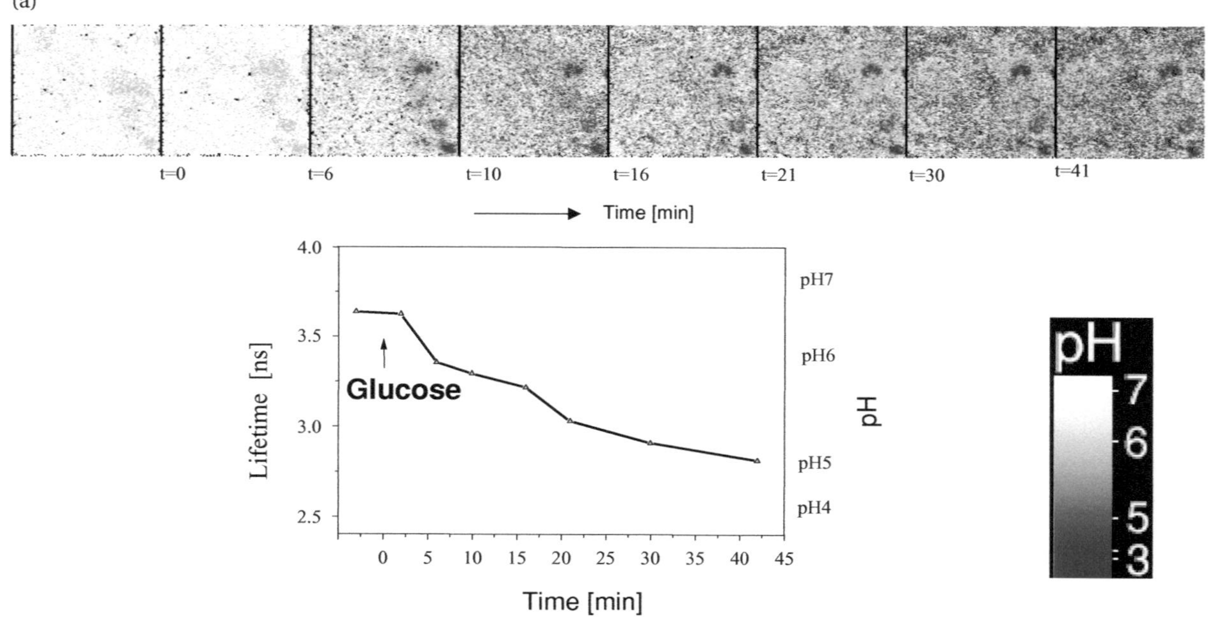
(a)
t=0
t=6
t=10
t=16
t=21
t=30
t=41
Time [min]
Lifetime [ns]
4.0
3.5
3.0
2.5
Glucose
0
5
10
15
20
25
30
35
40
45
Time [min]
pH7
pH6
pH5
pH4
pH
pH
7
6
5
3

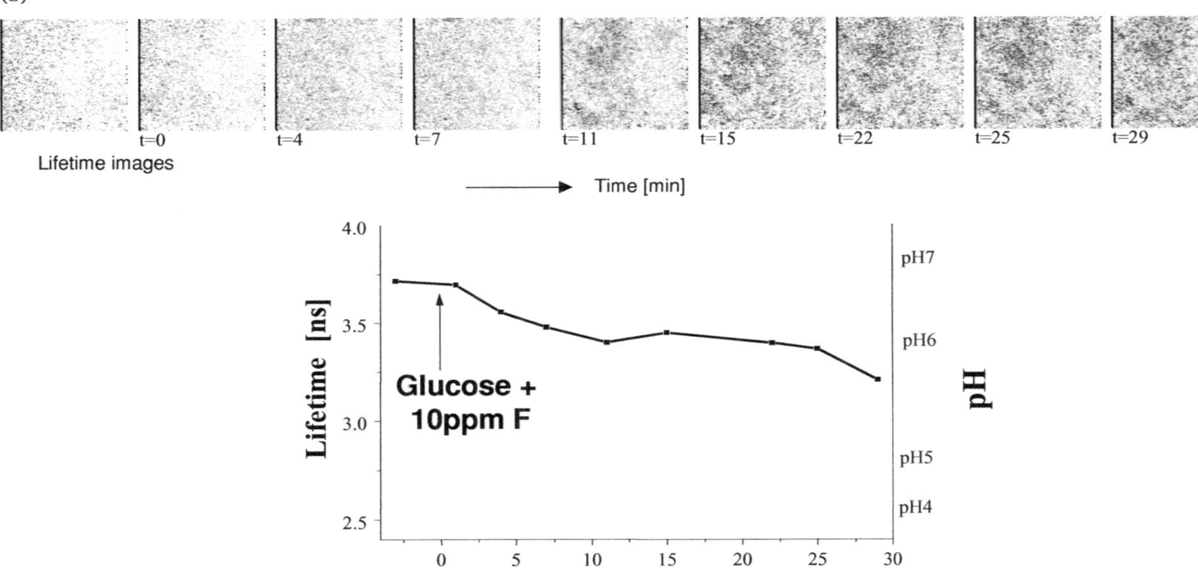

Figure 8.9: (a) Fall in pH in x-y section through biofilm following addition of glucose. (b) Fall in pH in x-y section through biofilm following addition of glucose + 10 ppm fluoride.

3.5 Measurement of Metabolic Inhibitor Effects in Biofilms Over Time

Fluoride is the most widely used chemical agent to control dental caries and, at least in part, acts by strengthening dental enamel against acidic attack. Fluoride can also interfere with bacterial metabolism, principally through effects on acid tolerance mechanisms and also by inhibiting various carbohydrate catabolism processes (Marquis, 1995). We examined whether we could visualise an inhibitory effect of a relatively low fluoride concentration on acid production within biofilms following glucose metabolism using TPEM-FLIM. Glucose solution was supplemented with 10 ppm F^- (0.53 m*M* NaF) and then overlaid onto the biofilms. The pH images and the data derived from them converted into a time series, before and following glucose or glucose/fluoride addition, are shown in Figure 8.9. A comparison of the data in Figures. 8.9a and 8.9b clearly shows that this low concentration of fluoride was able to inhibit the acidification of these mixed culture biofilms. Such a reduction of pH change in other microbial communities was previously shown to prevent enrichment of acid-tolerant species, such as mutans-streptococci (implicated in caries), and to maintain the proportions of species associated with health, many of which are acid sensitive (Bradshaw et al., 2002). In this way, even low concentrations of fluoride can help stabilise the beneficial properties of plaque biofilm.

4 CONCLUSIONS AND FUTURE OPPORTUNITIES

This chapter demonstrates the power of the nascent TPEM and FLIM techniques. These developments have been made possible by extraordinary advances in laser technology – a high-powered laser that can emit femtosecond (10^{-15} s) pulses of light, coupled to detectors with electronic 'shutters' which can resolve nanosecond time windows. In the dental field, further development of TPEM-FLIM could allow simultaneous detection of acid production, co-localisation of specific bacteria, and direct visualisation of calcium release as a tooth is demineralised. This would represent a similar 'holy grail' moment to the wonderful early images produced by Listgarten, Saxton, and others using electron microscopes to reveal the ultrastructure and interactions of dental plaque, discussed at the beginning of this chapter.

In the wider world, biofilm research is at a crossroads. Molecular biological techniques are beginning to reveal bacterial gene expression *in situ* in these systems. The TPEM-FLIM methodology offers the possibility to visualise the chemical microenvironments which individual cells experience within a biofilm matrix. The coupling of these techniques promises to reveal many of the mysteries of why biofilms behave in the way they do. As we begin to understand this, great new opportunities for biofilm control and exploitation will become apparent.

5 ACKNOWLEDGEMENTS

The authors wish to acknowledge Drs. Jurrien Vroom, Kees de Grauw, and Hans Gerritsen of the University of Utrecht, the Netherlands for the TPEM-FLIM work presented in this chapter. We would also like to thank Unilever Dental Research for providing generous funding for these studies.

REFERENCES

Bradshaw, D. J., Marsh, P. D., Hodgson, R. J. and Visser, J. M. (2002). Effects of glucose and fluoride on competition and metabolism within *in vitro* dental bacterial communities. *Caries Research*, 36, 81–86.

Centonze, V. E. and White, J. (1998). Multiphoton excitation provides optical sectioning from deeper within scattering specimen than confocal imaging. *Biophysical Journal*, 75, 2015–2024.

Costerton, J. W., Lewandowski, Z., Caldwell, D. E., Korber, D. R. and Lappin-Scott, H. M. (1995). Microbial biofilms. *Annual Review of Microbiology*, 49, 711–745.

Cummins, D., Moss, M. C., Jones, C. L., Howard, C. V. and Cummins, P. G. (1992). Confocal microscopy of dental plaque development. *Binary*, 4, 86–91.

Denk, W., Strickler, J. and Webb, W. W. (1990). Two-photon laser scanning fluorescence microscopy. *Science*, 248, 73–76.

Kolenbrander, P. E. (1988). Intergeneric coaggregation among human oral bacteria and ecology of dental plaque. *Annual Reviews of Microbiology*, 42, 627–656.

Lawrence, J. R., Korber, D. R., Hoyle, B. D., Costerton, J. W. and Caldwell, D. E. (1991). Optical sectioning of microbial biofilms. *Journal of Bacteriology*, 173, 6558–6567.

Listgarten, M. A. (1976). Structure of the microbial flora associated with periodontal health and disease in man. A light and electron microscopic study. *Journal of Periodontology*, 47, 1–18.

Listgarten, M. A. (1999). Formation of dental plaque and other oral biofilms. In *Dental Plaque Revisited*, eds. H. N. Newman and M. Wilson, pp. 187–210. Bioline, Cardiff.

McHugh, W. D. (1999). Dental plaque: thirty years on. In *Dental Plaque Revisited*, eds. H. N. Newman and M. Wilson, pp. 1–4. Bioline, Cardiff.

Marquis, R. E. (1995). Antimicrobial actions of fluoride for oral bacteria. *Canadian Journal of Microbiology*, 41, 955–964.

Nyvad, B. and Fejerskov, O. (1989). Structure of dental plaque and the plaque-enamel interface in human experimental caries. *Caries Research*, 23, 151–158.

Saxton, C. A. (1973). Scanning electron microscope study of the formation of dental plaque. *Caries Research*, 7, 102–119.

Vroom, J. M., De Grauw, K. J., Gerritsen, H. C., Bradshaw, D. J., Marsh, P. D., Watson, G. K., Birmingham, J. J. and Allison, C. (1999). Depth penetration and detection of pH gradients in biofilms by two-photon excitation microscopy. *Applied and Environmental Microbiology*, 65, 3502–3511.

Wimpenny, J. W. (1982). Responses of microorganisms to physical and chemical gradients. *Philosophical Transactions of the Royal Society of London Series B: Biological Sciences*, 297, 497–515.

Wood, S. R., Kirkham, J., Marsh, P. D., Shore, R. C., Nattress, B. and Robinson, C. (2000). Architecture of intact natural human plaque biofilms studied by confocal laser scanning microscopy. *Journal of Dental Research*, 79, 21–27.

CHAPTER NINE

Oral Streptococcal Genes That Encode Biofilm Formation

C. Y. Loo

1 INTRODUCTION

1.1 Oral Biofilms

Biofilms are surface-attached bacterial communities formed by unicellular organisms of single or multiple species. Most bacteria in their natural ecosystems colonise surfaces and are found in biofilm communities, rather than as planktonic cells. Biofilm formation is a highly structured process that occurs for numerous reasons, including protection from host immune systems, nutrient availability, and protection from harsh changes in the environment (Costerton et al., 1995; Costerton, Stewart, and Greenberg, 1999). A number of studies examining planktonic cells grown in batch culture have usually treated bacteria as unicellular species, even though they often exist in biofilms.

Although biofilm formation has been recognised and documented for approximately 100 years, we are just beginning to understand this process at the molecular level. Increasingly, bacteria have been studied as multicellular populations and, in some cases, viewed as interactive multicellular organisms (Shapiro, 1998), partly due to the fact that biofilm cells exist in a physical and physiological state that can increase their resistance to antimicrobials and mechanical forces (Costerton et al., 1995, 1999). Bacteria in a biofilm often display a dramatically different phenotype when compared with their counterparts in liquid culture. For example, biofilm cells often display higher resistance to antimicrobial agents, and they often exist in localised anoxic microenvironments and/or microenvironments that vary significantly in pH and ionic strength (Costerton et al., 1995, 1999). They may coordinate their activities, interact and communicate with other cells, and build complex structures of mature biofilms that protect them from the environment (Shapiro, 1998; Davey and O'Toole, 2000). These cells undergo changes during the highly regulated transition process from planktonic organisms to cells that are part of a complex biofilm community. These changes are reflected in the new phenotypic characteristics that develop and respond to a variety of environmental signals. This altered physiological state may result from gene activation initiated by contact with surfaces (Zhang and Normark, 1996) or in response to regulation by bacterial signal molecules, which allow sensing of cell density (Davies et al., 1998; Pringent-Combaret et al., 1999).

The structure, bacterial composition, adherence mechanisms, immunogenicity, and interspecies interactions in oral biofilms have been reviewed by Bowden and Hamilton (1998). Numerous studies have elucidated the mechanisms of initial streptococcal adhesion (Ganeshkumar, Song, and McBride, 1988; Fenno, LeBlanc, and Fives-Taylor, 1989; Rosan et al., 1989; Bleiweis, Oyston, and Brady, 1992; Clemans and Kolenbrander, 1995; Demuth et al., 1996; Gong and Herzberg, 1997; Jenkinson and Lamont, 1997; Rogers et al., 1998), but the process of bacterial accumulation and proliferation leading to functionally heterogeneously organised communities known as oral biofilms is poorly understood. Identification of virulence factors in oral bacteria (Kuramitsu, 2001) and understanding the molecular basis for biofilm formation will facilitate the design and development of vaccines, replacement therapy strategies, and inhibitors of biofilm formation.

1.2 Formation of Oral Biofilms

Oral biofilms, also referred to as dental plaque, play a role in the pathogenesis of dental caries (Loesche, 1986; Listgarten, 1994; Marsh, 1999) and periodontal disease (Socransky and Haffajee, 1994), two of the most common diseases in humans. They are complex systems that can contain more than 500 different bacterial taxa (Whittaker, Klier, and Kolenbrander, 1996). Recent studies suggest that oral biofilms may also play a role in systemic disease, including cardiovascular disease, respiratory infections, diabetes, and low-birthweight complications (Scannapieco and Genco, 1999; Li et al., 2000; Wu et al., 2000).

Oral biofilms are non-mineralised microbial accumulations that adhere to different oral surfaces, including enamel, cementum, restorations, prosthetic appliances, and epithelial cells of oral mucosa (Kolenbrander, 2000). They are composed of organic matrices derived from salivary glycoproteins and extracellular microbial products (Listgarten, 1994) and form in two distinct sequential steps: adhesion of early colonisers to host tissue components (Gibbons, 1989; Scannapieco, 1994), followed by time-dependent accumulation of multilayered cell clusters embedded in a matrix of bacterial and host constituents (Vickerman, Clewell, and Jones, 1991; Schilling and Bowen, 1992; Bloomquist, Reilly, and Liljemark, 1996). Biofilms on hard surfaces are usually several bacterial cell layers thick and can reach several hundred micrometers in thickness, whereas those on soft tissues often occur as a monolayer, as the epithelial cells are constantly being sloughed and replenished by the host (Kolenbrander, 2000).

Within minutes after a tooth surface is freshly cleaned, an acquired pellicle, composed primarily of salivary proteins, is adsorbed to the exposed hydroxyapatite crystallites (Listgarten, 1994). The surface is gradually covered by colonies of dividing bacteria that initially spread laterally along the tooth surface. Once the available surface is covered, the proliferating bacteria begin to grow away from the tooth in the form of columnar microbial colonies that are closely packed and compete for space and nutrients with neighbouring colonies. The first bacteria to colonise the surface are mostly Gram-positive, facultative cocci, mainly streptococci, and coccobacilli, mainly actinomyces. Initial colonisation is characterised by a transient, reversible adhesion to the tooth (Listgarten, 1994). The adhesion becomes stronger over time, mediated though proteoglycans and surface proteins in fimbriae and pili. Adhesins of oral bacteria are proteins that react with receptors on various oral surfaces and have been reviewed comprehensively by Jenkinson and Lamont (1997) and Whittaker et al. (1996).

Early oral biofilms consist of morphologically distinct palisading columns of cocci. These streptococci are the dominant 'pioneer' bacteria that initiate biofilm formation and constitute approximately 20 per cent of the normal human oral flora. The adhesion of these streptococci to oral surfaces has been studied extensively (Ganeshkumar et al., 1988; Fenno et al., 1989; Rosan et al., 1989; Bleiweis et al., 1992; Clemans and Kolenbrander, 1995; Demuth et al., 1996; Gong and Herzberg, 1997; Jenkinson and Lamont, 1997; Rogers et al., 1998). Viridans streptococci, which include *Streptococcus gordonii*, are ubiquitous initial colonisers that constitute a majority of the cultivable bacteria found in oral biofilms (Nyvad and Kilian, 1990) and are the most frequent aetiologic agents of bacterial endocarditis (Tao and Herzberg, 1999). Viridans streptococci are also opportunistic pathogens and a major cause of bacteraemia in immunocompromised patients, accounting for 40 per cent of infections in neutropenic patients (Tao and Herzberg, 1999).

The streptococci are followed by increasing proportions of actinomyces and the eventual conversion of the biofilm to a 'mature' community with high levels of Gram-negative anaerobic filamentous organisms (Rosan and Lamont, 2000). Physical interaction among distinct types of bacteria that live in close juxtaposition on oral surfaces occur through coadhesion, which occurs between suspended and already attached cells in biofilms, and coaggregation, which occurs between cells that are suspended (Kolenbrander, 2000). The suspended coaggregates, as well as single cells, contribute to bacterial accumulation to the existing biofilm. These interactions may provide additional benefits beyond facilitating adhesion and bacterial succession. For example, *Fusobacterium nucleatum* facilitated the survival of obligate anaerobes *Porphyromonas gingivalis* and *Prevotella nigrescens* in biofilms under aerated conditions (Bradshaw et al., 1998).

1.3 Resistance of Oral Biofilms to Antimicrobial Agents

Common infections such as urinary tract infections, catheter infections, middle-ear infections, and gingivitis are all caused by biofilms and are hard to treat or relapse frequently (Lewis, 2001). Factors which may contribute to an increase in resistance of cells in a biofilm to antimicrobial agents include restricted penetration of antimicrobials into a biofilm, decreased growth rate, altered gene expression, and microenvironmental factors (Wilson, 1996).

Bacteria living in oral biofilms are generally more resistant to antimicrobial agents than planktonic cells. Resistance of *S. sanguinis* biofilm cells to amoxicillin, doxycycline, and chlorhexidine increased considerably [up to 500 times the minimum inhibitory concentration (MIC)] when compared to planktonic cells (Larsen and Fiehn, 1996). Growth in the presence of sucrose offered further protection to biofilms (Embleton, Newman, and Wilson, 1998). A six-member oral biofilm (which included three streptococcal species) was refractory to chlorhexidine, and efficacy was shown to decrease with an increase in cell density (Pratten, Barnett, and Wilson, 1998), suggesting that resistance to antimicrobials may be modulated by population-density–dependent gene regulation, which can induce adaptive stress responses.

2 GENETICS OF ORAL BIOFILM FORMATION

The strategy used to screen for biofilm-defective mutants using abiotic surfaces in conjunction with transposon mutagenesis was pioneered by Genevaux, Muller, and Bauda (1996). Random transposon insertion libraries were used for high-throughput screens of thousands of mutants for genes associated with biofilm formation by detecting the ability of the mutant clones to form biofilms.

A simple macroscopic assay that facilitates rapid screening of thousands of mutants for the desired phenotype is used. In this assay, microtitre plates are used to test potential biofilm-defective mutants, and the presence or absence of a biofilm within a well of the plate is detected by staining the wells with a dye. Potentially interesting mutants are then analysed in more detail using molecular approaches and direct microscopic observation. Studies using this method have identified novel genes involved in biofilm formation by *Staphylococcus epidermidis, Escherichia coli, Pseudomonas fluorescens, P. aeruginosa,* and *Vibrio cholerae* (Davey and O'Toole, 2000). The biofilm formation of *Streptococcus gordonii,* a non-motile Gram-positive oral bacterium, was examined using a similar *in vitro* microtitre plate biofilm formation assay together with Tn*916* transposon mutagenesis (Loo, Corliss, and Ganeshkumar, 2000).

One limitation of this approach is that biofilm formation occurs in a stationary environment in the absence of hydrodynamic influences, which are present *in vivo.* Another limitation of the transposon insertion screening studies is that they test the mass of the cells making up the biofilm, which cannot provide any information on possible defects in biofilm architecture. For example, a *P. aeruginosa lasI* mutant defective in production of the quorum-sensing factor *N*-(3-oxododecanoyl)-L-homoserine lactone (HSL) formed biofilms on a glass surface that were easily dislodged by sodium dodecyl sulphate (SDS), whereas biofilms formed on polystyrene were not affected (Brooun, Liu, and Lewis, 2000).

2.1 Genes Important in Biofilm Formation by *S. gordonii*

Fully developed, surface-attached oral biofilms are highly structured, with distinct architectural and physiochemical properties commonly observed in other biofilm communities (Costerton et al., 1995, 1999). Oral biofilms are subject to harsh environmental conditions such as nutrient availability (feast or famine), aerobic-to-anaerobic transitions, pH changes, as well as exposure to antimicrobials and other constituents of dentifrices and mouthwashes, all of which may contribute to the regulation of biofilm formation. As sessile populations reflect conditions *in vivo* more accurately than planktonic bacteria, the genes expressed by biofilm bacteria are likely to play a role in colonisation of tooth surfaces and the virulence of viridans streptococci in susceptible hosts (Tao and Herzberg, 1999). Much of the progress that has been made in understanding the development of bacterial biofilms is due to the recent focus on analysing biofilms using genetic and molecular biological approaches (Davey and O'Toole, 2000). Analyses of biofilm formation by the viridans streptococci species *S. gordonii* have identified a number of genetic components required for the formation of single-species bacterial biofilms by this early coloniser of oral biofilms (Loo et al., 2000; Rogers et al., 2001).

2.1.1 Specific Adhesins

Amylase is the most abundant enzyme in saliva, binding specifically and with high affinity to several species of oral streptococci (Whittaker et al., 1996). AbpA is an α-amylase-binding protein in *S. gordonii* (Rogers et al., 1998), which mediates the adhesion of these bacteria to amylase in oral salivary pellicles on tooth surfaces. The *abpA* gene appears to be specific to *S. gordonii* and differs from genes encoding amylase-binding proteins in other amylase-binding streptococci, such as *Streptococcus mitis, Streptococcus crista, Streptococcus anginosus,* and *Streptococcus parasanguinis* (Brown et al., 1999). A recent study using a flowcell system demonstrated that AbpA functions in human saliva-supported biofilm formation by *S. gordonii* (Table 9.1), in addition to adhesion to amylase-coated hydroxyapatite and salivary-amylase-mediated catabolism of dietary starches (Rogers et al., 2001). The α-amylase enzyme binds with high affinity to oral streptococci (Scannapieco et al., 1989) via AbpA (Rogers et al., 1998) to promote biofilm formation on nutritionally poor, saliva–coated tooth surfaces.

Table 9.1: Genes of oral streptococci involved in biofilm formation

Protein Function	Gene	Species	Reference
Adhesin	*abpA*	*Streptococcus gordonii*	Loo et al., 2000; Rogers et al., 2001
	sspA	*S. gordonii*	Egland et al., 2001
	sspB	*S. gordonii*	Demuth et al., 2001; Egland et al., 2001
	fap1	*S. parasanguinis*	Froeliger and Fives-Taylor, 2001
	gbpA	*S. mutans*	Hazlett et al., 1999
	?gbpB	*S. mutans*	Mattos-Graner et al., 2001
Cell-to-cell signalling	*comD*	*S. gordonii*	Loo et al., 2000
Cell wall formation	*bacA*	*S. gordonii*	Loo et al., 2000
	glmM	*S. gordonii*	Loo et al., 2000
	mutT	*S. gordonii*	Loo et al., 2000
	pbp2B	*S. gordonii*	Loo et al., 2000

The initial stages of oral biofilm formation involve the adhesion of early colonisers such as *S. gordonii* and *Actinomyces naeslundii* to saliva-coated tooth surfaces and to each other. *S. gordonii* has been shown to bind to the salivary pellicle and outcompete the late coloniser, *Porphyromonas gingivalis*, for adhesion (Cook et al., 1998). During the maturation of oral biofilms, *S. gordonii* may provide an attachment substrate for colonisation and accretion of the periodontal pathogen *P. gingivalis* onto oral biofilms because biofilm formation was observed only subsequent to coadhesion of *P. gingivalis* to *S. gordonii* cells deposited on the salivary pellicle (Cook et al., 1998).

S. gordonii surface proteins SspA and SspB are members of the antigen I/II family, which bind to immobilized human salivary agglutinin glycoprotein and to collagen (Demuth et al., 1996). A differential display-PCR method also indicated that the syntheses of SspA and SspB are induced by saliva (Du and Kolenbrander, 2000). Expression of *sspB* was also positively regulated by the *sspA* gene product (El-Sabaeny et al., 2001). Transcription of *sspA* and *sspB* is regulated differentially in response to oral environmental cues, suggesting that they have distinct functional roles in adhesion (El-Sabaeny et al., 2000). SspA and SspB contribute to both the initiation and the maturation of oral biofilms (Table 9.1) by interacting with the early coloniser *A. naeslundii* (Jenkinson and Demuth, 1997; Egland, Du, and Kolenbrander, 2001) and the late coloniser *P. gingivalis* (Lamont et al., 1994) via coaggregation or coadhesion.

The adherence of *P. gingivalis* to *S. gordonii* appears to occur through a protein–protein interaction requiring *S. gordonii* SspB and the minor fimbrial component of *P. gingivalis* (Demuth et al., 2001). A 26-amino-acid region of SspB mediates the adherence of *P. gingivalis* to *S. gordonii* in the development of mature oral biofilms (Table 9.1). This species specificity of adherence arises from the interaction involving a discrete structural determinant of SspB that is not conserved in SpaP, its counterpart in *S. mutans* (Bleiweis et al., 1992). Within the oral cavity, the specificity of *P. gingivalis* interactions with streptococci may be important for identifying a suitable environmental niche in the growing oral biofilm. Demuth et al. (2001) has proposed that the specific adherence to *S. gordonii* may represent a mechanism by which *P. gingivalis* avoids colonising sites that are rich in acid-tolerant bacteria such as *S. mutans*, as an acidic local environment is not physiologically favourable for *P. gingivalis*. In contrast, the metabolism of arginine by *S. gordonii* plays a role in maintaining a neutral or slightly basic environment (Demuth et al., 2001). Preferential adherence to *S. gordonii* may also explain, in part, the clinical observation that adult periodontitis does not usually occur simultaneously with coronal dental caries caused by *S. mutans.*

In another oral streptococcal species, *S. parasanguinis,* Fap1 is a protein required for the assembly of fimbriae (Wu and Fives-Taylor, 1999). *S. parasanguinis* containing a disrupted *fap1* gene was deficient in biofilm formation in an *in vitro* microtitre plate biofilm assay (Froeliger and Fives-Taylor, 2001), suggesting that long, peritrichous fimbriae of *S. parasanguinis* are critical for the formation of biofilms on solid surfaces (Table 9.1). This finding is of particular interest, as many Gram-negative bacteria also require fimbriae or pili for initial interactions with a surface (Davey and O'Toole, 2000). The predicted Fap1 protein contains an unusually long signal sequence, a cell wall sorting signal and two repeat regions with a similar highly conserved dipeptide composition of (E/V/I)S, composed of 28 and 1000 repeats, respectively (Wu and Fives-Taylor, 1999). The repeat regions combined account for 80 per cent of the *fap1* coding region and are highly conserved, suggesting that repetitive blocks are structurally important and may be virulence determinants (Froeliger and Fives-Taylor, 2001). In *Staphylococcus aureus* clumping factor A, serine and aspartate dipeptide repeats (DS) are required for spanning the cell wall peptidoglycan and presenting the ligand-binding domain distal from the cell surface, while the non-repetitive region possesses a fibrinogen-binding domain, which facilitates the binding of *S. aureus* to fibrinogen. It has been proposed that *Streptococcus parasanguinis* Fap1 may function as an adhesin in a similar manner (Froeliger and Fives-Taylor, 2001).

A number of bacterial adhesins have been identified as important in biofilm formation by large-scale genetic screening (Loo et al., 2000) and by the traditional approach of examining the ability of mutants that lack specific adhesins to form biofilms, either by using an *in vitro* microtitre plate assay (Froeliger and Fives-Taylor, 2001; Mattos-Graner et al., 2001) or in a flow cell system (Rogers et al., 2001). Both approaches lend support to the importance of interactions of specific bacterial adhesins with oral surfaces or other bacteria in the formation of biofilms. There may be an overlap in factors required for biofilm formation and those for adhesion and/or pathogenesis *in vivo.* Alternatively, as bacterial adhesins may promote adhesion to non-nutritive, abiotic surfaces, sessile growth on solid surfaces in a nutritionally poor ecological niche may represent a survival strategy utilised by certain bacteria, including pioneer oral biofilm bacteria.

In other bacteria, structural components of the cell wall are also associated with initial biofilm formation: holdfast in the stalked *Caulobacter* (Yun, Ely, and Smit, 1994); type I fimbriae, lipopolysaccharide, and curli in *E. coli*;

type IV fimbriae in *Pseudomonas aeruginosa*; flagella in *E. coli, P. aeruginosa, P. fluorescens*, and *V. cholerae*; an autolysin in *S. epidermidis*; and MSHA pili (also type IV pili) in *V. cholerae* (Davey and O'Toole, 2000). All these studies demonstrate that cell structures and other surface proteins play important roles in early biofilm formation.

2.1.2 Cell-to-Cell Signalling

In order to regulate a variety of physiological functions, bacteria sense environmental signals and process this information into specific transcriptional responses. In Gram-positive bacteria, cell-density-dependent gene expression regulatory models appear to follow a common theme, in which the signal molecule is a posttranslationally processed peptide that is secreted by a dedicated ATP-binding cassette (ABC) exporter protein. This peptide pheromone functions as the input signal which is recognised by the sensor of a two-component signal transduction system (Dunny and Leonard, 1997), which then interacts with cytoplasmic response regulator proteins. The peptide pheromone accumulates extracellularly in proportion to the total number of cells, providing an index of population densities (Dunny and Leonard, 1997). Therefore, these bacterial autoinduction systems represent both intra- and interspecies cell-to-cell communication, which is also referred to as quorum sensing.

The functions of histidine kinases, the sensor component of the two-component signal transduction systems, are signal recognition and transduction. In *S. gordonii*, the sensing component is ComD, an autophosphorylating histidine kinase. ComD is the receptor for the *comC*-encoded competence-stimulating peptide (Havarstein et al., 1996; Lunsford and London, 1996; Pestova, Havarstein, and Morrison, 1996). The second component of the two-component signal transduction system in *S. gordonii*, the cognate response regulator ComE, becomes activated after receiving the phosphoryl group from ComD and undergoes conformational changes that elicit a response by binding to specific promoter regions of appropriate target genes, thereby acting as a transcriptional factor (Pestova et al., 1996). In *S. gordonii, comD* from the competence locus (*comCDE* operon) (Havarstein et al., 1996) is involved in biofilm formation (Loo et al., 2000). As ComD is a histidine kinase that can act as an environmental sensor and its activity is regulated in response to specific stimuli, this suggests that that cell-to-cell signalling is involved in biofilm formation (Table 9.1).

Although little is known about competence in biofilms per se, competence may be affected by the accumulation of bacterial populations. Cell-to-cell signalling may modulate acid adaptation, which is important to the survival of certain streptococci in the oral cavity. A study by Li et al. (2001a) showed that disruption of the *comC*, -*D*, or -*E* competence genes in *S. mutans* resulted in a diminished log-phase acid tolerance response, whereas the addition of synthetic competence stimulating peptide to a *comC* mutant restored the acid tolerance response.

Another oral streptococcal cell-to-cell signalling mechanism is mediated by lantibiotics, antimicrobial peptides that are produced by and are active against closely related Gram-positive organisms. *S. salivarius* produces a 22-amino-acid residue lantibiotic, SalA, which inhibits the growth of a range of streptococci, including all strains of *S. pyogenes*. *S. pyogenes* SalA1 is able to induce *S. salivarius salA* transcription, whereas production of both SalA1 and SalA is autoregulated. These peptides modulate lantibiotic production and possibly influence streptococcal population ecology in the oral cavity (Upton et al., 2001).

Interspecies cell-to-cell signalling has also been observed between streptococci and other oral bacteria. A 59-kDa surface protein of *S. cristatus* with specificity for the *fimA* gene caused repression of the *Porphyromonas gingivalis* fimbrial gene *fimA* (Xie et al., 2000). This third streptococcal cell-to-cell signalling system, found in *S. cristatus*, may modulate virulence gene expression in *P. gingivalis*, inhibit the ability of *P. gingivalis* to form biofilm microcolonies with *S. cristatus*, and, consequently, influence the development of pathogenic oral biofilms (Xie et al., 2000).

In addition to these Gram-positive signalling systems, a *luxS* homologue has been discovered in the *S. mutans* chromosome (Cvitkovitch, 2001). A comparison with the luxS-based systems in Gram-negative bacteria needs to be made in order to understand the similarities and differences between them and the role of this system in Gram-positive oral streptococci.

Signalling systems have also been found in a number of periodontal pathogens. *Prevotella intermedia*, *F. nucleatum*, *Porphyromonas gingivalis* (Frias, Olle, and Alsina, 2001), and *Actinobacillus actinomycetemcomitans* (Fong et al., 2001) produce extracellular signalling molecules related to the type 2 autoinducer in *V. harveyi* that stimulate the expression of luminescence genes. These molecules may play a role in the development of mature and subgingival biofilms and in the virulence of these oral bacteria. The luxS-based signalling system in *P. gingivalis* may control the expression of genes involved in haemin acquisition, although it does not seem to be involved in *P. gingivalis* biofilm formation with *S. gordonii* or in invasion of epithelial cells (Chung et al., 2001). The *luxS* homologue in *A. actinomycetemcomitans* may modulate iron uptake and aspects of virulence and may induce responses in other periodontal pathogens in mixed-species biofilms (Fong et al., 2001).

In oral streptococci, a connection between signalling and bacterial cell growth has been discovered (Liljemark et al., 1997). A low molecular weight cell-to-cell signalling molecule, known as START, may act as a quorum-sensing signalling molecule to accelerate the growth or cell division of biofilm cells (Liljemark et al., 1997) and may play a role in rapid accumulation of biofilms on oral surfaces.

A similar connection between signalling and growth has been found in *Pseudomonas aeruginosa*. A molecular link between the environment and the transition from a planktonic to a biofilm mode of growth has been identified in the Crc protein. Crc plays a role in sensing carbon source availability and affects expression of the type IV *pilA* structural gene. Crc is also required for biofilm formation (O'Toole et al., 2000a). As type IV pili-mediated twitching motility is required for *P. aeruginosa* biofilm formation, Crc links carbon availability to the decision whether or not to enter a biofilm mode of growth (O'Toole et al., 2000a).

Some of the other regulatory functions that are involved in biofilm formation of other bacteria include the two-component sensor systems CpxA/CpxR and EnvZ/OmpR (Dorel et al., 1999; Davey and O'Toole, 2000), which are involved in regulation of curli synthesis in *E. coli*. Also, ClpP in *P. fluorescens*, which is normally exposed at the cell surface, is involved in Clp protease synthesis, acting at a posttranslational level (Davey and O'Toole, 2000).

2.1.3 Cell Wall Formation

The bacterial cell wall consists of a wide variety of molecules and serves a multitude of functions. Although its primary function is to provide a rigid exoskeleton for protection against both high osmotic pressures and severe external environments, it also serves as an attachment site for proteins that interact with the environment. The cell wall of Gram-positive bacteria is a peptidoglycan macromolecule, a covalently closed, netlike polymer of glycan strands linked by peptides. The cell wall peptides are cross-linked with other peptides that are attached to a neighbouring glycan strand, thereby generating a three-dimensional molecular network. Most cell walls consist of the same basic type of disaccharide-, tri-(or tetra-)peptide subunits in their peptidoglycan, and the peptidoglycan structure is determined by many biosynthetic and autolytic reactions (Massidda et al., 1998). Four of the genes associated with biofilm formation are involved in cell wall formation of *S. gordonii* (Loo et al., 2000).

The *bacA* gene, which encodes an undecaprenyl kinase involved in lipid phosphorylation that confers bacitracin resistance in *E. coli,* is involved in *S. gordonii* biofilm formation (Table 9.1). The antibiotic resistance is due to the tight binding of bacitracin to a complex of undecaprenyl diphosphate (UDP) and a metal ion. This tight binding prevents undecaprenyl diphosphate from functioning as a membrane-associated carrier of intermediates during peptidoglycan biosynthesis (Cain et al., 1993). A *Staphylococcus aureus bacA* mutant showed slightly reduced virulence in a mouse model of infection, whereas a *Streptococcus pneumoniae bacA* mutant was highly attenuated in a mouse model of infection and had considerably increased susceptibility to bacitracin (Chalker et al., 2000). These *bacA* mutants showed no detectable changes in morphology or growth parameters, suggesting that although they were healthy *in vitro,* the absence of undecaprenyl kinase caused metabolic changes sufficient to compromise infectivity.

A phosphoglucomutase gene in *E. coli, glm,* also has a role in biofilm formation (Table 9.1). GlmM in *S. gordonii* is a member of the family of bacterial phosphoglucosamine mutases, which have been identified in *E. coli* (Mengin-Lecreulx and Heijenoort, 1996), *Staphylococcus aureus* (Jolly et al., 1997), *Helicobacter pylori,* and *P. aeruginosa* (Tavares et al., 2000). GlmM catalyses formation of glucosamine-1-phosphate from glucosamine-6-phosphate, the first step in the biosynthetic pathway leading to UDP-*N*-acetyl glucosamine, a common precursor to essential cell envelope components such as peptidoglycan and lipopolysaccharides (Mengin-Lecreulx and Heijenoort, 1996). Inactivation of *glmM* in *E. coli* led to the progressive depletion of the pools of precursors located downstream from glucosamine-1-phosphate in the pathway for peptidoglycan synthesis, followed by various alterations of cell shape and finally cells lysis, when cellular peptidoglycan content decreased to a critical value corresponding to about 60 per cent of its normal level (Mengin-Lecreulx and Heijenoort, 1996). Inactivation of *femD,* which encodes the phosphoglucosamine mutase of *S. aureus,* caused a drastic reduction in methicilllin resistance (Jolly et al., 1997; Glanzmann et al., 1999), an abnormal cell wall composition characterised by the complete disappearance of unsubstituted disaccharide-pentapeptides from the peptidoglycan (Jolly et al., 1997) and hypersusceptibility to the glycopeptide teicoplanin (Glanzmann et al., 1999).

Another gene associated with *Streptococcus gordonii* biofilm formation was *mutT* (Table 9.1) (Loo et al., 2000). This codes for a nucleoside triphosphatase which prevents A·T --> C·G tranversions during DNA replication by removing an oxidised form of guanine, 8-oxodGTP, from the nucleotide pool, thus preventing it from mis-pairing with template A and maintaining replication fidelity (Fowler and Schaaper, 1997). More importantly, the *mutT* gene is one of a cluster of genes in *S. pneumoniae* responsible for cell wall biosynthesis and cell division known as the *dcw* (division and cell wall) cluster (Massidda et al., 1998). Although many different bacteria, including Gram-negative and Gram-positive rods or cocci, possess this region, which is similar to that found at the 2-minute region of the *E. coli* chromosome, the genetic organisation of the *S. pneumoniae dcw* cluster differed significantly from that of other bacteria reported to date. The genes were distributed in three separate regions of the chromosome in *S. pneumoniae* strains G54, R6, and TIGR4 (Massidda et al., 1998). In all three strains, the first region (*dcw1*) contained *pbp2b* (which encodes a penicillin-binding protein PBP2b), *ddl, murF,* and *mutT* at one end and *ftsA* and *ftsZ* cell division genes at the other end. The second region (*dcw2*) consisted of *murD, murG,* and *divlB* genes, whereas the third region contained *yllC, yllD, pbp2x,* and *mraY* genes (Massidda et al., 1998). The genetic organisation of the *dcw* clusters in these three *S. pneumoniae* strains is consistent, and this organisation is substantially different from that of the *dcw* cluster of other eubacterial species (Massidda et al., 1998).

Penicillin-binding proteins (PBPs) comprise one of the many sets of enzymes which play a role in the construction, maintenance, and regulation of the rigid peptidoglycan. Their functions include recycling old peptidoglycan, functioning as zippers during cell division, and making room for the insertion of new peptidoglycan material. β-lactams act as antimicrobials by inactivating PBPs, weakening the cell wall peptidoglycan structure (Dowson et al., 1990).

Peptidoglycan determines the cell shape of bacteria, and its synthesis requires a transglycosylase activity to polymerise the glycan strand and a transpeptidase activity to cross-link these strands via their peptide side chains. PBPs can be grouped into high molecular weight Class A and B and low molecular weight PBPs (Stingele and Bollet, 1996). PBP2B, a transpeptidase from the Class B high molecular weight PBPs, has higher similarity to PBPs involved in lateral elongation and plays a role in cell wall elongation, which occurs in some Gram-positive cocci (Massidda et al., 1998). This may explain the absence of biofilm formation when *pbp2B* was inactivated (Loo et al., 2000). Interestingly, the *pbp2B* gene is present in the same end of *dcw1* of *S. pneumoniae* as another biofilm gene, *mutT* (Massida et al., 1998). The *pbp2B* of *S. thermophilus* is involved in exopolysaccharide production (Stingele and Bollet, 1996). *S. thermophilus* mutants with disrupted *pbp2B* genes also displayed altered cell morphology (Stingele and Bollet, 1996). The cells did not grow in linear chains typical of streptococci, but grew in chains that were curled and twisted. Most cells displayed a rather asymmetrical, nearly wedge-shaped morphology and not the regular ovoid shape, whereas some mutants also showed impaired growth with the same abnormal cell morphology (Stingele and Bollet, 1996).

2.2 *S. mutans* Biofilms

Similarly, surface components of the cariogenic oral streptococci, *S. mutans,* also appear to play a critical role in biofilm formation, as demonstrated by the importance of the ratio of glucan to glucan-binding protein (GBP) present (Hazlett, Mazurkiewicz, and Banas, 1999). The sucrose-dependent adhesion of *S. mutans* is mediated by glucans, produced by polymerisation of glucose moieties of sucrose by extracellular glucosyltransferases (GTFs). *S. mutans* synthesises three GBPs: GbpA, GbpB, and GbpC. GbpA is a secreted protein associated with the cell surface and in the extracellular medium. The carboxyl-terminal three-quarters of GbpA, which has homology to the putative glucan-binding domain of GTFs, mediates binding to α-1,6-glucosidic linkages present in water-soluble and, to a lesser extent, water-insoluble glucans and undergoes a conformational shift upon binding to dextran. When *S. mutans* were grown in the presence of sucrose within hydroxyapatite-coated wells, the biofilm consisted primarily of large aggregates that did not completely coat the hydroxyapatite surface, whereas disruption of *gbpA* produced a biofilm which consisted of a uniform layer of smaller aggregates which almost entirely coated the hydroxyapatite (Hazlett et al., 1999). These changes in biofilm structure correlated with differences in susceptibility to ampicillin, with the *gbpA* biofilm organisms being more susceptible. The changes in biofilm structure also correlated with changes in virulence, as inactivation of *gbpA* increased virulence in a gnotobiotic rat model and promoted *in vivo* accumulation of organisms. GbpA contributes to *S. mutans* biofilm development (Table 9.1), and it seems likely that *S. mutans* biofilm structure influences virulence (Hazlett et al., 1999).

The *gbpB* gene of *S. mutans* may also have a role in biofilm formation (Table 9.1), as the amounts of GbpB produced by individual strains correlated positively with their ability to grow as biofilms in an *in vitro* microtitre plate biofilm assay (Mattos-Graner et al., 2001). GbpB has homology to a putative peptidoglycan hydrolase from group B streptococci, suggesting that GbpB is involved in peptidoglycan synthesis. *S. mutans gbpB* is also stress responsive, displaying increased expression under conditions of high osmolarity and temperature (Chia et al., 2001). These properties may contribute the underlying mechanisms of biofilm formation of *S. mutans.*

Proteome analysis to compare the protein profiles of two bacterial populations was carried out in a recent study to examine differences between planktonic and biofilm cells of *S. mutans* grown in a chemostat (Svensater et al., 2001). Up-regulation (1.3-fold or more) was observed in expression of proteins involved in elongation-translation, ribosomal complex, transcription-RNA elongation, amino acid biosynthesis, cell division, translation, AMP phosphorylation, fatty acid biosynthesis, protein folding, and secretion. Biosynthetic processes were important in *S. mutans*, while glycolytic enzymes involved in acid formation were repressed. This study also showed that novel proteins of as yet unknown function were expressed by biofilm cells, but not by their planktonic counterparts (Svensater et al., 2001).

3 GENETIC EXCHANGE IN ORAL BIOFILMS

S. mutans biofilm cells have been shown to be 10- to 600-fold more efficient at incorporating plasmid DNA than their planktonic counterparts (Li et al., 2001b). Intergeneric transfer of a conjugative transposon, Tn*5397*, in a mixed-species biofilm demonstrated the ability of conjugative transposons to disseminate antibiotic resistance genes from a non-oral *Bacillus subtilis* donor to *S. acidominimus*, an oral commensal (Roberts et al., 1999). A Tn*916*-like element from a tetracycline-resistant *S. salivarius* was successfully transferred to another streptococcal species, *S. parasanguinis*, within an eleven-membered oral biofilm consortium consisting of ten tetracycline-sensitive oral bacteria and the tetracycline-resistant donor, *S. salivarius* (Roberts et al., 2001).

PBP2B, which is needed for biofilm formation (Loo et al., 2000), is a major killing target for penicillin in pneumococci because PBPs covalently bind β-lactam antibiotics. Penicillin-resistant strains of *S. pneumoniae* possess altered forms of PBPs, and the *pbp2B* genes of these strains have a mosaic structure, consisting of regions that are very similar to those in penicillin-sensitive strains, alternating with regions that are highly diverged (Dowson et al., 1990). The *pbp2B* genes of penicillin-resistant isolates of *S. sanguinis* were identical in sequence to the mosaic Class B *pbp2B* genes of *S. pneumoniae*, whereas the *pbp2B* genes of penicillin-resistant *S. oralis* strains were similar to the mosaic Class B *pbp2B* gene of penicillin-resistant strains of *S. pneumoniae*, but possessed an additional block of diverged sequence. These observations indicated that in penicillin-resistant strains of viridans streptococci, horizontal gene transfer occurred between *S. pneumoniae* and *S. sanguinis* and probably *S. oralis* (Dowson et al., 1990).

These studies lend support to the notion that oral biofilms can act as genotypic reservoirs that harbour promiscuous mobile elements and genes that can undergo horizontal transfer, allowing the persistence, transfer, and selection of genetic elements that confer resistance to antimicrobial compounds. Since the mouth is the primary point of entry for most bacteria that colonise humans, genetic exchange may occur between indigenous and transient bacteria in the mouth and genetic transfer is highly likely among oral microflora living in dense biofilms on oral surfaces.

4 CONCLUSIONS

Oral biofilms have important effects detrimental to our oral health and may also contribute to systemic disease. Studies performed with planktonic oral bacteria since the early 1970s have yielded tremendous amounts of information about the physiology, genetics, and interactions of oral bacteria. This pioneering work has provided a solid foundation to begin to apply the knowledge and technologies developed using planktonic cells for studying oral bacteria in biofilms, which more closely mimic conditions in the oral cavity. Studies on oral biofilms may provide an ideal system for linking a broad ecological perspective of biofilms with the power of molecular genetics to identify and dissect genetic determinants required for the formation of sessile bacterial communities (Davey and O'Toole, 2000).

The development of a biofilm proceeds from an early stage through maturation and maintenance and finally to dissolution. Bacteria initiate biofilm development in response to environmental cues such as nutrient availability. For example, *S. gordonii* (Loo et al., 2000) and *E. coli* (O'Toole, Kaplan, and Kolter, 2000b) seem to form biofilms in a nutritionally limited environment, leading to the proposal that the starvation response pathway may be a part of the overall biofilm developmental cycle (O'Toole et al., 2000b). Novel aspects of bacterial physiology may also play a part in biofilm formation, as genes that have no homology in the databases, together with a number of known genes, have been identified as being associated with the initial biofilm formation of *S. gordonii* (Loo et al., 2000).

Further studies are needed to identify the genetic determinants and bacterial elements necessary for the development of oral biofilms into maturation, maintenance of the biofilms, and subsequent detachment of biofilm cells. Mature biofilms may develop complex architectural features, increased resistance to antimicrobials, and increased rates of genetic exchange, which may create a protective environment. Cell-to-cell signalling plays a role in the differentiation of *P. aeruginosa* biofilms into complex structures that are resistant to the biocide SDS (Davies et al., 1998). Therefore, in addition to its role in early biofilm formation, the possible role of cell-to-cell signalling in the maturation of *S. gordonii* biofilms needs to be investigated. Little is known about the functions or regulatory pathways involved in detachment of bacteria from biofilms. This may involve an active process, as *S. mutans* has a surface protease that cleaves its own surface proteins (Lee, Li, and Bowden, 1996), which may serve as a mechanism to release cells from the biofilm.

To date, a number of genetic components required for the formation of a single-species bacterial biofilm by the viridans streptococci species *S. gordonii* Challis have been identified (Loo et al., 2000; Demuth et al., 2001; Egland, Du, and Kolenbrander, 2001), indicating that a signalling system and a number of bacterial cell wall components modulate biofilm formation. Further investigation of these genes will lead to a better understanding of the steps involved in initiating biofilm formation and the cellular components required to accomplish these steps. Bacteria must be able to adhere to surfaces, to sense their cell density and ultimately form a three-dimensional structure of biofilms. New potential targets to control dental and other infections are needed, due to the increasing incidence of persistent and chronic bacterial diseases and the emergence of bacterial resistance to multiple antibiotics. Understanding the genetic basis of biofilm formation will provide insight into this unique process and may facilitate the development of therapeutic agents and strategies that target the biofilm phenotype and cell-to-cell signalling for the control of biofilm-mediated infections.

5 ACKNOWLEDGEMENTS

I gratefully acknowledge my principal investigator, Dr. Nadarajah Ganeshkumar, for his valuable advice and critical review of the manuscript and the National Institute of Dental and Craniofacial Research for financial support by grant IR01 DE13328-01A1.

REFERENCES

Bleiweis, A. S., Oyston, P. C. and Brady, L. J. (1992). Molecular, immunological and functional characterization of the major surface adhesin of *Streptococcus mutans*. *Advances in Experimental Medical Biology*, 327, 229–241.

Bloomquist, C. G., Reilly, B. E. and Liljemark, W. F. (1996). Adherence, accumulation, and cell division of a natural adherent bacterial population. *Journal of Bacteriology*, 178, 1172–1177.

Bowden, G. H. and Hamilton, I. R. (1998). Survival of oral bacteria. *Critical Reviews in Oral Biology and Medicine*, 9, 54–85.

Bradshaw, D. J., Marsh, P. D., Watson, G. K. and Allison, C. (1998). Role of *Fusobacterium nucleatum* and coaggregation in anaerobe survival in planktonic

and biofilm oral microbial communities during aeration. *Infection and Immunity*, 66, 4729–4732.
Brooun, A., Liu, S. and Lewis, K. (2000). A dose-response study of antibiotic resistance in *Pseudomonas aeruginosa* biofilms. *Antimicrobial Agents and Chemotherapy*, 44, 640–646.
Brown, A. E., Rogers, J. D., Hassse, E. M., Zelasko, P. M. and Scannapieco, F. A. (1999). Prevalence of the amylase-binding protein A gene (*abpA*) in oral streptococci. *Journal of Clinical Microbiology*, 37, 4081–4085.
Cain, B. D., Norton, P. J., Eubanks, W., Nick, H. S. and Allen, C. M. (1993). Amplification of the *bacA* gene confers bacitracin resistance to *Escherichia coli*. *Journal of Bacteriology*, 175, 3784–3789.
Chalker, A. F., Ingraham, K. A., Lundsford, R. D., Bryant, A. P., Bryant, J., Wallis, N. G., Broskey, J. P., Pearson, S. C. and Holmes, D. J. (2000). The *bacA* gene, which determines bacitracin susceptibility in *Streptococcus pneumoniae* and *Staphylococcus aureus*, is also required for virulence. *Microbiology*, 146, 1547–1553.
Chia, J.-S., Lee, Y.-Y., Huang, P.-T. and Chen, J.-Y. (2001). Identification of stress-responsive genes in *Streptococcus mutans* by differential display reverse transcription-PCR. *Infection and Immunity*, 69, 2493–2501.
Chung, W. O., Park, Y., Lamont, R. J., McNab, R., Barbieri, B. and Demuth, D. R. (2001). Signalling system in *Porphyromonas gingivalis* based on a LuxS protein. *Journal of Bacteriology*, 183, 3903–3909.
Clemans, D. L. and Kolenbrander, P. E. (1995). Identification of a 100-kDa putative coaggregation-mediating adhesin of *Streptococcus gordonii* DL1 (Challis). *Infection and Immunity*, 63, 4890–4893.
Costerton, J. W., Lewandowski, Z., Caldwell, D. E., Korber D. R. and Lappin-Scott, H. M. (1995). Microbial biofilms. *Annual Review of Microbiology*, 49, 711–745.
Costerton, J. W., Stewart, P. S. and Greenberg, E. P. (1999). Bacterial biofilms: a common cause of persistent infections. *Science*, 284, 1318–1322.
Cvitkovitch, D. G. (2001). Genetic competence and transformation in oral streptococci. *Critical Reviews in Oral Biology and Medicine*, 12, 217–243.
Davey, M. E. and O'Toole, G. A. (2000). Microbial biofilms: from ecology to molecular genetics. *Microbiology and Molecular Biology Reviews*, 64, 847–867.
Davies, D. G., Parsek, M. R., Pearson, J. P., Iglewski, B. H., Costerton, J. W. and Greenberg, E. P. (1998). The involvement of cell-to-cell signals in the development of a bacterial biofilm. *Science*, 280, 295–298.
Demuth, D. R., Duan, Y., Brooks, W., Holmes, A. R., McNab, R. and Jenkinson, H. F. (1996). Tandem genes encode cell-surface polypeptides SspA and SspB which mediate adhesion of the oral bacterium *Streptococcus gordonii* to human and bacterial receptors. *Molecular Microbiology*, 20, 403–413.
Demuth, D. R., Irvine, D. C., Costerton, J. W., Cook, G. S. and Lamont, R. J. (2001). Discrete protein determinant directs the species-specific adherence of *Porphyromonas gingivalis* to oral streptococci. *Infection and Immunity*, 69, 5736–5741.
Dorel, C., Vidal, O., Prigent-Combaret, C., Vallet, I. and Lejeune, P. (1999). Involvement of the Cpx signal transduction pathway of *Escherichia coli* in biofilm formation. *FEMS Microbiology Letters*, 178, 169–175.

Dowson, C. G., Hutchison, A., Woodford, N., Johnson, A. P., George, R. C. and Spratt, B. G. (1990). Penicillin-resistant viridans streptococci have obtained altered penicillin-binding protein genes from penicillin-resistant strains of *Streptococcus pneumoniae*. *Proceedings of the National Academy of Sciences USA*, 87, 5858–5862.

Du, L. D. and Kolenbrander, P. E. (2000). Identification of saliva-regulated genes of *Streptococcus gordonii* DL1 by differential display using random arbitrarily primed PCR. *Infection and Immunity*, 68, 4834–4837.

Dunny, G. M. and Leonard, B. A. B. (1997). Cell-cell communication in Gram-positive bacteria. *Annual Review of Microbiology*, 51, 527–564.

Egland, P. G., Du, L. D. and Kolenbrander, P. E. (2001). Identification of independent *Streptococcus gordonii* SspA and SspB functions in coaggregation with *Actinomyces naeslundii*. *Infection and Immunity*, 69, 7512–7516.

El-Sabaeny, A., Demuth, D. R. and Lamont, R. J. (2001). Regulation of *Streptococcus gordonii sspB* by the *sspA* gene product. *Infection and Immunity*, 69, 6520–6522.

El-Sabaeny, A., Demuth, D. R., Park, Y. and Lamont, R. J. (2000). Environmental conditions modulate the expression of the *sspA* and *ssB* genes in *Streptococcus gordonii*. *Microbial Pathogenesis*, 29, 101–113.

Embleton, J. V., Newman, H. M. and Wilson, M. (1998). Influence of growth mode and sucrose on susceptibility of *Streptococcus sanguis* to amine fluorides and amine fluoride-inorganic fluoride combinations. *Applied and Environmental Microbiology*, 64, 3503–3506.

Fenno, J. C., LeBlanc, D. J. and Fives-Taylor, P. (1989). Nucleotide sequence analysis of a type 1 fimbrial gene of *Streptococcus sanguis* FW213. *Infection and Immunity*, 57, 3527–3533.

Fong, K. P., Chung, W. O., Lamont, R. J. and Demuth, D. R. (2001). Intra- and interspecies regulation of gene expression by *Actinobacillus actinomycetemcomitans* LuxS. *Infection and Immunity*, 69, 7625–7634.

Fowler, R. G. and Schaaper, R. M. (1997). The role of *mutT* gene of *Escherichia coli* in maintaining replication fidelity. *FEMS Microbiology Reviews*, 21, 43–54.

Frias, J., Olle, E. and Alsina, M. (2001). Periodontal pathogens produce quorum sensing signal molecules. *Infection and Immunity*, 69, 3431–3434.

Froeliger, E. H. and Fives-Taylor, P. (2001). *Streptococcus parasanguis* fimbria-associated adhesin Fap1 is required for biofilm formation. *Infection and Immunity*, 69, 2512–2519.

Ganeshkumar, N., Song, M. and McBride, B. C. (1988). Cloning of a *Streptococcus sanguis* adhesin which mediates binding to a saliva-coated hydroxyapatite. *Infection and Immunity*, 56, 1150–1157.

Genevaux, P., Muller, S. and Bauda, P. (1996). A rapid screening procedure to identify mini-Tn*10* insertion mutants of *Escherichia coli* K-12 with altered adhesion properties. *FEMS Microbiology Letters*, 142, 27–30.

Gibbons, R. J. (1989). Bacterial adhesion to oral tissues: a model for infectious diseases. *Journal of Dental Research*, 68, 750–760.

Glanzmann, P., Gustafson, J., Komatsuzawa, H., Ohta, K. and Berger-Bachi, B. (1999). *glmM* operon and methicillin-resistant *glmM* supressor mutants in *Staphylococcus aureus*. *Antimicrobial Agents and Chemotherapy*, 43, 240–245.

Gong, K. and Herzberg, M. C. (1997). *Streptococcus sanguis* expresses a 150-kilodalton two-domain adhesin: characterization of several independent adhesin epitopes. *Infection and Immunity*, 65, 3815–3821.

Havarstein, L. S., Gaustad, P., Nes, I. F. and Morrison, D. A. (1996). Identification of the streptococcal competence-pheromone receptor. *Molecular Microbiology*, 21, 863–869.

Hazlett, K. O., Mazurkiewicz, J. E. and Banas, J. A. (1999). Inactivation of the *gbpA* gene of *Streptococcus mutans* alters structural and functional aspects of plaque biofilms, which are compensated by recombination of the *gtfB* and *gtfC* genes. *Infection and Immunity*, 67, 3909–3914.

Jenkinson, H. F. and Demuth, D. R. (1997). Structure, function and immunogenicity of streptococcal antigen I/II polypeptides. *Molecular Microbiology*, 23, 183–190.

Jenkinson, H. F. and Lamont, R. J. (1997). Streptococcal adhesion and colonization. *Critical Reviews in Oral Biology and Medicine*, 8, 175–200.

Jolly, L., Wu, S., van Heijenoort, J., de Lencastre, H., Mengin-Lecreulx, D. and Tomasz, A. (1997). The *femR315* gene from *Staphylococcus aureus*, the interruption of which results in reduced methicillin resistance, encodes a phoshoglucosamine mutase. *Journal of Bacteriology*, 179, 5321–5325.

Kolenbrander, P. E. (2000). Oral microbial communities: biofilms, interactions and genetic systems. *Annual Review of Microbiology*, 54, 413–437.

Kuramitsu, H. K. (2001). Virulence properties of oral bacteria: impact of molecular biology. *Current Issues in Molecular Biology*, 3, 35–36.

Lamont, R. J., Gil, S., Demuth, D. R., Malamud, D. and Rosan, B. (1994). Molecules of *Streptococcus gordonii* that bind to *Porphyromonas gingivalis*. *Microbiology*, 140, 867–872.

Larsen, T. and Fiehn, N.-E. (1996). Resistance of *Streptococcus sanguis* biofilms to antimicrobial agents. *APMIS*, 104, 280–284.

Lee, S. F., Li, Y. H. and Bowden, G. H. (1996). Detachment of *Streptococcus mutans* biofilm cells by an endogenous enzymatic activity. *Infection and Immunity*, 64, 1035–1038.

Lewis, K. (2001). Riddle of biofilm resistance. *Antimicrobial Agents and Chemotherapy*, 45, 999–1007.

Li, X., Kolliveit, K. M., Tronstad, L. and Olsen, I. (2000). Systemic diseases caused by oral infection. *Clinical Microbiology Reviews*, 13, 547–558.

Li, Y.-H., Hanna, M. N., Svensater, G., Ellen, R. P. and Cvitkovitch, D. G. (2001a). Cell density modulates acid adaptation in *Streptococcus mutans*: implications for survival in biofilms. *Journal of Bacteriology*, 183, 6875–6844.

Li, Y.-H., Lau, P. C. Y., Lee, J. H., Ellen, R. P. and Cvitkovitch, D. G. (2001b). Natural genetic transformation of *Streptococcus mutans* growing in biofilms. *Journal of Bacteriology*, 193, 897–908.

Liljemark, W. F., Bloomquist, C. G., Reilly, B. E., Bernards, C. J., Townsend, D. W., Pennock, A. T. and LeMoine, J. L. (1997). Growth dynamics in a natural biofilm and its impact on oral disease management. *Advances in Dental Research*, 11, 14–23.

Listgarten, M. A. (1994). The structure of dental plaque. *Periodontology, 2000* 5, 52–65.

Loesche, W. J. (1986). Role of *Streptococcus mutans* in human dental decay. *Microbiology Reviews*, 50, 353–380.

Loo, C. Y., Corliss, D. A. and Ganeshkumar, N. (2000). *Streptococcus gordonii* biofilm formation: identification of genes that code for biofilm phenotypes. *Journal of Bacteriology*, 182, 1374–1382.

Lunsford, R. D. and London, J. (1996). Natural genetic transformation in *Streptococcus gordonii*: *comX* imparts spontaneous competence on strain Wicky. *Journal of Bacteriology*, 178, 5831–5835.

Marsh, P. D. (1999). Microbiologic aspects of dental plaque and dental caries. *Dental Clinics of North America*, 43, 599–614.

Massidda, O., Anderluzzi, D., Friedli, L. and Feger, G. (1998). Unconventional organization of the division and cell wall gene cluster of *Streptococcus pneumoniae*. *Microbiology*, 144, 3069–3078.

Mattos-Graner, R. O., Jin, S., King, W. F., Chen, T., Smith, D. J. and Duncan, M. J. (2001). Cloning of the *Streptococcus mutans* gene encoding glucan binding protein B and analysis of genetic diversity and protein production in clinical isolates. *Infection and Immunity*, 69, 6931–6941.

Mengin-Lecreulx, D. and van Heijenoort, J. (1996). Characterization of the essential gene *glmM* encoding phosphoglucosamine mutase in *Escherichia coli*. *Journal of Biological Chemistry*, 271, 32–39.

Nyvad, B. and Kilian, M. (1990). Comparison of the initial streptococcal microflora on dental enamel in caries-active and in caries-inactive individuals. *Caries Research*, 24, 267–272.

O'Toole, G. A., Gibbs, K. A., Hager, P. W., Phibbs, P. V., Jr. and Kolter, R. (2000a). The global carbon metabolism regulator Crc is a component of a signal transduction pathway required for biofilm development by *Pseudomonas aeruginosa*. *Journal of Bacteriology*, 182, 425–431.

O'Toole, G. A., Kaplan, H. B. and Kolter, R. (2000b). Biofilm formation as microbial development. *Annual Review of Microbiology*, 54, 49–79.

Pestova, E. V., Havarstein, L. S. and Morrison, D. A. (1996). Regulation of competence for genetic transformation in *Streptococcus pneumoniae* by an auto-induced peptide pheromone and two-component regulatory system. *Molecular Microbiology*, 21, 853–862.

Pratten, J., Barnett, P. and Wilson, M. (1998). Composition and susceptibility to chlorhexidine of multispecies biofilms of oral bacteria. *Applied and Environmental Microbiology*, 64, 3515–3519.

Pringent-Combaret, C., Vidal, O., Dorel, C. and Lejeune, P. (1999). Abiotic surface sensing and biofilm-dependent regulation of gene expression in *Escherichia coli*. *Journal of Bacteriology*, 181, 5993–6002.

Roberts, A. P., Cheah, C., Ready, D., Pratten, J., Wilson, M. and Mullany, P. (2001). Transfer of Tn*916*-like elements in microcosm dental plaques. *Antimicrobial Agents and Chemotherapy*, 45, 2943–2946.

Roberts, A. P., Pratten, J., Wilson, M. and Mullany, P. (1999). Transfer of a conjugative transposon, Tn*5397* in a model oral biofilm. *FEMS Microbiology Letters*, 177, 63–66.

Rogers, J. D., Haase, E. M., Brown, A. E., Douglas, C. W. I., Gwynn, J. P. and Scannapieco, F. A. (1998). Identification and analysis of a gene (*abpA*) encoding a major amylase-binding protein in *Streptococcus gordonii*. *Microbiology*, 144, 1223–1233.

Rogers, J. D., Palmer, R. J., Jr., Kolenbrander, P. E. and Scannapieco, F. A. (2001). Role of *Streptococcus gordonii* amylase-binding protein A in adhesion to hydroxyapatite, starch metabolism and biofilm formation. *Infection and Immunity*, 69, 7046–7056.

Rosan, B. C. T., Baker Nelson, G. M., Berman, R., Lamont, R. J. and Demuth, D. R. (1989). Cloning and expression of an adhesin antigen of *Streptococcus sanguis* G9B in *Escherichia coli*. *Journal of General Microbiology*, 135, 531–538.

Rosan, B. and Lamont, R. J. (2000). Dental plaque formation. *Microbes and Infection*, 2, 1599–1607.

Scannapieco, F. A. (1994). Saliva-bacterium interactions in the oral microbial ecology. *Critical Reviews in Oral Biology and Medicine*, 5, 203–248.

Scannapieco, F. A., Bergey, E. J., Reddy, M. S. and Levine, M. J. (1989). Characterization of salivary α-amylase binding to *Streptococcus sanguis*. *Infection and Immunity*, 57, 2853–2863.

Scannapieco, F. A. and Genco, R. J. (1999). Association of periodontal infections with atherosclerotic and pulmonary diseases. *Journal of Periodontal Research*, 34, 340–345.

Schilling, K. M. and Bowen, W. H. (1992). Glucans synthesized *in situ* in experimental salivary pellicle function as specific binding sites for *Streptococcus mutans*. *Infection and Immunity*, 60, 284–295.

Shapiro, J. A. (1998). Thinking about bacterial populations as multicellular organisms. *Annual Review of Microbiology*, 52, 81–104.

Socransky, S. S. and Haffajee, A. D. (1994). Evidence of bacterial etiology: a historical perspective. *Periodontology 2000*, 5, 7–25.

Stingele, F. and Bollet, B. (1996). Disruption of the gene encoding penicillin-binding protein 2b *(pbp2b)* causes altered cell morphology and cease in exopolysaccharide production in *Streptococcus thermophilus* Sfi6. *Molecular Microbiology*, 22, 357–366.

Svensater, G., Welin, J., Wilkins, J. C., Beighton, D. and Hamilton, I. R. (2001). Protein expression by planktonic and biofilm cells of *Streptococcus mutans*. *FEMS Microbiology Letters*, 205, 139–146.

Tao, L. and Herzberg, M. C. (1999). Identifying *in vivo* expressed streptococcal genes in endocarditis. *Methods in Enzymology*, 310, 109–116.

Tavares, I. M., Jolly, L., Pompeo, F., Leitao, J. H., Fialho, A. M., Sa-Correia, I. and Mengin-Lecreulx, D. (2000). Identification of the *Pseudomonas aeruginosa glmM* gene, encoding phosphoglucosamine mutase. *Journal of Bacteriology*, 182, 4453–4457.

Upton, M., Tagg, J. R., Wescombe, P. and Jenkinson, H. F. (2001). Intra- and interspecies signalling between *Streptococcus salivarius* and *Streptococcus pyogenes* mediated by SalA and SalA1 lantibiotic peptides. *Journal of Bacteriology*, 183, 3931–3938.

Vickerman, M. M. D., Clewell, B. and Jones, G. W. (1991). Sucrose-promoted accumulation of growing glucosyltransferase variants of *Streptococcus gordonii* on hydroxyapatite surfaces. *Infection and Immunity*, 59, 3523–3530.

Whittaker, C. J., Klier, C. M. and Kolenbrander, P. E. (1996). Mechanisms of adhesion by oral bacteria. *Annual Review of Microbiology*, 50, 513–552.

Wilson, M. (1996). Susceptibility of oral bacterial biofilms to antimicrobial agents. *Journal of Medical Microbiology*, 44, 79–87.

Wu, H. and Fives-Taylor, P. M. (1999). Identification of dipeptide repeats and a cell wall sorting signal in the fimbriae-associated adhesin, Fap1, of *Streptococcus parasanguis*. *Molecular Microbiology*, 34, 1070–1081.

Wu, T., Trevisan, M., Genco, R. J., Dorn, J. P., Falkner, K. L. and Sempos, C. T. (2000). Periodontal disease and risk of cerebrovascular disease: the first national health and nutrition examination survey and its follow-up study. *Archives of Internal Medicine*, 160, 2749–2755.

Xie, H., Cook, G. S., Costerton, J. W., Bruce, G., Rose, T. M. and Lamont, R. J. (2000). Intergeneric communication in dental plaque biofilms. *Journal of Bacteriology*, 182, 7067–7069.

Yun, C., Ely, B. and Smit, J. (1994). Identification of genes affecting production of the adhesive holdfast of a marine caulobacter. *Journal of Bacteriology*, 176, 796–803.

Zhang, J. P. and Normark, S. (1996). Induction of gene expression in *Escherichia coli* after pilus-mediated adherence. *Science*, 273, 1234–1236.

CHAPTER TEN

Gene Expression in Oral Biofilms

Robert A. Burne, Yi-Ywan M. Chen, Yunghua Li,
Samir Bhagwat, and Zezhang Wen

1 INTRODUCTION

The only natural habitat for the aetiological agents of dental caries and periodontal diseases, as well as the overwhelming majority of oral bacteria, is the mouth. These organisms exist almost exclusively as constituents of the biofilms that form on the soft and hard tissues of the oral cavity and, therefore, can be thought of as 'obligate biofilm organisms' (Burne, 1998a). Bacteria capable of sustaining a free-living existence have sensing and differentiation systems that allow the organisms to detect acquisition of a suitable host, to alter motility and capsule expression, and to transition from oligotrophic systems to relatively nutrient-rich systems where they must confront the innate and acquired defences of a host. In contrast, oral bacteria do not have to transition between radically different environments, such as moving from the gut of a mammal to a mountain stream or decaying forest material and *vice versa*. Therefore, it is reasonable to predict that, through normal evolutionary processes, oral biofilm bacteria may have sloughed off many of the elaborate genetic systems for responding to the transition from the free-living state to the host-associated or biofilm state. In fact, with the completion or near completion of a number of genomes of oral bacteria, it is clear that oral bacteria have comparatively small chromosomes and lack many of the sensing and differentiation pathways that are present in organisms that can exist in a free-living state, such as *Pseudomonas aeruginosa*, *Escherichia coli*, and *Yersinia pestis*. Notwithstanding, intermittent eating by the host, the host's diurnal rhythms, and the use of antimicrobials in oral health care products, among other factors, expose oral bacteria to relatively large and very rapid changes in their environment. The bacteria in the mouth must respond rapidly and efficiently to these stresses to optimise their growth and persistence (Burne, 1998a,b). Since these organisms do persist and have become very well adapted to the oral environment, these bacteria must have evolved very effective genetic, physiologic, and biochemical circuitry to deal with the rapid and significant changes that occur in oral biofilms. Not surprisingly then, as investigators begin to explore environmental regulation of gene expression in oral bacteria, the studies reveal complex networks for control of virulence expression, as well as important differences between oral bacteria and other organisms. Dissecting how the pathogenic constituents of oral biofilms adapt to environmental changes to capitalise on conditions that are conducive to the development of disease is an essential first step along the path to novel and more effective therapies and preventive measures to combat oral diseases.

In many ways, diseases of the oral cavity are fundamentally different from those studied by traditional medical microbiologists. Caries and periodontal diseases are not caused by overt pathogens, and rarely, if ever, are these diseases elicited by individual organisms. Under conditions of health, the host tissues and the biofilms formed on these tissues exist in a dynamic equilibrium that is compatible with maintenance of the integrity of the host tissues. Caries and periodontal diseases can develop when there are environmental changes that foster an increase in the proportions of suspected pathogens and that induce changes in the biochemical activities and virulence of the biofilms colonising teeth and soft tissues (Bowden, Ellwood, and Hamilton, 1979; Burne, 1998a,b). The development of cariogenic or periodontopathic biofilms is complex and undoubtedly results from changes in the environment that are more favourable to the growth of the putative pathogens. Such changes include, but are not limited to, sustained and repeated acidification of biofilms, reduction in the oxygen content or redox potential of plaque, enhanced nutrient availability, a change in the flow rate or composition of saliva or gingival fluids, injury or foreign body presence, and alterations in the immune responses to particular species. Organisms in the biofilm, and thus the biofilm as a whole, adapt to these changes, and these adaptations can enhance the pathogenic potential of the biofilms. Of the environmental factors known to have major influences on the composition and pathogenic potential of tooth biofilms, pH and carbohydrate availability seem to be the most critical, at least for supragingival plaque (Bowden et al., 1979; Burne, 1998a). Studying population dynamics of oral biofilms in response to external stimuli and dissecting how individual species respond to new environmental conditions is a major challenge, but is absolutely necessary if a complete understanding of disease development is to be grasped. In this chapter, we discuss studies we have initiated to begin to understand how oral streptococci behave when growing in three-dimensional biofilms on surfaces.

2 GENE EXPRESSION IN ORAL BIOFILMS

2.1 Carbohydrate Availability and pH as Modulators of Oral Biofilm Gene Expression – the *Streptococcus salivarius* Urease Genes as a Model System

The genes encoding the urease of *S. salivarius* 57.I are regulated in a manner that is different from other urease genes of eubacteria (Chen and Burne, 1996). Whereas other urease genes are either constitutive or regulated by urea induction or nitrogen limitation (Collins and Dorazio, 1993), the urease genes of *S. salivarius* are almost completely repressed at neutral pH. As the pH of the growth medium is lowered to 6 and then to 5.5, transcription of urease genes is observed (Table 10.1). Once urease becomes derepressed at low pH, transcription of the urease genes can be dramatically enhanced by cultivation of the cells under carbohydrate excess conditions, even though there is no

Table 10.1: The CAT specific activity of recombinant *S. salivarius* strains growing in *in vitro* biofilms at quasi-steady state and following a 25 m*M* glucose pulse

Strain/Time after Glucose Pulse	pH Control[a]		No pH Control[b]	
	pH after Glucose Pulse	CAT Activity (U/mg Protein)[c]	pH after Glucose Pulse	CAT Activity (U/mg Protein)
PureI				
T_0	6.70 ± 0.20	0.86 ± 0.32	5.12 ± 0.20	9.58 ± 1.40
T_{15}	5.76 ± 0.11	8.04 ± 4.40	4.86 ± 0.12	12.6 ± 1.00
T_{30}	5.21 ± 0.12	17.6 ± 4.28	4.81 ± 0.10	13.3 ± 1.10
T_{60}	4.98 ± 0.13	22.0 ± 1.95	4.72 ± 0.21	16.2 ± 0.60
PureIΔ100				
T_0	6.67 ± 0.11	19.9 ± 4.85	5.32 ± 0.10	18.5 ± 3.62
T_{15}	6.30 ± 0.15	21.1 ± 2.61	5.02 ± 0.05	18.0 ± 1.64
T_{30}	5.87 ± 0.24	21.8 ± 3.65	4.88 ± 0.12	18.9 ± 1.92
T_{60}	5.27 ± 0.19	22.1 ± 1.96	4.86 ± 0.15	20.0 ± 1.28

Note: Data from Li, Chen, and Burne (2000). Reproduced with permission from Blackwell Publishing.
[a] pH control: 50 m*M* KPO_4, pH 7.8.
[b] No pH control: no buffering system, but supplemented with 90 m*M* KCl.
[c] CAT (chloramphenicol acetyltransferase) activity is expressed as nmols of chloramphenicol acetylated min^{-1} mg^{-1} protein. The values shown here are averages and standard deviations from at least three independent runs of the Rototorque, and all assays were performed in triplicate.

apparent effect of carbohydrate availability when cells are growing at neutral pH values. Thus, pH is the dominant factor, but carbohydrate acts as a key modulator of urease expression. Because of the manner in which urease is regulated in *S. salivarius*, we reasoned that this gene cluster could serve as a good model system to begin to study effects of growth pH and carbohydrate source and availability on the regulation of genes in biofilm bacteria.

To study urease gene expression in adherent populations of *S. salivarius* 57.I, promoter fusions of a chloramphenicol acetyltransferase gene (*cat*) to the intact, pH- and carbohydrate-sensitive urease promoter, and to a deletion derivative of the urease promoter that is not repressed at neutral pH (Chen et al., 1998), were constructed and integrated into the *S. salivarius* chromosome. The latter fusion served as a control to be certain that observed effects of pH and carbohydrate on gene expression were, in fact, due to the environmental variables of interest and that the effects were being exerted through the cognate *cis*-acting elements. Single-species biofilms of *S. salivarius* were established in a modified Rototorque, continuous-flow biofilm fermenter (Characklis, 1990; Burne and Chen, 1998). Cells were cultivated to quasi-steady state in a tryptone-yeast extract medium with sucrose as the limiting carbohydrate for an equivalent of ten mean generation times, as previously described (Li, Chen, and Burne, 2000). Control of the pH of the culture was maintained by buffering the medium with potassium phosphate, such that the pH of the liquid phase of the culture was 6.70 ± 0.20 and the biofilm pH was measured at 6.12 ± 0.15 (Li et al., 2000). Alternatively, cultures grown without pH control had a suspended phase pH of 5.12 ± 0.20 and a biofilm pH of 5.35 ± 0.16 (Li et al., 2000). Thus, the use of buffers proved to be a reasonable method to create populations of biofilms that were growing at significantly different pH values. The biofilms formed in the Rototorque were heterogeneous and ranged in thicknesses up to about 100 μm (Li et al., 2000). Chemical analysis of the biofilm revealed that the biofilms grown with pH control were about 46 per cent carbohydrate, whereas biofilms formed without pH control were composed of a significantly greater proportion of carbohydrate (60.1 per cent). Not surprisingly, the percentage of viable cells as measured by plate counting and dividing by the number of cells enumerated in a Petroff–Hauser chamber was significantly greater in cultures grown with pH control (68 per cent) than without (15 per cent). However, total viable counts of bacteria did not differ significantly because absolute biomass of the low pH biofilms was substantially greater than in the pH controlled culture (Li et al., 2000).

Once cultures were established at quasi-steady state, three slides were removed from the Rototorque and the CAT activity expressed in the biofilms was measured as previously described. As can be seen in Table 10.1 (Li et al., 2000), the strain PureI, which carries the wild-type, pH-responsive urease promoter fusion, had very low levels of CAT activity due to repression of urease at neutral pH. This behaviour can be contrasted with PureI growing without pH control, which had higher activity arising from enhanced transcription of the promoter at the lower pH values in the biofilms. Eliminating the *cis*-acting elements involved in pH control in strain PureIΔ100 resulted in constitutively high expression, providing support for the idea that pH is, in fact, being sensed by the cells in the biofilms. Immediately after the initial three slides were removed, glucose was added to the vessel to a final concentration of 0.025 *M*, the pH of the vessel was monitored, and CAT activity was assayed in the biofilm cells at 15, 30, and 60 minutes. As expected, the pH of the culture fluid of PureI maintained at neutral pH fell rapidly to just below 5 as the cells metabolised the added carbohydrate. This pH fall was accompanied by a rapid induction of transcription of about 25-fold from the wild-type urease promoter. The pH of the culture growing without added buffer fell from 5.1 to 4.7. Since the operon was largely derepressed at pH 5.1, only about a 60 per cent increase in CAT activity expressed from the wild-type promoter was observed following the carbohydrate pulse. No significant additional induction of the pH-unresponsive promoter fusion strain (PureIΔ100) was observed. Considering that the low pH biofilms were grown under carbohydrate-limiting conditions and that addition of excess carbohydrate did not result in as dramatic an increase as was seen in the cultures growing with pH control, the primary control point for induction appears to be pH.

Urease plays two primary roles in *S. salivarius* (Chen et al., 2000). The production of ammonia from urea can alkalinise the cytoplasm of the organism and can raise the pH of the environment. In this manner, ureolysis protects *S. salivarius* from acid killing. Also, urea provides *S. salivarius* with a bioenergetically efficient source of nitrogen. The biofilm studies with gene fusions to the pH-regulated urease promoter demonstrated that environmental pH is a major factor governing the expression of genes important for the persistence and survival of oral bacteria growing as adherent populations and that large changes in gene expression levels could be induced rapidly by pH changes in biofilms. Thus, these studies established the feasibility of examining virulence gene expression in adherent oral streptococci in response to pH, carbohydrate source, and carbohydrate availability. We next moved on to studying the expression of the expolysaccharide machinery of the cariogenic bacterium *S. mutans* in biofilms.

2.2 Regulation of the Exopolysaccharide Machinery of *S. mutans* in Biofilms

One of the only virulence attributes to be unequivocally shown to be essential for *S. mutans* to initiate caries in experimental animals is the ability to produce and degrade extracellular homopolysaccharides of glucose and fructose (Yamashita et al., 1993; Burne et al., 1996). *S. mutans* produces three glucosyltransferases (Gtfs) that convert sucrose to high-molecular-mass polysaccharides composed almost entirely of α1,3- and α1,6-linked glucose moieties (Kuramitsu, 1987). These largely water-insoluble polysaccharides serve as an adhesive scaffolding for biofilm formation by these organisms. *S. mutans* also produces a fructosyltransferase (ftf) enzyme, which catalyzes the synthesis of homopolymers of β2,1- and β2,6-linked fructose monomers. Unlike glucans, plaque fructans are short lived and appear to serve mainly as extracellular reserves of carbohydrate that can be utilised when exogenous sources are exhausted (Burne, 1998a,b).

Using continuous chemostat cultivation of strains carrying *cat* gene fusions to the *gtfBC* and *ftf* promoters, we previously reported that the expression of these genes could be influenced by growth rate and pH and that the genes could be induced by adding sucrose to carbohydrate-limited cultures. In a later study, we explored how growth in biofilms might influence expression of *gtfBC-cat* and *ftf-cat* fusions (Wexler, Hudson, and Burne, 1993). Using a Rototorque to grow monospecies biofilms of *S. mutans*, we showed that expression of the genes for exopolysaccharide production in biofilms that were allowed to form after 48 hours – fairly thin films that had been established for less than ten mean generation times – behaved similarly to the planktonic cells we examined in our previous chemostat experiments (Burne, Chen, and Penders, 1997). However, cultivation of the biofilms for 7 days caused dramatic changes in expression of both *gtfBC* and *ftf*. In the 7-day biofilms, *gtf* expression was some 10- to 70-fold higher than in planktonic cells, whereas *ftf* expression was nearly completely repressed. Consistent with our hypothesis that *S. mutans* has developed mechanisms to sense its environment and alter virulence expression, we reasoned that perhaps conditions in microenvironments created in the mature biofilms induced changes in gene expression. Since pH and carbohydrate availability are major environmental factors dealt with by natural dental plaque, we used the experimental platform we established with the urease gene expression work to explore the regulation of *S. mutans* genes for exopolysaccharide synthesis (Li and Burne, 2001).

S. mutans strains SMS101 (*ftf-cat*) and SMS102 (*gtfBC-cat*) (Hudson and Curtiss, 1990) were cultivated to quasi-steady state (ten mean generations) in the Rototorque using a modified tryptone-yeast extract medium (Li et al.,

2000) supplemented with 10 m*M* sucrose or, in specific cases, with 5 m*M* glucose. Similar to the aforementioned experiments, three slides were removed (T_0) and CAT activity was measured. We performed several different series of experiments. First, we explored expression of the genes in response to pH by growing the biofilms in buffered or unbuffered medium followed by addition of sucrose. Whereas pH appeared to have no substantive effect on *ftf* expression (Table 10.2), CAT activity expressed from the *gtfBC* promoter in biofilms in buffered medium (actual biofilm pH = 6.1) was roughly half that of cells growing in biofilms without pH control (actual biofilm pH = 5.3) (Li and Burne, 2001). As might be predicted from previous studies, the expression of both operons was rapidly induced about threefold following the addition of sucrose to the vessel. Of note, the absolute levels of CAT expressed from each of the constructs in the biofilms prior to, and after, sucrose pulsing were generally similar to that which was observed in carbohydrate-limited chemostat cultures before and after sucrose pulsing (Wexler et al., 1993). Thus, the biofilm bacteria were behaving largely like planktonic cells, and there appeared to be no specific alterations in exopolysaccharide gene expression arising simply as a result of growing the organisms on a solid surface.

Table 10.2: The CAT specific activity of recombinant *S. mutans* strains growing in *in vitro* biofilms following a 25 m*M* sucrose pulse

	Strain SMS101 (*ftf::cat*)		Strain SMS102 (*gtfBC::cat*)	
Time (min) after Sucrose Pulse	pH after Sucrose Pulse	CAT Activity (U/mg Protein)[a]	pH after Sucrose Pulse	CAT Activity (U/mg Protein)
With pH control[b]				
T_0	6.68 ± 0.15	0.112 ± 0.08	6.72 ± 0.12	0.054 ± 0.03
T_{15}	6.34 ± 0.13	0.265 ± 0.10	5.76 ± 0.10	0.112 ± 0.04
T_{30}	5.87 ± 0.12	0.290 ± 0.08	5.21 ± 0.11	0.185 ± 0.06
T_{60}	5.51 ± 0.08	0.269 ± 0.05	4.98 ± 0.07	0.232 ± 0.12
Without pH control[c]				
T_0	5.12 ± 0.25	0.091 ± 0.05	5.14 ± 0.16	0.106 ± 0.04
T_{15}	4.90 ± 0.33	0.298 ± 0.08	4.91 ± 0.14	0.224 ± 0.08
T_{30}	4.74 ± 0.28	0.262 ± 0.11	4.70 ± 0.04	0.318 ± 0.07
T_{60}	4.63 ± 0.35	0.183 ± 0.12	4.58 ± 0.06	0.382 ± 0.10

Note: Data from Li and Burne (2001).

[a] CAT activity is expressed as μmols of chloramphenicol acetylated min^{-1} mg^{-1} protein. The values shown here are averages and standard deviations from triplicate independent experiments.

[b] pH control: 50 m*M* KPO_4, pH 7.8.

[c] No pH control: no buffering system but supplemented with 90 m*M* KCl.

It has been suggested that sucrose may be a specific inducer of *ftf* and *gtfBC*, so it was of interest to examine expression of these genes in biofilms that were limited for a carbohydrate other than sucrose (Li and Burne, 2001). To accomplish this, biofilms were formed for the equivalent of ten generations and then a new quasi-steady state was allowed to establish for ten more generations with 5 m*M* glucose substituted for 10 m*M* sucrose in the base medium. If sucrose was a specific inducer of these genes, then one might expect that limiting the cells for glucose, instead of sucrose, would result in lower levels of CAT activity at T_0. Instead, the levels of CAT were no different than in quasi-steady state cells growing on sucrose (Table 10.3). Similar to the results obtained with sucrose-limited cultures, pH had a significant effect on *gtfBC* expression, with about a twofold increase noted in the cells growing at lower pH values (Table 10.3). Again, no differences in *ftf* expression as a function of pH were noted. Sucrose was found to be an efficient inducer of *gtfBC* and *ftf* expression in glucose-limited cells, as it was for cells growing with sucrose as the limiting carbohydrate (Table 10.3). As previously stated, the levels and patterns of CAT expressed by the recombinant strains were essentially those

Table 10.3: The CAT specific activity of recombinant *S. mutans* strains growing in *in vitro* biofilms[a] following a 25 m*M* sucrose pulse with and without pH control[b]

	Strain SMS101 (*ftf::cat*)		Strain SMS102 (*gtfBC::cat*)	
Time (min) after Sucrose Pulse	pH after Sucrose Pulse	CAT Activity[c] (U/mg Protein)	pH after Sucrose Pulse	CAT Activity (U/mg Protein)
With pH control				
T_0	6.62 ± 0.12	0.088 ± 0.07	6.64 ± 0.15	0.042 ± 0.04
T_{15}	6.38 ± 0.18	0.226 ± 0.12	6.20 ± 0.22	0.176 ± 0.08
T_{30}	5.92 ± 0.13	0.287 ± 0.10	5.80 ± 0.21	0.258 ± 0.05
T_{60}	5.52 ± 0.24	0.254 ± 0.09	5.49 ± 0.18	0.295 ± 0.06
Without pH control				
T_0	5.08 ± 0.22	0.098 ± 0.07	5.11 ± 0.25	0.082 ± 0.03
T_{15}	4.89 ± 0.28	0.326 ± 0.14	4.92 ± 0.22	0.214 ± 0.07
T_{30}	4.82 ± 0.33	0.268 ± 0.12	4.80 ± 0.19	0.342 ± 0.12
T_{60}	4.72 ± 0.24	0.154 ± 0.12	4.79 ± 0.23	0.395 ± 0.14

Note: Data from Li and Burne (2001).

[a] The biofilms were grown in sucrose-based medium for ten mean generation times and followed by growing in glucose-based medium for another ten mean generation times before sucrose pulse.

[b] pH control: 50 m*M* KPO_4, pH 7.8; without pH control: no buffering system, but supplemented with 90 m*M* KCl.

[c] The numbers are expressed as μmol of chloramphenicol acetylated min^{-1} mg^{-1} protein.

observed with suspended populations of chemostat-grown cells cultivated under similar, albeit not identical, conditions.

Because there was little difference in the levels of CAT expressed from the *gtfBC* or *ftf* promoters in cells growing with sucrose or glucose as the limiting carbohydrate, it seemed possible that sucrose was, in fact, not a specific inducer of exopolysaccharide gene expression. Based on our results, we postulated that the availability of carbohydrate was a key factor governing induction of the exopolysaccharide machinery of *S. mutans*. To test this idea, *S. mutans* gene fusion strains were cultivated for the equivalent of ten mean generations in biofilms under sucrose-limiting conditions, with or without pH control. Three slides were removed for the quasi-steady state time point, then the vessel was pulsed with glucose at a final concentration of 25 m*M* and slides were removed at 15, 30, and 60 minutes for measurement of CAT activity (Table 10.4). Both the *ftf-cat* and *gtfBC-cat* fusion strains showed evidence for induction by glucose, although the magnitude of the induction was not quite as large as when sucrose was added to the vessel in excess. Nevertheless, it was clear that simply increasing the amount of carbohydrate available to the cells was enough to stimulate transcription of both the *ftf* and the *gtfBC* gene

Table 10.4: The CAT specific activity of recombinant *S. mutans* strains growing in *in vitro* biofilms following a 25 m*M* glucose pulse

	Strain SMS101 (*ftf::cat*)		Strain SMS102 (*gtfBC::cat*)	
Time (min) after Glucose Pulse	pH after Glucose Pulse	CAT Activity (U/mg Protein)[a]	pH after Glucose Pulse	CAT Activity (U/mg Protein)
With pH control[b]				
T_0	6.70 ± 0.15	0.103 ± 0.03	6.65 ± 0.15	0.038 ± 0.03
T_{15}	6.20 ± 0.20	0.124 ± 0.05	6.20 ± 0.12	0.076 ± 0.04
T_{30}	5.80 ± 0.15	0.118 ± 0.04	5.75 ± 0.05	0.115 ± 0.04
T_{30}	5.50 ± 0.20	0.125 ± 0.06	5.50 ± 0.10	0.126 ± 0.06
Without pH control[c]				
T_0	5.04 ± 0.20	0.084 ± 0.02	5.12 ± 0.18	0.114 ± 0.03
T_{15}	4.82 ± 0.25	0.188 ± 0.05	4.84 ± 0.22	0.167 ± 0.05
T_{30}	4.65 ± 0.26	0.175 ± 0.04	4.66 ± 0.20	0.198 ± 0.07
T_{30}	4.60 ± 0.24	0.116 ± 0.03	4.62 ± 0.21	0.224 ± 0.06

Note: Data from Li and Burne (2001).

[a] The numbers are expressed as μmol of chloramphenicol acetylated min^{-1} mg^{-1} protein.

[b] pH control: 50 m*M* KPO_4, pH 7.8.

[c] No pH control: no buffering system, but supplemented with 90 m*M* KCl.

fusions. We have followed these studies with chemostat studies that explore effects of pH and carbohydrate availability on expression of the *gtf* and *ftf* genes and have found that these environmental stimuli are major signals regulating these genes (Li and Burne, 2001). It also appears reasonable to speculate that microenvironments created in the biofilms were, in part, responsible for the differences we observed in previous experiments in which we found that 7-day biofilms had altered exopolysaccharide gene expression.

3 A GENOMIC APPROACH TO IDENTIFYING GENES REQUIRED FOR BIOFILM MATURATION

Although the actual stages of biofilm development are less clearly defined than the conceptual stages defined by many biofilm researchers, most agree that it is at least convenient to categorise biofilm formation into four distinct events: (1) initial attachment, (2) microcolony formation, (3) a growth and recruitment phase where the biofilms develop and extracellular polysaccharide synthesis becomes very evident, and (4) formation of a mature biofilm, sometimes referred to as a climax community (Costerton et al., 1995). In an effort to understand what justifiably can be thought of as a complex form of multicellular developmental process (Danese, Pratt, and Kolter, 2001), many laboratories have initiated studies to define the gene products necessary for a single organism to initiate formation of a biofilm (for example, see Danese et al., 2001). A wide variety of molecules, including adhesins, flagella, intercellular adhesins, and exopolysaccharides, as well as signal transduction pathways, have been shown to participate in the various phases of biofilm formation. Most of this work has been done with organisms that have both a free-living, usually motile, existence as well as a host-associated or biofilm existence. Although much is now known about how oral bacteria adhere to their respective target tissues, relatively little has been done to dissect oral biofilm development and maturation or to understand how bacteria in oral biofilms respond to their environment.

One approach that has been taken in an effort to understand oral biofilm formation has been to use random mutagenesis of oral streptococci to identify genes necessary for biofilm formation (Loo, Corliss, and Ganeshkumar, 1999). Many of the identified gene products were predicted to be important for envelope biogenesis, and thus, mutations in these genes probably interfere with adhesion to abiotic surfaces or intercellular interactions. At least one signalling system, that being the peptide-based quorum sensing system regulating competence development, was also identified in the screen for biofilm genes conducted by Ganeshkumar (Loo et al., 1999). Recently, Cvitkovitch and co-workers (Li et al., 2001) confirmed that competence and signalling may be part of the global communication and differentiation network involved in transitioning from single microcolonies to large adherent biofilms, although it remains to be determined whether these *in vitro* observations are relevant to biofilm formation in the mouth. Still, it is important to understand which pathways are essential for stable biofilm formation by oral bacteria *in vitro* as a prelude to determining whether biofilm formation *in vivo* is affected by the identified gene products and whether the gene products can be useful targets with which to interfere and inhibit initiation of oral diseases. We have elected to take a functional genomics approach to understanding biofilm formation, with a particular interest in defining gene regulation pathways that are critical specifically for maturation of biofilms of oral bacteria.

Our approach to find genes in *S. mutans* that were necessary for formation of stable biofilms was fairly straightforward. In general, we had less interest in identifying structural and enzymatic components that underpin biofilm formation than in understanding the regulation of the maturation process. Consequently, we chose to focus on DNA and RNA binding proteins, two-component signal transduction systems, and gene products that might be involved in cell:cell communication. We initially selected around fifteen genes, isolated the genes by polymerase chain reaction (PCR), mutated the genes by allelic exchange, and tested the ability of the various mutants to form biofilms in an *in vitro* assay. As might be expected, inactivation of most of these genes had no discernable effect on biofilm formation as assessed in a modified microtitre assay based on that developed for *Pseudomonas* (O'Toole and Kolter, 1998). However, three of the genes we selected were found to be required for biofilm formation (Figure 10.1). In particular, we found that inactivation of the *ccpA* orthologue of *S. mutans* UA159 resulted in about a 60 per cent decrease in biofilm-forming capability (Wen and Burne, 2002). CcpA, a member of the GalR/LacI family of transcriptional repressors, has been shown to be a global regulator of carbon catabolite repression (CCR) in Gram-positive bacteria (Henkin, 1996). The role of CcpA in *S. mutans* is as yet undefined, and mutation of *ccpA* does not profoundly affect CCR in this organism (Simpson and Russell, 1998). Thus, CcpA may regulate other genes, some of which are needed for efficient biofilm maturation. Interestingly, the loss of function of a *luxS* orthologue of *S. mutans* UA159, which is predicted to synthesise a small signalling molecule of the AI-2 family of autoinducers (Schauder et al., 2001), had no effect on biofilm-forming capacity. Recently, some evidence has emerged that indicates that AI-2 signals may function in

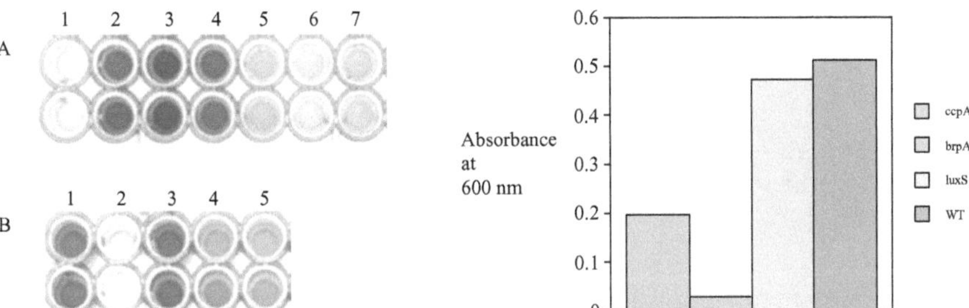

Figure 10.1: Biofilm formation of *S. mutans* UA159 and its derivatives in BM medium. Panels A and B show crystal violet-stained 24-hour biofilms of *brpA* (A5-7), *ccpA* (B4-5), and $luxS_{sm}$ (B3) mutants and their parental strain UA159 (A2-4, B1). Rows A1 and B2 are un-inoculated BM medium as negative control. The graph shows quantitation of the biofilms formed after 24 hours by *ccpA* (ccpA), *brpA* (brpA), and $luxS_{sm}$ (luxS) mutants and the wild-type (WT) strains. See text for more details. Data are representative of no fewer than three separate experiments.

interspecies communication (Bassler, 1999) and that LuxS may regulate virulence in streptococci (Lyon et al., 2001). Therefore, LuxS could be important in determining the biological activities, composition, or architecture of complex, multispecies oral biofilms. Since environmental sensing is likely to be an important factor in persistence of oral biofilms, we also targeted two-component signal transduction systems of *S. mutans*. We were able to inactivate five of six predicted response regulators of these systems – inactivation of the sixth appeared to be a lethal event. Four of the mutants behaved indistinguishably from the wild-type strain in the biofilm assay system. The fifth mutant, which was defective in production of a response regulator for a two-component system with the greatest homology to a peptide sensing system in *Streptococcus pneumoniae*, was almost completely unable to form stable biofilms (data not shown; Bhagwat et al., 2001). None of the aforementioned mutants had any detectable defects in growth or initial adhesion to polystyrene surfaces, so we interpret these data to suggest that the mutants are defective in the regulation of gene products that are necessary for biofilm maturation, which presumably would require normal envelope biogenesis and production of entities that mediate intercellular adherence.

One additional mutant was found to display a biofilm-negative phenotype. In this case, the gene of interest was identified by searching for DNA binding proteins using the CcpC gene of *Bacillus subtilis* as a virtual probe. CcpC is a member of the LysR family of transcriptional regulators that is involved in catabolite repression in *B. subtilis*. The primary structure of BrpA is shown

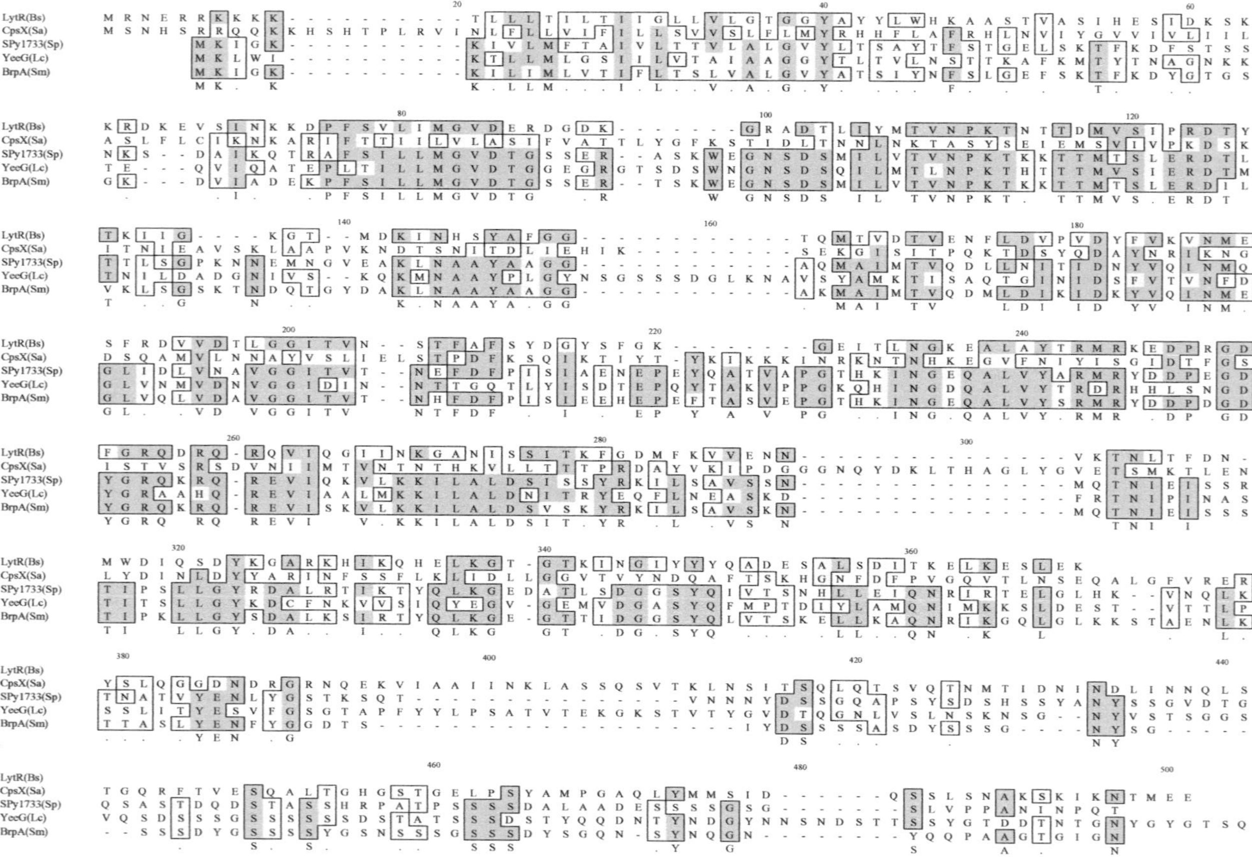
LytR(Bs)
CpsX(Sa)
SPy1733(Sp)
YeeG(Lc)
BrpA(Sm)

in Figure 10.2 (Wen and Burne, 2002). BrpA is predicted to be a membrane-spanning protein with a large, serine-rich extracellular domain. There are no apparent helix-turn-helix domains, nor are there any phosphorylation sites typical of two-component systems. The only discernable differences between BrpA-deficient strains and *S. mutans* UA159 are that the mutant tends to form aberrantly long chains in stationary phase cultures, but not exponentially growing cells; the mutant aggregates more rapidly than the wild-type strain; and the mutant undergoes autolysis more readily than *S. mutans* UA159. We are exploring the idea that defects in envelope biogenesis or failure to properly display surface structures needed for intercellular interactions could lead to the autolytic phenotype or *vice versa.*

4 SUMMARY AND CONCLUSIONS

The use of *in vitro* models to study biofilm formation and the behaviour of adherent populations of microorganisms is becoming more widely accepted as a technique to study bacteria in an environment that more closely mimics that of the native state of the organisms. Using single-species biofilms of oral streptococci, we have found that there are potentially important differences in the behaviours of suspended and sessile populations of these organisms. In large part, these differences are probably attributable to changes in the microenvironment of the bacteria that lead to changes in growth rates, growth domains, and alterations in gene expression patterns. Using a functional genomics approach, we have identified candidate regulatory molecules that may be intimately involved in sensing the state of the bacterial cells and the environment and converting these signals into important changes in gene expression that allow oral pathogens to become dominant members of pathogenic oral biofilms. Perturbing these signalling pathways may offer a unique opportunity to disrupt the development of pathogenic biofilms in the oral cavity and elsewhere in and on the body.

5 ACKNOWLEDGEMENTS

This work was supported by Grants DE12236, DE10362, and DE13239 from the National Institute of Dental and Craniofacial Research.

Figure 10.2: Alignment of predicted amino acid of BrpA of *S. mutans* (Sm) with transcriptional regulator LytR of *B. subtilis* (Bs), and putative regulatory proteins CpsX of *S. agalactiae* (Sa), YeeG of *L. lactis* ssp. *lactis* (Lc), and Spy1733 of *S. pyogenes* (Sp). See text for more detail.

REFERENCES

Bassler, B. L. (1999). How bacteria talk to each other: regulation of gene expression by quorum sensing. *Current Opinions Microbiology*, 2, 582–587.

Bhagwat, S. P., Nary, J. and Burne, R. A. (2001). Effects of mutating putative two-component systems on biofilm formation by *Streptococcus mutans* UA159. *FEMS Microbiology Letters*, 205(2), 225–230.

Bowden, G. H., Ellwood, D. C. and Hamilton, I. R. (1979). Microbial ecology of the oral cavity. In *Advances in Microbial Ecology*, ed. M. Alexander, pp. 135–217. New York: Plenum Press.

Burne, R. A. (1998a). Oral streptococci: products of their environment. *Journal of Dental Research*, 77, 445–452.

Burne, R. A. (1998b). Regulation of gene expression in adherent populations of oral streptococci. In *Microbial Pathogenesis: Current and Emerging Issues*, eds. D. J. LeBlanc, M. S. Lantz and L. M. Switalski, pp. 55–70. Indianapolis, IN: Indiana University Press.

Burne, R. A. and Chen, Y.-Y. M. (1998). The use of continuous flow bioreactors to explore gene expression and physiology of suspended and adherent populations of oral streptococci. *Methods in Cell Science*, 20, 181–190.

Burne, R. A., Chen, Y. M., Wexler, D. W., Kuramitsu, H. and Bowen, W. H. (1996). Cariogenicity of *Streptococcus mutans* strains with defects in fructan metabolism assessed in a program-fed specific pathogen free rat model. *Journal of Dental Research*, 75, 1572–1577.

Burne, R. A., Chen, Y. Y. and Penders, J. E. (1997). Analysis of gene expression in *Streptococcus mutans* in biofilms *in vitro*. *Advances in Dental Research*, 11, 100–109.

Characklis, W. G. (1990). Laboratory biofilm reactors. In *Biofilms*, eds. W. G. Charachlis and K. C. Marshall, pp. 55–89. New York: John Wiley & Sons.

Chen, Y. M. and Burne, R. A. (1996). Analysis of *Streptococcus salivarius* urease expression using continuous chemostat culture. *FEMS Microbiology Letters*, 135, 223–229.

Chen, Y. Y., Weaver, C. A. and Burne, R. A. (2000). Dual functions of *Streptococcus salivarius* urease. *Journal of Bacteriology*, 182, 4667–4669.

Chen, Y.-Y. M., Weaver, C. A., Mendelsohn, D. R. and Burne, R. A. (1998). Transcriptional regulation of the *Streptococcus salivarius* 57.I urease operon. *Journal of Bacteriology*, 180, 5769–5775.

Collins, C. M. and Dorazio, S. E. F. (1993). Bacterial ureases – structure, regulation of expression and role in pathogenesis. *Molecular Microbiology*, 9, 907–913.

Costerton, J. W., Lewandowski, Z., Caldwell, D. E., Korber, D. R. and Lappin-Scott, H. M. (1995). Microbial biofilms. *Annual Review of Microbiology*, 49, 711–745.

Danese, P. N., Pratt, L. A. and Kolter, R. (2001). Biofilm formation as a developmental process. *Methods in Enzymology*, 336, 19–26.

Henkin, T. M. (1996). The role of the CcpA transcriptional regulator in carbon metabolism in *Bacillus subtilis* [review]. *FEMS Microbiology Letters*, 135, 9–15.

Hudson, M. C. and Curtiss, R. I. (1990). Regulation of expression of *Streptococcus mutans* genes important to virulence. *Infection and Immunity*, 58, 464–470.

Kuramitsu, H. K. (1987). Recent advances in defining the cariogenicity of mutans streptococci: molecular genetic approaches. *European Journal of Epidemiology*, 3, 257–260.

Li, Y. and Burne, R. A. (2001). Regulation of the *gtfBC* and *ftf* genes of *Streptococcus mutans* in biofilms in response to pH and carbohydrate. *Microbiology*, 147, 2841–2848.

Li, Y. H., Chen, Y. M. and Burne, R. A. (2000). Regulation of urease gene expression by *Streptococcus salivarius* growing in biofilms. *Environmental Microbiology*, 2, 169–177.

Li, Y. H., Hanna, M. N., Svensater, G., Ellen, R. P. and Cvitkovitch, D. G. (2001). Cell density modulates acid adaptation in *Streptococcus mutans*: implications for survival in biofilms. *Journal of Bacteriology*, 183, 6875–6884.

Loo, C. Y., Corliss, D. A. and Ganeshkumar, N. (1999). *Streptococcus gordonii* biofilm formation: identification of genes that code for biofilm phenotypes. *Journal of Bacteriology*, 182, 1374–1382.

Lyon, W. R., Madden, J. C., Levin, J. C., Stein, J. L. and Caparon, M. G. (2001). Mutation of luxS affects growth and virulence factor expression in *Streptococcus pyogenes*. *Molecular Microbiology*, 42, 145–157.

O'Toole, G. A. and Kolter, R. (1998). Initiation of biofilm formation in *Pseudomonas fluorescens* WCS365 proceeds via multiple, convergent signalling pathways: a genetic analysis. *Molecular Microbiology*, 28, 449–461.

Schauder, S., Shokat, K., Surette, M. G. and Bassler, B. L. (2001). The LuxS family of bacterial autoinducers: biosynthesis of a novel quorum-sensing signal molecule. *Molecular Microbiology*, 41, 463–476.

Simpson, C. L. and Russell, R. R. B. (1998). Identification of a homolog of CcpA catabolite repressor protein in *Streptococcus mutans*. *Infection and Immunity*, 66, 2085–2092.

Wen, Z. T. and Burne, R. A. (2002). Functional genomics approach to identifying genes required for biofilm development by *Streptococcus mutans*. *Applied Environmental Biology*, 68(3), 1196–1203.

Wexler, D. L., Hudson, M. C. and Burne, R. A. (1993). *Streptococcus mutans* fructosyltransferase (*ftf*) and glucosyltransferase (*gtfBC*) operon fusion strains in continuous culture. *Infection and Immunity*, 61, 1259–1267.

Yamashita, Y., Bowen, W. H., Burne, R. A. and Kuramitsu, H. K. (1993). Role of the *Streptococcus mutans gtf* genes in caries induction in the specific-pathogen-free rat model. *Infection and Immunity*, 61, 3811–3817.

PART FOUR

BIOFILMS ON SHEDDING SURFACES

CHAPTER ELEVEN

Dissection of the Genetic Pathway Leading to Multicellular Behaviour in *Salmonella enterica* Serotype Typhimurium and Other Enterobacteriaceae

Ute Römling, Werner Bokranz, Ulrich Gerstel, Heinrich Lünsdorf, Manfred Nimtz, Wolfgang Rabsch, Helmuth Tschäpe, and Xhavit Zogaj

1 INTRODUCTION

Many environmental as well as host-associated microorganisms not only live as single independent cells, but are able to interact with each other and to build multicellular communities, whose architecture is determined by a self-produced extracellular matrix. Also, in the family Enterobacteriaceae, different types of multicellular behaviour have been identified, for example, in *Salmonella enterica* serotype Typhimurium (*S. typhimurium*) and *Escherichia coli* (Harshey and Matsuyama, 1994; Romling et al., 1998a). The rdar (red, dry and rough colony morphology) morphotype (Figure 11.1), first identified in *S. enterica* serotype Typhimurium (Romling et al., 1998a), represents a characteristic multicellular behaviour of *S. enterica* serotypes (*Salmonella* spp.) and *E. coli* in the late stationary phase of growth (Zogaj et al., 2001). Life within a community, compared with planktonic growth, is significantly different for *S. enterica* serotype Typhimurium and *E. coli* with respect to, for instance, cell density, nutrient supply, and production of architectural components (extracellular matrix formation); cell–cell communication, gene expression, and regulation are altered on various levels (Pringent-Combaret et al., 1999; Zogaj et al., 2001). An additional regulatory network, under tight control by environmental conditions, is required to coordinate the transition from the multicellular to the free-living form (Gerstel and Romling, 2001). Novel regulatory pathways involving genes with signalling domains of unknown function are part of this network (Romling et al., 2000). The impact of multicellular behaviour on the life cycle of *S. enterica* and *E. coli* has not been unambiguously proven. However, aside from its original function outside the host, a switch

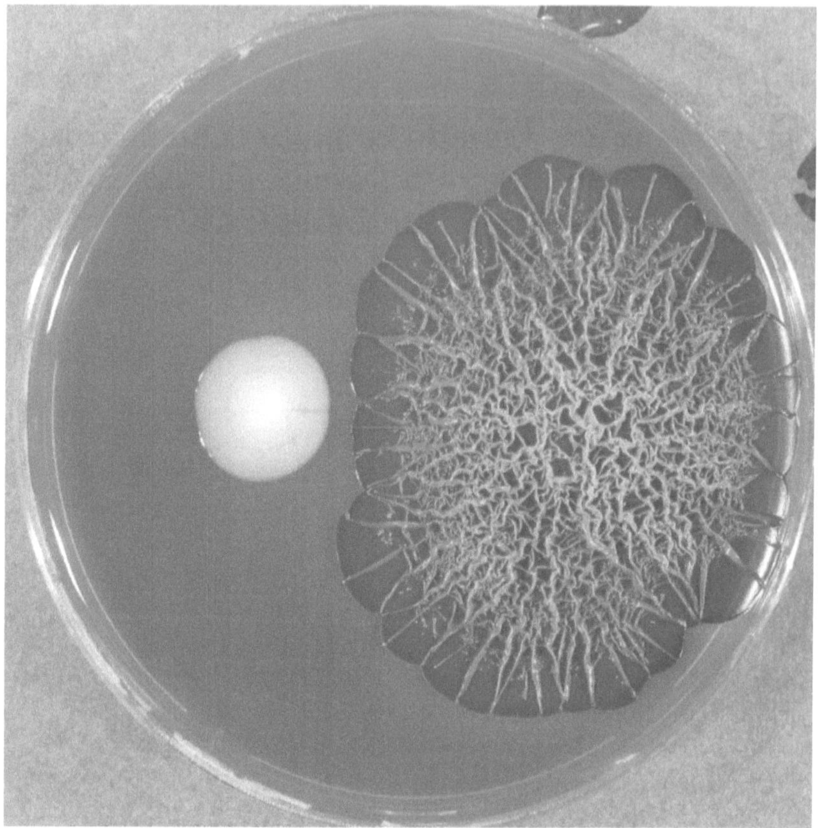

Figure 11.1: The rdar morphotype of *S. enterica* serotype Typhimurium. Left: *S. enterica* serotype Typhimurium ATCC14028 colony with regulated rdar morphotype expression. Right: *S. enterica* serotype Typhimurium ATCC 14028 colony with semi-constitutive rdar morphotype expression. The two strains are isogenic mutants and differ by the insertion of one nucleotide in the promoter region of *agfD*. Strains were grown on Luria Bertani (LB) agar medium without salt at 37°C for 4.5 days. To enhance the appearance of the rdar morphotype, Congo red was added to the agar. (See colour section.)

in functionality is proposed, which accompanies the association of these microorganisms with a vertebrate host giving multicellular behaviour a role in transmission, colonisation, and/or bacterial–host interactions.

2 REGULATION OF THE rdar MORPHOTYPE

The rdar morphotypes of *S. enterica* serotype Typhimurium and *E. coli* exhibit multicellular behaviour which is expressed as a network of spreading cells on agar plates (Figure 11.1), cell aggregation in liquid culture, pellicle formation at the air–liquid interface in standing culture, and biofilm formation on abiotic

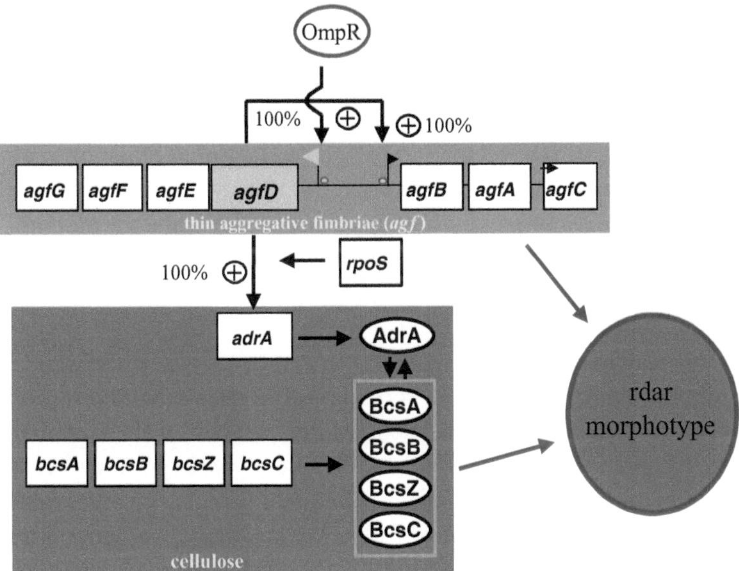

Figure 11.2: Regulatory network leading to expression of the rdar morphotype. Two functional modules, thin aggregative fimbriae (*agf*) and cellulose (*bcs*), are required for this expression. AgfD regulates the activity of both modules. *AgfD* itself requires the transcriptional regulator *ompR* for expression.

surfaces (Romling et al., 1998b; Romling and Rohde, 1999). The rdar morphotype is regulated by a complex network, which is far from being completely resolved (Figure 11.2).

2.1 Regulation by Mutations

The positive transcriptional regulator, AgfD in *S. enterica* serotype Typhimurium or CsgD in *E. coli*, plays a central role in regulation of the rdar morphotype, and its expression turns on multicellular behaviour (Romling et al., 1998b). The level of expression of multicellular behaviour (the ratio of planktonic to multicellular cells) can be modified by mutations in the *agfD* promoter (P*agfD*) region, which convert a rdar morphotype highly regulated by environmental conditions to a semi-constitutive rdar morphotype. Two commonly used virulent wild-type strains of *S. enterica* serotype Typhimurium, ATCC14028 and SR-11, display regulated multicellular behaviour and have an identical sequence in the *agfD-agfBA* intergenic region. From each strain, one semi-constitutive mutant could be independently isolated, which showed an individual point mutation in the P*agfD* region (Romling et al., 1998b). In ATCC14028, the mutation was the insertion of an A at position –17 upstream of the transcriptional start site, which alters the spacer between the –10 and –35 regions from 16 to 17 base pairs (bp). In strain SR-11, a G → T transversion occurred in the region of the OmpR binding site, but not in the highly conserved residues (Huang and Igo, 1996). Whatever the primary consequences of these mutations, possibly enhanced binding of RNA polymerase (RNAP) and OmpR to their respective sites, both mutations have the same effect; they significantly enhance the activity of P*agfD* (Gerstel and Romling, 2001). As a consequence, the expression of P*agfD* and the rdar morphotype is not restricted to ambient temperature, but also occurs at 37°C and under a variety of environmental conditions (Romling et al., 1998b; Gerstel and Romling, 2001).

While the regulated expression of P*agfD* (or P*csgD* in *E. coli*) is abolished in an *rpoS* mutant (Hammar et al., 1995; Romling et al., 1998a), expression at both semi-constitutive *agfD* promoters becomes independent of the stationary phase sigma factor, σ^S (Romling et al., 1998b). Alternative hypotheses can be postulated to explain *rpoS* independence. The single base pair changes might enhance binding of the RNAP loaded with σ^D or an alternative sigma factor to P*agfD*. As a consequence, the activator(s) encoded by an *rpoS*-dependent pathway is not necessary for transcription of P*agfD*. An alternative possibility is a sigma factor switch in which the two promoter mutations reversibly alter the binding specificity for RNAP loaded with σ^S and σ^D.

There are alternative ways to achieve *rpoS* independence of P*agfD*/P*csgD*. The transcriptional regulator OmpR binds immediately upstream of P*agfD*/P*csgD* and is required for their expression (Romling et al., 1998a; Pringent-Combaret et al., 2001). In the *E. coli* K-12 derivative MG1655, OmpR with the L43R mutation binds more strongly to its binding site, conferring an *rpoS*-independent *csgD* transcription (Pringent-Combaret et al., 2001). The mutated OmpR might induce structural changes in P*csgD*, improve binding of RNAP, or recruit an RNAP loaded with another sigma factor. The mutation in OmpR led to enhanced biofilm formation in minimal medium, but it was not reported whether temperature regulation of *csgD* expression still occurred. *RpoS*-independent transcription of *csgD* was also achieved by H-NS deficiency, but the regulation by temperature, growth phase, and osmolarity remained (Olsen et al., 1993).

Therefore, at the *agfD*/*csgD* promoter there is a labile balance of promoter recognition/transcription initiation by RNAPs loaded with different sigma factors, which leads to different levels of expression of the rdar morphotype. Obviously, the recruitment of RNAP loaded with different sigma subunits is mediated by several factors, which may contribute by various mechanisms to sigma factor selection.

2.2 Regulation by Transcriptional Regulators

Besides being one of the largest regions without coding capacity in *S. enterica* serotype Typhimurium and *E. coli*, the 521-bp intergenic region between *agfD-agfBA* (*csgD-csgBA*) is also unique with respect to its low AT content, high curvature, and other structural parameters (Pedersen et al., 2000). Although there is the potential for a whole range of regulatory interactions, binding sites for only two transcriptional regulators, OmpR and CpxR, have been identified immediately upstream of the *agfD*/*csgD* promoter. AgfD does not affect its own expression by autoregulation (Romling et al., 2000).

At the top of the signalling hierarchy stands *ompR*, the response regulator of the two-component system *ompR*/*envZ* (Liljestrom et al., 1988) which is necessary and sufficient for *agfD*/*csgD* promoter activation (Romling et al., 1998a) (Figure 11.2). OmpR is a global transcriptional regulator, the role of which has been extensively studied in the osmoregulation of two outer membrane proteins, OmpC and OmpF, in the logarithmic growth phase (Pratt et al., 1996). In *S. enterica* serotype Typhimurium, OmpR regulates stationary phase genes and is involved in pathogen–host interaction by causing macrophage late cell death through regulation of the horizontally acquired transcriptional regulator, *ssrA* (Bang et al., 2000; Lee et al., 2000). A common feature of the protein products of all genes regulated by OmpR is their extracytoplasmic location.

At the *agfD*/*csgD* promoter, phosphorylated OmpR binds to a consensus sequence located −49.5 bp upstream of the transcriptional start site and, subsequently, activates expression of the *agfDEFG*/*csgDEFG* operon (Romling et al., 1998a; Pringent-Combaret et al., 2001). Phosphorylation of OmpR *in vivo* can occur via the cognate sensor kinase EnvZ, but also via the small phosphate-donor molecules, acetyl phosphate and presumably other phosphate donors (Heyde et al., 2000). While EnvZ phosphorylates OmpR during osmoregulation of *ompC*/*ompF* in the logarithmic phase of growth, the phosphorylating agent of OmpR in the stationary phase has not been identified. EnvZ is not involved in phosphorylating OmpR when it is required for the acid-tolerance response and in regulation of the *agfD* promoter in the stationary phase (Bang et al., 2000; our unpublished results). Indirectly, it was also shown that acetyl phosphate does not contribute to the phosphorylation of OmpR in the stationary phase (Dorel et al., 1999).

The CpxA/CpxR two-component system is suggested to be involved in the negative regulation of *agfD*/*csgD* expression under certain environmental conditions (Figure 11.3). In *E. coli*, phosphorylated CpxR activates extracytoplasmic stress-induced genes such as *degP*, *dsbA*, *rotA*, and *cpxP*. (Danese and

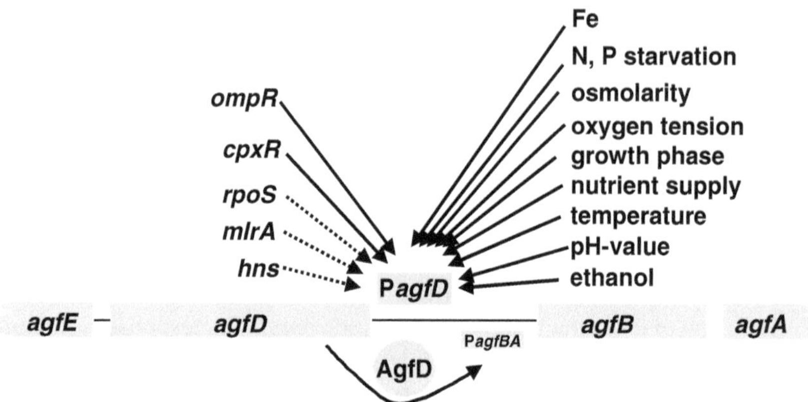

Figure 11.3: Environmental conditions and genes that influence transcription from P*agfD*. Dotted arrows indicate genes that have an effect only at rP*agfD*, but not at scP*agfD*.

Silvahy, 1997; Raivio and Silhavy, 1997) and expression of *pap* pili by favouring the phase variation switch to the ON state (Hung et al., 2001); however, motility and chemotaxis are repressed (De Wulf et al., 1999). *In vivo* experiments showed that phosphorylated CpxR represses *csgA* expression in *E. coli* (Dorel et al., 1999). This effect was observed in minimal medium, but not in rich medium, consistent with the activation profile of CpxR when it regulates other systems (Danese et al., 1995). In addition, it was proposed that the Cpx pathway is activated when the bacteria encounter high-osmolarity conditions (Pringent-Combaret et al., 2001). In *S. enterica* serotype Typhimurium and *E. coli*, the binding site for CpxR in the *agfD*/*csgD* promoter region overlaps the OmpR binding site (Pringent-Combaret et al., 2001; our unpublished results). In electrophoretic shift assays, CpxR has been shown to bind to the *csgD* promoter region; however, whether or not it competes with OmpR for binding to the DNA is an open question. Another CpxR binding site is located just downstream of the transcriptional start site of *agfBA*/*csgBA* (Pringent-Combaret et al., 2001), adding another level of control by CpxR to the expression of thin aggregative fimbriae.

Although OmpR has been shown to activate expression of both the regulated and the semi-constitutive *agfD* promoter (rP*agfD* and scP*agfD*), other regulatory proteins were shown to influence only expression of rP*agfD*/*csgD*. Therefore, yet another level of complexity is observed at rP*agfD*.

Under all environmental conditions examined, rP*agfD*/*csgD* has been shown to be completely dependent on *rpoS* (Hammar et al., 1995; Romling et al., 1998a). The first suggestion was that σ^S directly recognises P*agfD*/P*csgD*, but this idea still remains to be proven by *in vitro* experiments. The *agfD*/*csgD* promoter sequence only partially employs sequence characteristics of a σ^S-dependent promoter (Becker and Hengge-Aronis, 2001). However, it was suggested that σ^S and σ^D actually have the same consensus sequence, but that the tolerance level of σ^S is higher (Gaal et al., 2001).

Recently, the transcriptional regulator *mlrA* (formerly called *yehV*) was shown to be required for expression of the regulated rdar morphotype by targeting the *rpoS*-dependent *agfD*/*csgD* promoter (Brown et al., 2001) (Figure 11.3). MlrA is a member of the MerR family, which regulate responses to a wide variety of stresses such as exposure to toxic compounds, including oxygen radicals, and to nutrient starvation (Heldwein and Brennan, 2001). The homologous N-terminal domains of MerR family proteins contain a DNA binding helix-turn-helix motif, whereas the diverse C-terminal domains interact with small indicator molecules. The C-terminal part of MlrA is not homologous to any of the characterised MerR family members, suggesting that MlrA interacts with a novel kind of indicator molecule (Brown et al., 2001).

MlrA is transcriptionally regulated by *rpoS*. However, when *mlrA* was expressed from a plasmid in an *rpoS* background, no rdar morphotype was exhibited. Also, expression of *rpoS* in an *mlrA* background did not lead to rdar morphotype expression (Brown et al., 2001). These data indicate that, although *rpoS* is required for *mlrA* expression, *rpoS* has yet another role in the expression of the regulated rdar morphotype.

As far as has been characterised, MerR family members target promoters with a 19-bp spacer between the –10 and –35 consensus sequences. Since rP*agfD*/*csgD* has a 16-bp spacer, it would represent an atypical target for the MerR family member MlrA.

2.3 Regulation by Environmental Conditions

A wide variety of environmental conditions regulate expression of the rdar morphotype through P*agfD*/P*csgD* (Romling et al., 1998b; Gerstel and Romling, 2001) (Figure 11.3). It was originally suggested that the regulated and the semi-constitutive rdar morphotype follow different regulatory patterns, since the only condition where the regulated rdar morphotype was expressed was on agar plates composed of rich medium of low osmolarity, whereas the semi-constitutive morphotype was expressed under a variety of environmental conditions (Romling et al., 1998b). However, detailed expression studies using transcriptional fusions to the *agfD* promoter showed that expression of the regulated and semi-constitutive rdar morphotypes were similarly regulated by environmental conditions (Gerstel and Romling, 2001).

Oxygen tension is a major factor, which influences the expression of rP*agfD* and scP*agfD*. When *S. enterica* serotype Typhimurium cells were grown in rich medium, expression reached a maximum in microaerophilic conditions and decreased to 30 per cent in aerobic and anaerobic conditions. In minimal medium, oxygen tension also strongly affected expression. Maximal expression occurred in aerobic conditions, whereas it decreased 30 per cent to a minimum in microaerophilic conditions. While rP*agfD* and scP*agfD* showed the same dependency on oxygen tension in both growth media, the level of expression varied significantly. As a consequence, the rP*agfD*-regulated rdar morphotype is expressed at maximal levels only in optimal conditions of aerobic atmosphere in minimal medium and in microaerophilic conditions in rich medium. Since the rdar morphotype confers biofilm formation, surface induction was considered a possible signal for the expression of the regulated rdar morphotype. However, conditions on plates probably resemble microaerophilic conditions as the formation of huge cell aggregates would prevent oxygen diffusion and, therefore, enable rP*agfD*-mediated rdar morphotype expression.

Another environmental factor which strongly influences expression of the rdar morphotype is osmolarity (Olsen et al., 1993; Sukupolvi et al., 1997a; Romling et al., 1998b; Pringent-Combaret et al., 2001). With increasing salt concentration, the activity of P*agfD*/P*csgD* decreases (Romling et al., 1998b). The relative change in expression of rP*agfD* in response to increasing osmolarity is much bigger than that of scP*agfD* (Romling et al., 1998b; our unpublished results). When sucrose was used as an osmotic agent, no effect on *agfD*/*csgD* and rdar morphotype expression was noticed, showing that the high ion content is sensed (our unpublished results).

It has been suggested that in minimal medium the Cpx pathway is activated by high osmolarity in an *rpoS*-dependent fashion (Pringent-Combaret et al., 2001). Somewhat contradictory to this, we found that in rich medium the decrease of scP*agfD* expression in response to high osmolarity was more dramatic in an *rpoS* mutant when compared to the wild type (our unpublished results). It remains to be seen if different pathways in different growth media mediate the downregulation of PagfD expression in response to osmolarity.

Iron deficiency is another factor that regulates expression of P*agfD*. Whereas iron abundance decreases expression of P*agfD*, iron deficiency enhances expression at rP*agfD* and scP*agfD* (Romling et al., 1998b). Elevation of rP*agfD* expression at 37°C is far above the threshold value which is required for the expression of the rdar morphotype (Romling et al., 1998b).

A key step in activating the rdar morphotype is the significant enhancement of expression of P*agfD*/P*csgD* in the stationary phase of growth (Gerstel and Romling, 2001). Quorum sensing, cell-to-cell signalling by the density-dependent expression of low molecular weight substances, has been shown to be required in at least some steps of attachment and biofilm formation (Davies et al., 1998). In *S. enterica* serotype Typhimurium, quorum sensing is not a signal to trigger expression from P*agfD*. On the other hand, nutrient deficiency, but not deficiency of the energy source, is a signal for elevated expression of P*agfD* (Gerstel and Romling, 2001). Nitrogen and phosphate were identified as factors, the depletion of which triggered *agfD* expression. But the lack of additional components is sensed by signal integration, since addition of the nitrogen and phosphate source to conditioned medium only partially suppressed the elevated expression of P*agfD* (Gerstel and Romling, 2001).

Ethanol is a stress factor, which enhances P*agfD* and AgfD expression in the logarithmic growth phase. As a consequence, isolates prone to express the rdar morphotype survived treatment with 4 per cent ethanol in the logarithmic phase of growth better than the respective *agfD* mutant (Gerstel and Romling, 2001).

In conclusion, several environmental conditions but few regulators have been identified which modulate expression of both *agfD*/*csgD* promoters. The challenge of the future is to identify the whole network of regulators and to establish the signal transduction network that transforms environmental signals into molecular events, which finally regulate *agfD*/*csgD* promoter expression and, subsequently, the rdar morphotype.

3 EXTRACELLULAR MATRIX COMPONENTS

The extracellular matrix is a major determinant of multicellular behaviour. It mediates the spatial distribution of cells, which determines the three-dimensional architecture of the community of microorganisms and other features such as stability. The extracellular matrix is also responsible for anchoring cells to abiotic surfaces in biofilms and it interacts with host components.

3.1 Thin Aggregative Fimbriae

Thin aggregative fimbriae in *S. enterica* serotype Typhimurium (or curli, as the homologous fibres are called in *E. coli*) are one component of the extracellular matrix. Thin aggregative fimbriae appear as thin (2–4 nm), long, curled filaments in electron micrographs of negatively stained cells (Figure 11.4) (Olsen,

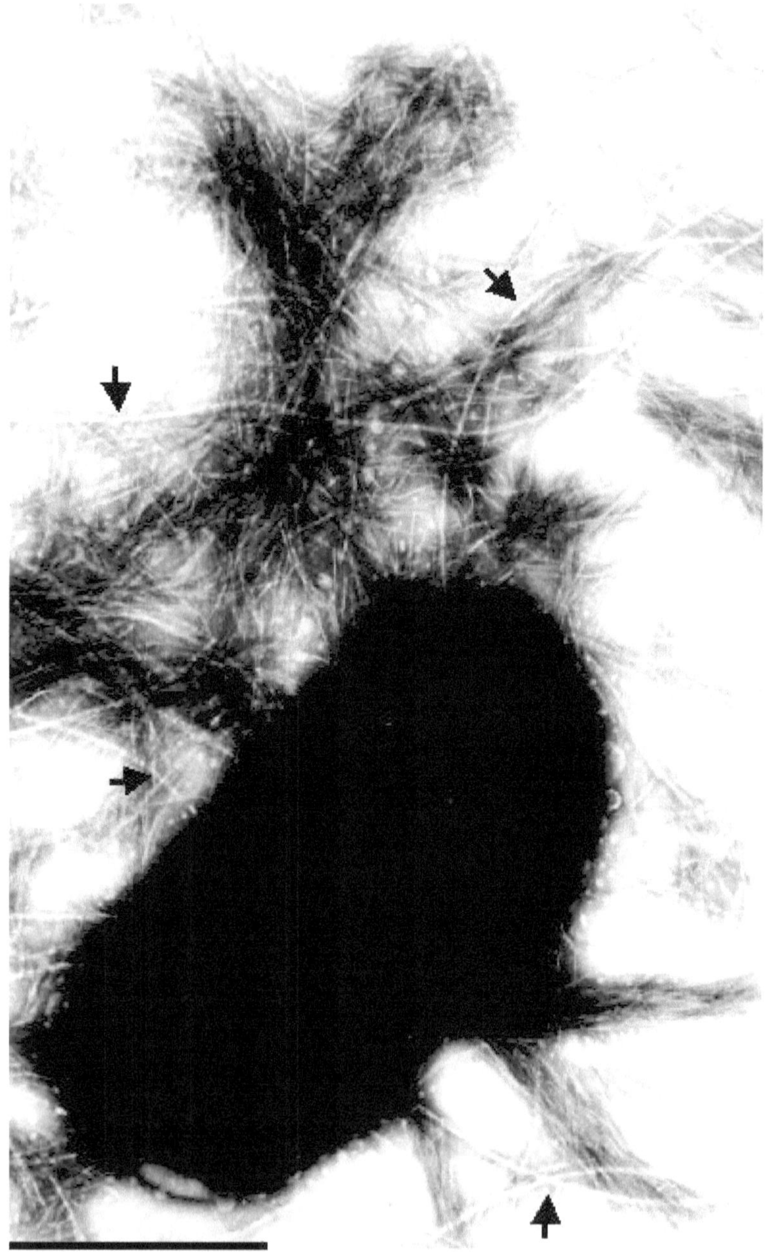

Figure 11.4: Electron micrograph of a negatively stained *S. enterica* serotype Typhimurium SR11-b derivative expressing thin aggregative fimbriae, but no cellulose. Arrows point to flagella. Bar, 1 μm.

Jonsson, and Normark, 1989). Thin aggregative fimbriae and curli are highly homologous at the biochemical, serological, genetic, regulatory, and functional levels (Olsen et al., 1989; Collinson et al., 1992, 1993; Austin et al., 1998; Romling et al., 1998a,b; Vidal et al., 1998). To follow the conventions for gene designation, these fibres should bear the same name in both species.

Thin aggregative fimbriae/curli are encoded by the two divergently transcribed operons, *agfBAC*/*csgBAC* and *agfDEFG*/*csgDEFG*. *AgfA*/*csgA* encodes the fibre subunit, whereas AgfB/CsgB is considered to be anchored at the cell surface, where it acts as a nucleator for the polymerisation of AgfA/CsgA (Bian and Normark, 1997). In addition, AgfB/CsgB was found in the fibre in minor amounts (more than 20 AgfA per AgfB protein) (Bian and Normark, 1997; White et al., 2001). The particular roles of *agfF*/*csgF* and *agfE*/*csgE* are not clear, but for both genes phenotypes exist with respect to expression and polymerisation of thin aggregative fimbriae. Deletion of *csgF*, like that of *csgB*, led to secreted CsgA monomers (Hammar, Bian, and Normark, 1996; Chapman et al., 2002). Deletion of *csgE* reduced the amount of CsgA significantly, but did not affect polymerisation (Chapman et al., 2002). The lipoprotein agfG/CsgG localises to the inner leaflet of the outer membrane and may serve as an assembly platform for thin aggregative fimbriae (Loferer, Hammar, and Normark, 1997).

Regulation of the biosynthesis of thin aggregative fimbriae/curli occurs via *agfD*/*csgD*, which is considered to directly activate expression at the *agfBA*(*C*) promoter. AgfD, a UphA family member of transcriptional regulators, has the conserved aspartate at position 59, whose phosphorylation by a cognate histidine kinase is considered to be essential for activity (Volz, 1993). However, the mechanism of activation of *agfD*/*csgD* needs to be investigated.

Expression of thin aggregative fimbriae is regulated by *rpoS* via *agfD* in the regulated rdar morphotype (Romling et al., 1998a). Since expression of *agfD* is independent of *rpoS* in the semi-constitutive rdar morphotype, expression of thin aggregative fimbriae is also not dependent on *rpoS* (Romling et al., 1998b). The only difference between the two regulatory morphotypes is the point mutation in the *agfD* promoter. Therefore, *rpoS* is not required for transcription of the *agfBA*(*C*) promoter in both rdar morphotypes. Also OmpR, the other global regulator of rdar morphotype expression, acts only at the *agfD* promoter level and is not required for *agfBA*(*C*) promoter expression (Pringent-Combaret et al., 2001).

A major property of thin aggregative fimbriae is their high stability; treatment with more than 90 per cent formic acid is required to disassemble polymerised AgfA and AgfB into monomers (Collinson et al., 1991; White et al.,

2001). A second important characteristic is the unique capability of polymerised AgfA to bind to a variety of biological macromolecules, such as tissue matrix proteins fibronectin, laminin, several types of collagens, and MHC-I molecules (Collinson et al., 1993; Olsen et al., 1998). Plasminogen is absorbed by thin aggregative fimbriae/curli and subsequently cleaved by simultaneously captured tissue-type plasminogen activator (t-PA) to plasmin, a fibrinolytic compound (Sjobring, Pohl, and Olsen, 1994). Although it has been suggested that the binding capabilities of thin aggregative fimbriae lead to enhanced adhesion and colonisation properties of strains expressing those surface structures, this effect still awaits to be proven in *in vivo* experiments. Thin aggregative fimbriae/curli also bind contact phase proteins, especially H-kininogen, and, subsequently, release the proinflammatory and vasoactive peptide bradykinin (Ben Nasr et al., 1996) which may contribute to the symptoms of sepsis and septic shock (Herwald et al., 1998). Indeed, more than half of *E. coli* isolates from sepsis patients expressed curli at 37°C (Bian et al., 2000). Curli have also been shown to contribute to reduced blood pressure, another symptom of septic shock, by the induction of nitric oxide (NO) which has been demonstrated *in vivo* and *in vitro* (Bian et al., 2001).

Due to the almost universal binding characteristics, cells expressing thin aggregative fimbriae/curli show enhanced interaction with a variety of cell types (e.g., macrophages and epithelial cells) of different organisms (Sukupolvi et al., 1997b; Dibb-Fuller et al., 1999; La Ragione, Cooley, and Woodward, 2000a; Gophna et al., 2001; Johansson et al., 2001). Through this cell association, thin aggregative fimbriae/curli mediate or enhance invasion (Dibb-Fuller et al., 1999; Gophna et al., 2001).

Besides interacting with the host, thin aggregative fimbriae may also play a role outside the host. It was shown independently by several groups that thin aggregative fimbriae/curli confer binding to a range of hydrophobic and hydrophilic abiotic surfaces such as stainless steel, glass, polystyrene, and Teflon, thereby enabling biofilm formation (Austin et al., 1998; Romling et al., 1998b; Vidal et al., 1998). Thin aggregative fimbriae/curli mediate binding to the substrate and establish interactions among the cells in the microcolonies (Austin et al., 1998; Romling and Rohde, 1999; Pringent-Combaret et al., 2000).

Thin aggregative fimbriae/curli also bind to the exopolysaccharide cellulose, which is co-expressed with the fibres (Zogaj et al., 2001). The interaction is apparent by the rigid matrix network around the cells and can be visualised by fluorescence microscopy (Figure 11.5). The cellulose fibres are tightly wrapped around cells when co-expressed with thin aggregative fimbriae, in

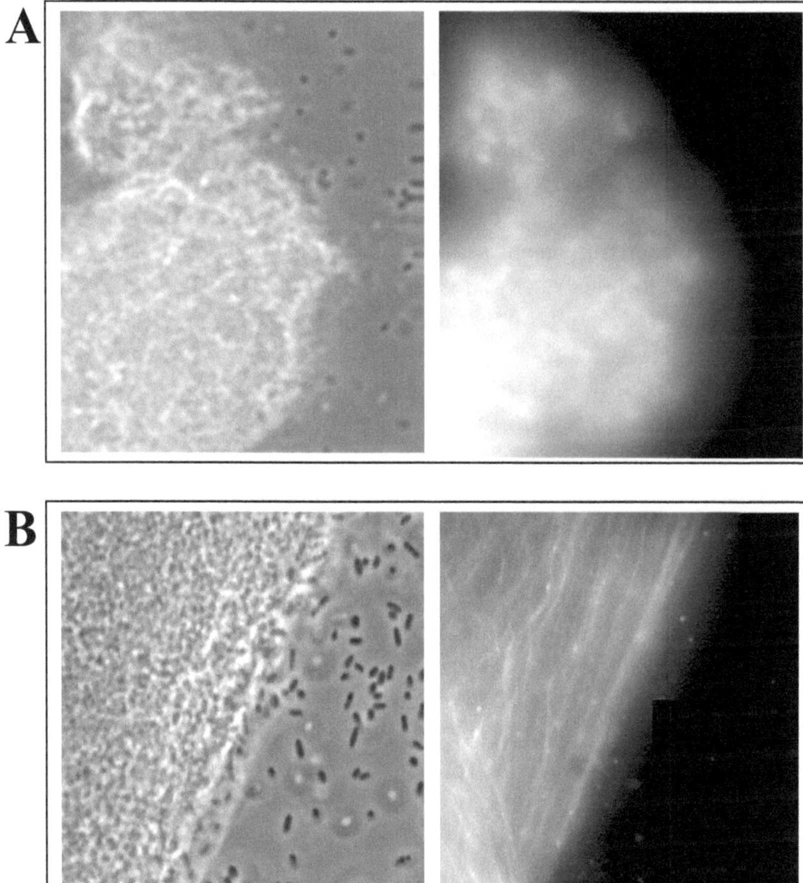

Figure 11.5: Interaction of thin aggregative fimbriae with cellulose as visualised by fluorescence microscopy. (A) Arrangement of cellulose fibres when co-expressed with thin aggregative fimbriae. Cellulose fibres are tightly wrapped around the cells. (B) Cellulose fibres, but not thin aggregative fimbriae, are expressed. The cellulose fibres run in long bands and are only loosely connected with the cell. Magnification × 400. (See colour section.)

contrast to the free-floating cellulose fibres. The strength of binding cannot be assessed at the moment, since, as the two substances are polymers, even a weak interaction might have a substantial effect. However, preliminary data suggest that by binding to cellulose, the binding of thin aggregative fimbriae to other macromolecules such as fibronectin is reduced (our unpublished data).

3.2 Cellulose

Cellulose is the second major component of the extracellular matrix of the rdar morphotype. A search specific for a second matrix component was performed, since it was noticed that after the deletion of thin aggregative fimbriae, *S. enterica* serotype Typhimurium cells still showed a specific adherence pattern to abiotic surfaces and produced an extracellular component which looked like chewing gum in scanning electron micrographs (Romling et al., 2000). Transposon mutagenesis identified several independent insertions in two adjacently located genes of unknown function (Zogaj et al., 2001). Those genes belonged to an operon that was renamed *bcsABZC* (bacterial cellulose synthesis, Figure 11.2). The genes of the *bcsABZC* operon have high similarity to the genes in the *bcsABCD* operon of *Gluconacetobacter xylinus* [formerly called *Acetobacter xylinus(m)*], which encodes for cellulose biosynthesis (Wong et al., 1990). Cellulose biosynthesis is well established in *G. xylinus*, which has long been the model organism for cellulose biosynthesis.

The highest similarity is displayed by BcsA, the catalytic subunit of the cellulose synthase, encoded by the first gene in the two operons. The central region of this protein has an amino acid similarity as high as 72 per cent and contains several conserved motifs, among them the D_3D_2D35QRXRWA motif. The D,D,D35Q(R,Q)XRW motif is characteristic for processive β-glycosyltransferases, whereas its D_3D_2D35QRXRWA submotif is found exclusively in bacterial cellulose synthases (Romling, 2002). The second gene in the cluster, *bcsB*, encodes for the regulatory subunit of cellulose synthase. The third gene, *bcsZ*, encoding an endo-1,4-β-glucanase (cellulase family D), was not encoded by the cellulose biosynthesis operon in *G. xylinus*. However, it is known that a cellulase adjacent to the *bcs* operon is needed for cellulose biosynthesis in *G. xylinus*. The protein product encoded by the fourth gene, *bcsC*, is the least conserved, but still shows a 36 per cent similarity over the entire length of the protein.

Where does the cellulose biosynthesis operon come from? This operon was already present in a common ancestor of *S. enterica* and *E. coli* 100 million years ago, since all the strains sequenced from those two species harbour the cellulose biosynthesis operon. Although the average GC content of *S. enterica* serotype Typhimurium is 53 per cent, the GC content of the *bcs* operon is 58 per cent, suggesting horizontal transfer from a species with higher GC content to a common enterobacterial ancestor. Indeed, the plant symbiont *Pseudomonas putida* KT2440 has the *bcsABZC* operon, which shows the highest homology on the protein as well as the nucleotide level. The BcsA protein is most conserved, but its homology reflects the similarity of the whole operon (Figure 11.6).

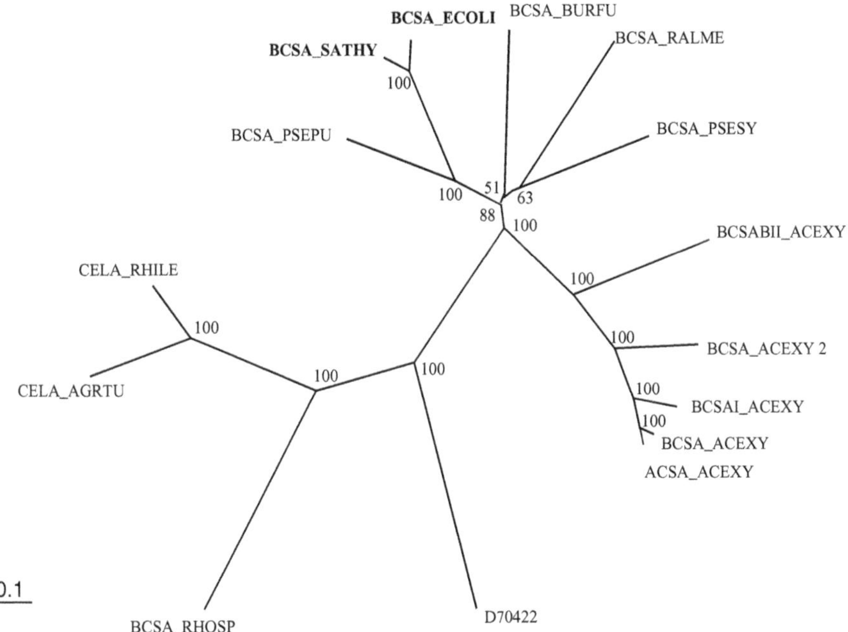

Figure 11.6: Phylogeny of bacterial cellulose synthases. Cellulose synthases from different bacterial species were aligned using PileUp (GCG package version 9, University of Wisconsin). The tree was created using the neighbourhood joining method and subjected to 10,000 bootstrap trials. The bootstrap value is shown as a percentage at the respective node of the tree. Protein sequences used: BCSA_ACEXY (BAA31463.1); ACSA_ACEXY (P19449); BCSAI_ACEXY (BAA77585.1); BCSA_ACEXY_2 (P21877); BCSABII_ACEXY (BAA77593.1); BCSA_SATHY (CAC44015.1); BCSA_ECOLI (P37653); CELA_AGRTU (NP_357298.1); CELA_RHILE (AAC41436.1). BCSA_PSEPU (*P. putida*) and BCSA_PSESY (*P. syringae*) are according to sequence data from http://www.tigr.org. BCSA_BURFU (*Burkholderia fungorum* LB400), BCSA_RALME (*Ralstonia metallidurans* CH34), and BCSA_RHOSP (*Rhodobacter sphaeroides* 2.4.1) are according to sequence data from www.jgi.doe.gov/JGI_microbial/html/.

Recently, a second operon, *bcsEFG*, adjacent to *bcsABZC*, was found to be required for cellulose synthesis (Solano et al., 2002). This operon encodes for a putative protease, a protein of unknown function and a second endo-1,4-β-glucanase gene of cellulase family D. The role of these genes in cellulose biosynthesis is unknown.

The production of cellulose by *Salmonella* spp. and *E. coli* strains was verified by chemical analysis (Zogaj et al., 2001). Due to the ordered alignment of the (1 → 4)-β-glucan chains to a crystalline tertiary structure, cellulose is a water insoluble and highly inert molecule, which requires specific isolation conditions. Whereas several (1 → 4)-glucans withstand hot alkaline solutions,

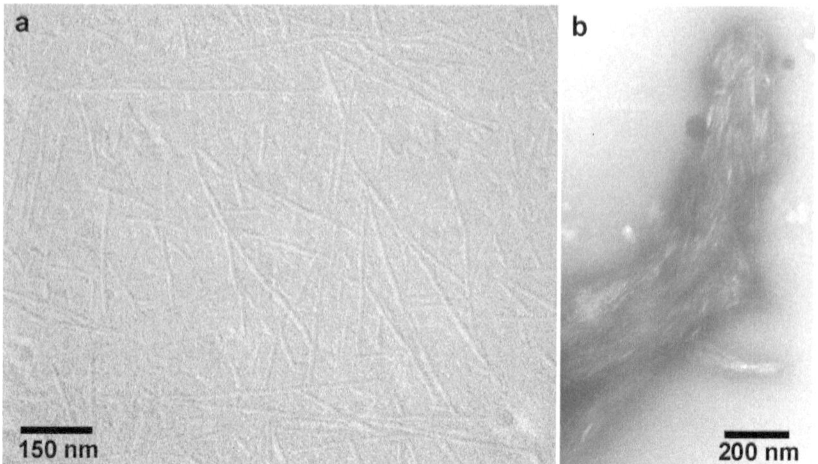

Figure 11.7: Electron microscopic images of cellulose isolated from *S. enterica* serotype Typhimurium MAE97 by different methods. (A) Shadowing of cellulose isolated by treatment of bacterial colonies with 1 *M* NaOH at 95°C. (B) Negative staining of cellulose isolated by treatment of bacterial colonies with the Updegraff method (58% acetic acid, 19% nitric acid) at 95°C.

crystalline cellulose is the only polysaccharide known to survive hot concentrated acids. After treatment of bacterial cells with the Updegraff reagent (Updegraff, 1969), a mixture of concentrated acetic and nitric acids, only crystalline cellulose is left over (Figure 11.7). The identity of the macromolecule was confirmed by cleavage of crystalline cellulose into the monomeric subunits that are exclusively glucose, the derivative of which can be detected by coupled gas chromatography/mass spectrometry (GC/MS) (Figure 11.8). Verification of the $(1 \rightarrow 4)$-glucan bond in the cellulose molecule was done by standard methods for the linkage analysis of polysaccharides (Figure 11.9).

Cellulose has a role in biofilm formation and in cell–cell interaction by providing elastic long-range cell bonds (Zogaj et al., 2001; Solano et al., 2002). In addition, cellulose protects cells from the bactericidal activity of chlorine (Solano et al., 2002).

3.3 Regulation of Cellulose Biosynthesis

AgfD regulates the rdar morphotype and, therefore, the biosynthesis of cellulose (Romling et al., 2000). However, regulation is not direct, but occurs via AdrA (*agfD* regulated). AdrA is a 371-amino-acid long protein consisting of two domains. The N-terminal domain contains four transmembrane helices, but has no homology to sequences in the database. The C-terminal part of

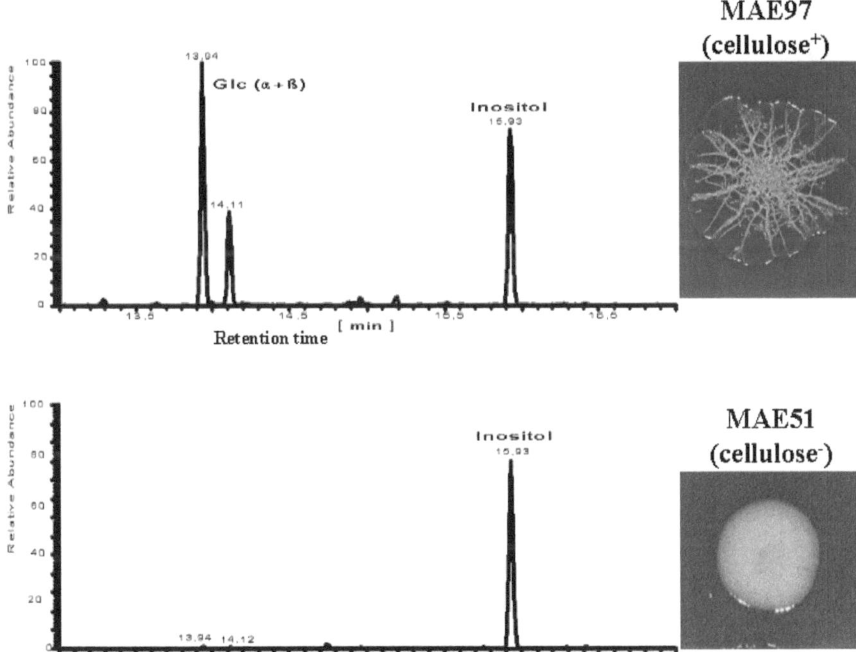

Figure 11.8: Detection of glucose monomers after isolation of crystalline cellulose. *S. enterica* serotype Typhimurium strain MAE97 expresses cellulose (upper right). When grown on LB medium without salt and containing Congo red, strain MAE97 absorbs the dye and expresses the pdar (pink, dry and rough) morphotype indicative of cellulose expression. After isolation of crystalline cellulose, the macromolecule was totally hydrolysed with 4 *N* Tri-fluoro-acetic acid at 100°C for 72 hours and pertrimethylsilylated. The glucose derivative was detected by GC/MS. MAE51, the *agfD* mutant, is cellulose negative. MAE51 remained white on CR medium (lower right); no sugar was detected by GC/MS. (See colour section.)

the protein encodes for a GGDEF domain (also called DUF1: domain of unknown function). The GGDEF domain was initially identified in the response regulator PleD that controls cell differentiation in the *Caulobacter crescentus* swarmer-to-stalked cell transition (Hecht and Newton, 1995). GGDEF domains are found in genomes of free-living bacteria and facultative pathogens, often in several copies (up to thirty-three in *P. aeruginosa*). *S. enterica* serotype Typhimurium has twelve proteins with this domain. A phenotype for the gene encoding the protein with the GGDEF domain is only known in a few cases (Wong et al., 1990; Hecht and Newton, 1995; Ausmees et al., 1999). Interestingly, several of them are involved in regulation of cellulose biosynthesis in various organisms, despite significantly different domain combinations

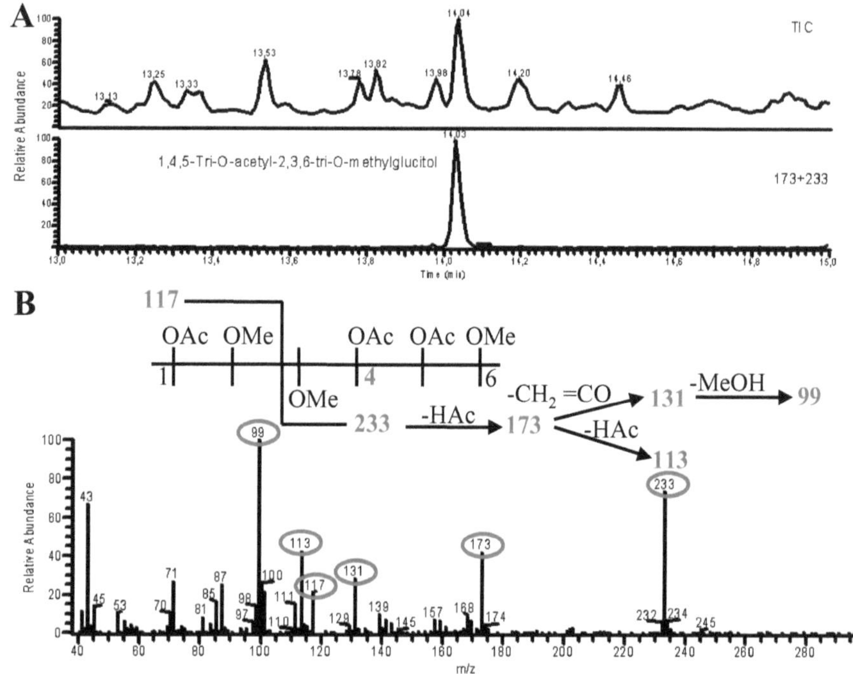

Figure 11.9: Methylation analysis of (1 → 4)-β-glucan cellulose isolated from *S. enterica* serotype Typhimurium MAE97. Cellulose was permethylated and subsequently hydrolysed. The sugar monomers were reduced, acetylated, and analysed by GC/MS. (A) The chromatogram of the GC/MS analysis indicated the presence of 1,4,5-Tri-*O*-acetyl-2,3,6-tri-*O*-methylglucitol, characteristic of 4-substituted glucose eluting at the expected retention time. Upper panel: total ion trace. Lower panel: trace of the sum of two characteristic fragment ions. No additional partially methylated alditol acetates were detected. (B) The mass spectrum confirmed the specific glucose derivative 1,4,5-Tri-*O*-acetyl-2,3,6-tri-*O*-methylglucitol.

(Wong et al., 1990; Ausmees et al., 1999; Zogaj et al., 2001). The function of the GGFED domain has not been experimentally proven, but sequence similarity suggests it confers nucleotide cyclisation activity (Pei and Grishin, 2001).

AdrA is positively regulated by AgfD on the transcriptional level under all growth conditions examined, such as rich medium, minimal medium, iron deficiency, and anaerobicity. The transcription rate of genes encoded by the *bcsABZC* operon, however, was not regulated by AgfD and, consequently, AdrA (Zogaj et al., 2001). Expression of the *bcsABZC* operon occurred under a variety of environmental conditions, including high osmolarity in which AgfD and the rdar morphotypes were not expressed. Activation of cellulose biosynthesis by AdrA could occur through direct interaction of AdrA with the Bcs proteins or, alternatively, and similar to the situation in *G. xylinus*, by the synthesis of the novel second messenger c-di-GMP (Romling, 2002).

No other AgfD-regulated genes are required for cellulose biosynthesis, since AdrA expressed constitutively from a plasmid is sufficient for cellulose biosynthesis in a Δ*agfD* strain (Zogaj et al., 2001). It has to be stated that, under conditions of sufficient carbon source, but otherwise strict nutrient deficiency, the biosynthesis of cellulose becomes independent of AgfD (Solano et al., 2002). Whether expression of *adrA* is independent of AgfD or whether cellulose biosynthesis is independent of AdrA under those conditions remains to be determined.

Cellulose biosynthesis is regulated by *rpoS* on several levels. In the regulated rdar morphotype, regulation of cellulose biosynthesis by *rpoS* occurs via regulation of *agfD* transcription (Romling et al., 1998a). In the semi-constitutive rdar morphotype, *rpoS* regulates the expression of *adrA* and, therefore, cellulose biosynthesis (Romling et al., 1998a, 2000).

4 OTHER FACTORS INVOLVED IN rdar MORPHOTYPE EXPRESSION AND BIOFILM FORMATION

In various organisms, flagella have been shown to contribute to biofilm formation. Flagella are proposed to bring the organism in close proximity to the surface and, through the motive force, override repelling mechanisms.

The appearance and spreading of the rdar morphotype on plates was not influenced by the expression of flagella (Romling and Rohde, 1999). Flagella are expressed by rdar morphotype colonies, but in amounts less than in the non-spreading standard colonies comprised of planktonic cells (Figure 11.1). In a steady state system of biofilm formation, flagella did not play a role in the quantity of biofilm formation. However, adherent cells tended to gather at the bottom of the tube, and pellicle formation at the air–liquid interface was delayed. Therefore, flagella play a role in the spatial and temporal setting of the biofilm, but not in its quantity. On the contrary, a mutation in *fliS*, which causes overproduction of flagellar structures, abolished pellicle formation and adherence (Solano et al., 2002). Timely regulation of flagella expression might be crucial for the development of biofilms. Indeed, in *E. coli* and other organisms, flagella and exopolysaccharide synthesis are inversely expressed in the biofilm (Garrett, Perlegas, and Wozniak, 1999; Pringent-Combaret et al., 1999, 2000). According to genetic and phenotypic analysis, the major extracellular matrix components of the rdar morphotype seem to be thin aggregative fimbriae and cellulose. Other fimbriae encoded by *S. enterica* serotype Typhimurium,

Type 1 fimbriae (*fim*), long polar fimbriae (*lpf*), and plasmid-encoded fimbriae (*pef*) did not influence the rdar colony morphology (our unpublished results). However, more components are involved in rdar colony morphology and biofilm formation. Colanic acid, another exopolysaccharide produced by *Salmonella* spp. and *E. coli*, contributes to the colony morphology of the rdar morphotype and biofilm formation in rich medium as determined in *S. enterica* serotype Enteritidis (*S. enteritidis*) (Solano et al., 2002). In *E. coli*, colanic acid has been shown to contribute to biofilm architecture, but does not influence the amount of cell adherence in minimal medium (Pringent-Combaret et al., 2000). Other surface polysaccharides such as lipopolysaccharide (LPS) and enterobacterial common antigen (ECA) are not matrix components, but they altered the colony morphology of the rdar morphotype and abolished biofilm formation (Solano et al., 2002).

5 EPIDEMIOLOGY OF rdar MORPHOTYPE EXPRESSION IN ENTEROBACTERIACEAE

Functionality studies suggested that both matrix components, thin aggregative fimbriae and cellulose, may have a function outside the host, but several roles for thin aggregative fimbriae in pathogen–host interactions have also been proposed. Another approach to elucidate the impact of the rdar morphotype in the life style of Enterobacteriaceae is epidemiological analysis of isolates from different habitats and diseases. Strictly speaking, a detailed molecular analysis of the expression of both extracellular matrix components has not been carried out. Binding of Congo Red (CR) dye and a wrinkled colonial morphology on agar media are strongly indicative of expression of the rdar morphotype. Therefore, for the evaluation of epidemiological data in this chapter, this colonial morphology will be considered as a marker for the expression of thin aggregative fimbriae and cellulose.

Early findings reported that CR binding at 28°C in *E. coli* was strictly associated with colisepticaemia, an invasive disease in poultry which starts with an upper respiratory tract infection caused by inhaled contaminated dust particles and progresses by infiltration of blood and internal organs. *E. coli* isolates from healthy birds and from the environment were variable in CR-binding capabilities (Berkhoff and Vinal, 1986). These results were confirmed with a collection of cloacae isolates from psittacine birds (Styles and Flammer, 1991), but not by Maurer et al. (1998), who found that 25 per cent of disease associated *E. coli* compared with 82 per cent of commensal *E. coli* from poultry showed CR-binding capabilities. Isolates from other animal infections have not been systematically investigated, but apart from bovine mastitis (55 per cent of isolates CR positive), no CR-positive strains have been reported (Olsen et al., 1989; Uhlich, Keen, and Elder, 2001).

In human disease, 55 per cent of *E. coli* isolates from urinary tract infections (UTIs) were positive for curli production at 28°C (Patri et al., 2000). The percentage of curli positive strains was confirmed in other investigations (our unpublished results). In intestinal infections, the majority of enterotoxigenic *E. coli* (ETEC) isolates, but not enteroinvasive or enteropathogenic *E. coli* (EIEC and EPEC, respectively), bound fibronectin, a strong indicator for the expression of curli (Olsen et al., 1993; Ben Nasr et al., 1996). The majority (70 per cent) of sepsis isolates was found to express curli at various levels (Bian et al., 2000). We have found that a similar percentage of commensal isolates expressed curli (our unpublished results). Co-expression of cellulose together with curli that results in formation of the rdar morphotype is not generally a trait of commensal isolates. Anecdotally, occurrence of curli expression has also been reported for *E. coli* strains causing diarrhoea and in enterohaemorrhagic *E. coli* (EHEC) (Collinson et al., 1992; Uhlich et al., 2001).

Taken together, the epidemiological data indicate no particular discrimination between disease and commensal isolates with respect to the expression of curli at 28°C or the rdar morphotype. Expression of thin aggregative fimbriae and cellulose at 37°C has only recently been investigated, so no conclusion about the association of the expression of the rdar morphotype with disease can be drawn. However, EPEC and EIEC isolates from gastrointestinal diseases (i.e., isolates showing a phenotype associated with epithelial invasion or destruction) did not express curli and presumably cellulose at 28°C (Olsen et al., 1993; Ben Nasr et al., 1996). Similarly, *Shigella* spp., which share the same pathogenic niche with EIEC, lacked expression of curli (Sakellaris et al., 2000). Loss of the expression of curli resulted mostly from insertions and deletions in the *csg* locus in all four *Shigella* spp.

Epidemiological data relevant to expression of the rdar morphotype have also been collected in *Salmonella* spp. The principal habitat of *Salmonella* is the intestinal tract of man and animals, where these microorganisms can live as commensals, colonise the host asymptomatically, or cause disease. Transmission almost always occurs by the oral route via contaminated food or water.

Over 800 *S. enterica* serotype Typhimurium and *S. enteritidis* strains received in the German National Reference Centre over a period of 2 months were screened for the rdar morphotype on CR plates. More than 90 per cent of strains expressed the morphotype (our unpublished data). Expression of solely thin aggregative fimbriae or cellulose was not generally observed as

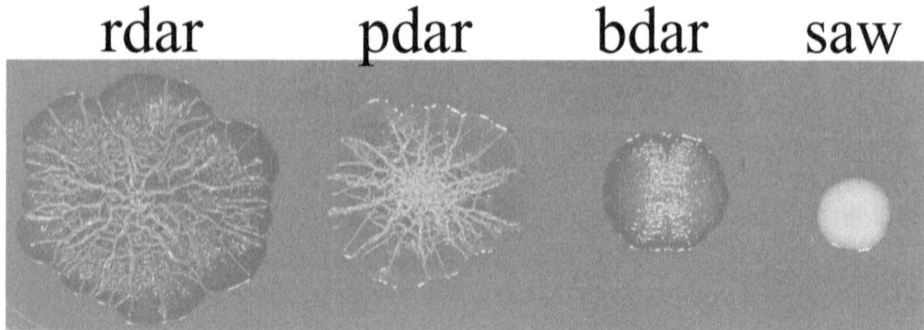

Figure 11.10: Morphotypes of *S. enterica* serotype Typhimurium expressing different matrix components. Rdar: thin aggregative fimbriae and cellulose are expressed; pdar: expression of cellulose; bdar: expression of thin aggregative fimbriae; saw: no expression of matrix components. For a pronounced visualisation, an ATCC14028 derivative with semi-constitutive expression of the rdar morphotype and respective mutants were used. Cells were incubated on LB plates without salt supplemented with Congo red dye at 37°C. (See colour section.)

determined by the appearance of characteristic bdar (brown, dry and rough) and pdar (pink, dry and rough) morphotypes, respectively (Figure 11.10). In the majority of strains, the rdar morphotype was only weakly expressed when examined after 2 days of growth at 28°C (Figure 11.11). However, strains from serovars with a restricted host range, *S. enterica* serotype Typhimurium var. copenhagen, DT2 and DT99 (Rabsch et al., 2002) isolated from pigeons, *S. enterica* serotype Choleraesuis (*S. choleraesuis*), and *S. enterica* serotype Typhi (*S. typhi*) consistently did not express the rdar morphotype. Those host-adapted serotypes cause invasive disease with septicaemia in their hosts. As in *E. coli*, invasive disease via the gastrointestinal tract is correlated with the loss of the rdar morphotype. However, the host-adapted serotypes display more features of niche adaptation, as they are auxotrophic and have lost immunogenic components such as flagella. Such visible host-adapted loss-of-function phenotypes are the tip of the iceberg, as sequencing of the *S. enterica* serotype Typhi genome has recently demonstrated more than 200 pseudogenes. Six of the twelve fimbrial operons assembled by the chaperone-usher pathway contained internal stop codons (Townsend et al., 2001). When the sequences of the *S. enterica* serotype Typhi *agfDEFG* and *agfBAC* operons, required for the biogenesis of thin aggregative fimbriae, were compared to the sequences of the rdar morphotype proficient *S. enterica* serotype Typhimurium ATCC14028, *agfD* from *S. enterica* serotype Typhi was found to harbour a stop codon that leads to a protein shorter by eight amino acids. The *bcsC* gene of *S. typhi* contains several stop codons early in the gene and a frame shift.

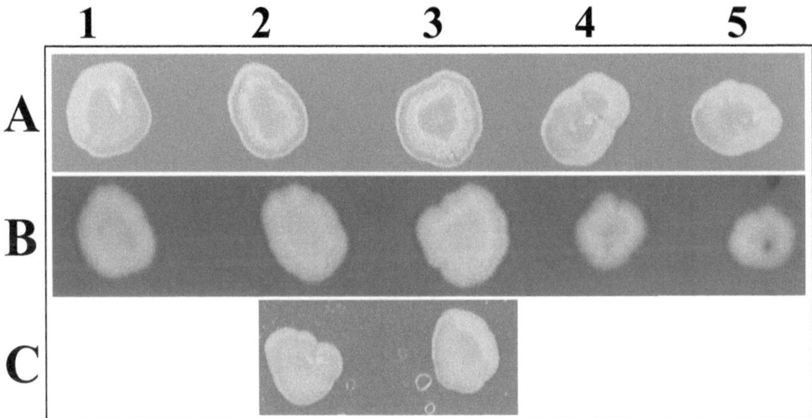

Figure 11.11: The most common regulatory pattern of *S. enterica* serotype Typhimurium/ Enteritidis strains expressing the rdar morphotype. (A) Shown are the rdar morphotype of *S. enterica* serotype Typhimurium strains incubated on LB agar plates without salt supplemented with Congo red dye at 28°C for 48 hours. The rdar morphotype is expressed weakly. The strains are 1, *S. enterica* serotype Enteritidis PT21/1; 2, *S. enterica* serotype Enteritidis PT4/6; 3, *S. enterica* serotype Enteritidis PT4/6; 4, *S. enterica* serotype Typhimurium DT 104; and 5, *S. enterica* serotype Typhimurium DT 104. (B) Shown are the same strains incubated on LB plates without salt substituted with Calcofluor at 28°C for 48 hours. Expression of cellulose is indicated by the binding of Calcofluor to the colonies, resulting in white fluorescence. (C) The strains do not show expression of the rdar morphotype when incubated on CR plates at 37°C for 24 hours. (See colour section.)

When more than 200 environmental, food, and disease (animal and human) isolates of *S. enteritidis* were screened, over 70 per cent of the disease isolates expressed the rdar morphotype, whereas only 50 per cent of environmental isolates expressed it on CR plates at 28°C (Solano et al., 2002). However, the actual capability of isolates to express cellulose was almost 90 per cent in a nutrient-deficient but carbon-source-rich medium (ATM medium: 60 mM NaCl, 30 mM NaHCO3, 20 mM KCl, 111 mM glucose at pH 8.4). Obviously, certain regulatory mechanisms are inoperative under specific environmental conditions. Thin aggregative fimbriae were not expressed in the ATM medium under those conditions.

An incubation condition that triggers expression of thin aggregative fimbriae is long-term incubation. When forty-three *Salmonella* serotypes were examined for the expression of thin aggregative fimbriae, a significant number expressed the fibres after incubation at 37°C for 5 days. The expression level varied within a serotype, but seventeen of twenty-two *S. enterica* serotype Enteritidis and six of twelve *S. enterica* serotype Typhimurium isolates showed expression of thin aggregative fimbriae (Doran et al., 1993). This finding is complemented by data showing that transcription of the *agfD* promoter plateaud only after 24 hours of growth and continued to increase up to the endpoint of measurement at 72 hours (our unpublished data).

In conclusion, the majority of the *S. enterica* serotype Typhimurium and *S. enterica* serotype Enteritidis isolates express the rdar morphotype under standard conditions at 28°C. However, the actual capacity of strains to express the rdar morphotype may be underestimated, since certain environmental conditions can override regulatory mechanisms.

6 THE rdar MORPHOTYPE IN PATHOGEN–HOST INTERACTION

The occurrence of the expression of the rdar morphotype or one of its matrix components in pathogenic *E. coli* and *Salmonella* spp. isolates of various origins suggests that the rdar morphotype has a role in the disease process. Indeed, various studies reported that thin aggregative fimbriae mediate invasion of epithelial cell lines by both *S. enterica* serotype Enteritidis and *E. coli* (Dibb-Fuller et al., 1999; Gophna et al., 2001; Uhlich, Keen, and Elder, 2002). However, there is also a contradictory report (Rajashekara et al., 2000). Cellulose had no effect on adherence or invasion of epithelial cells (Solano et al., 2002).

Studies on the virulence of *Salmonella* spp. and *E. coli* capable of producing extracellular matrix components gave variable results between the two species. *S. enterica* serotype Typhimurium causes an infection in mice resembling typhoid fever in humans. We showed that after oral inoculation, a strain with semi-constitutive expression of the rdar morphotype was slightly less virulent than its isogenic mutant that did not express the rdar morphotype (Romling et al., 2000). Mice also died later when the semi-constitutive rdar morphotype was expressed upon inoculation (our unpublished results). However, expression of the regulated rdar morphotype had no effect on virulence in comparison with a matrix-deficient mutant (Romling et al., 2000). Solano et al. (2002) found expression of cellulose had no effect on virulence of *S. enterica* serotype Enteritidis in mice.

Consistently, thin aggregative fimbriae or cellulose had no effect on the virulence of *S. enteritidis* in a 1-day-old chicken model (Allen-Vercoe, Sayers, and Woodward, 1999; Rajashekara et al., 2000; Solano et al., 2002). Thin aggregative fimbriae also had no effect on caecal colonisation and persistence of *S. enteritidis* in chickens (Allen-Vercoe and Woodward, 1999; Rajashekara et al., 2000). However, synergistic effects on virulence and persistence by the knock-out of several fimbrial genes have been reported (van der Velden et al., 1998; La Ragione et al., 2000b).

Another effect of the expression of thin aggregative fimbriae or the rdar morphotype was found in *E. coli.* In a 1-day-old chicken model, invasion of organs and persistence of bacteria in the cloacae was significantly diminished upon the knock-out of curli in an avian pathogenic strain (La Ragione, Sayers, and Woodword, 2000b). The virulence of EHEC isolates that expressed the rdar morphotype semi-constitutively was significantly higher when orally inoculated into streptomycin-treated mice (Uhlich et al., 2002).

Taking the virulence studies together, there is no consistent evidence concerning the role of the rdar morphotype in virulence. On the contrary, the rdar morphotype seems to confer almost opposite roles in pathogenicity in *S. enterica* serotype Typhimurium and *E. coli.* Certainly, more studies are needed to investigate this phenomenon.

7 CONCLUSIONS

The rdar morphotype was discovered at the beginning of the last century (Lingelsheim, 1913) and was only occasionally investigated subsequently (Jameson, 1996). Not much detailed attention was paid to this morphotype until recently (Olsen et al., 1989; Collinson et al., 1991; Romling et al., 1998a), although it is expressed by the majority of *S. enterica* serotypes and *E. coli* isolates of pathogenic and commensal origin. The rdar morphotype is characterised by an abundant extracellular matrix, which plays an architectural and protective role and modulates interaction with the inanimate and animate environment. Thin aggregative fimbriae and cellulose are the two major extracellular matrix components. Biosynthesis of both components is controlled by the transcriptional regulator AgfD. Expression of *agfD* is tightly regulated by mutations and environmental conditions, which is expected to involve a sophisticated regulatory network that still requires exploration. Therefore, investigation of the rdar morphotype spans the whole spectrum of microbiology ranging from biochemistry to epidemiology and host–pathogen interactions.

REFERENCES

Allen-Vercoe, E., Sayers, A. R. and Woodward, M. J. (1999). Virulence of *Salmonella enterica* serotype Enteritidis aflagellate and afimbriate mutants in a day-old chick model. *Epidemiology and Infection,* 122, 395–402.

Allen-Vercoe, E. and Woodward, M. J. (1999). Colonisation of the chicken caecum by afimbriate and aflagellate derivatives of *Salmonella enterica* serotype Enteritidis. *Veterinary Microbiology,* 69, 265–275.

Ausmees, N., Jonsson, H., Hoglund, S., Ljunggren, H. and Lindberg, M. (1999). Structural and putative regulatory genes involved in cellulose synthesis in *Rhizobium leguminosarum* bv. trifolii. *Microbiology,* 145, 1253–1262.

Austin, J. W., Sanders, G., Kay, W. W. and Collinson, S. K. (1998). Thin aggregative fimbriae enhance *Salmonella enteritidis* biofilm formation. *FEMS Microbiology Letters*, 162, 295–301.

Bang, I. S., Kim, B. H., Foster, J. W. and Park, Y. K. (2000). OmpR regulates the stationary-phase acid tolerance response of *Salmonella enterica* serovar Typhimurium. *Journal of Bacteriology*, 182, 2245–2252.

Becker, G. and Hengge-Aronis, R. (2001). What makes an *Escherichia coli* promoter sigma(S) dependent? Role of the -13/-14 nucleotide promoter positions and region 2.5 of sigma(S). *Molecular Microbiology*, 39, 1153–1165.

Ben Nasr, A., Olsen, A., Sjobring, U., Muller-Esterl, W. and Bjorck, L. (1996). Assembly of human contact phase proteins and release of bradykinin at the surface of curli-expressing *Escherichia coli. Molecular Microbiology*, 20, 927–935.

Berkhoff, H. A. and Vinal, A. C. (1986). Congo red medium to distinguish between invasive and non-invasive *Escherichia coli* pathogenic for poultry. *Avian Diseases*, 30, 117–121.

Bian, Z., Brauner, A., Li, Y. and Normark, S. (2000). Expression of and cytokine activation by *Escherichia coli* curli fibers in human sepsis. *Journal of Infectious Diseases*, 181, 602–612.

Bian, Z. and Normark, S. (1997). Nucleator function of CsgB for the assembly of adhesive surface organelles in *Escherichia coli. EMBO Journal*, 16, 5827–5836.

Bian, Z., Yan, Z. Q., Hansson, G. K., Thoren, P. and Normark, S. (2001). Activation of inducible nitric oxide synthase/nitric oxide by curli fibers leads to a fall in blood pressure during systemic *Escherichia coli* infection in mice. *Journal of Infectious Diseases*, 183, 612–619.

Brown, P. K., Dozois, C. M., Nickerson, C. A., Zuppardo, A., Terlonge, J. and Curtiss, R., III. (2001). MlrA, a novel regulator of curli (AgF) and extracellular matrix synthesis by *Escherichia coli* and *Salmonella enterica* serovar Typhimurium. *Molecular Microbiology*, 41, 349–363.

Chapman, M. R., Robinson, L. S., Pinkner, J. S., Roth, R., Heuser, J., Hammar, M., Normark, S. and Hultgren, S. J. (2002). Role of *Escherichia coli* curli operons in directing amyloid fiber formation. *Science*, 295, 851–855.

Collinson, S. K., Doig, P. C., Doran, J. L., Clouthier, S., Trust, T. J. and Kay, W. W. (1993). Thin, aggregative fimbriae mediate binding of *Salmonella enteritidis* to fibronectin. *Journal of Bacteriology*, 175, 12–18.

Collinson, S. K., Emody, L., Muller, K. H., Trust, T. J. and Kay, W. W. (1991). Purification and characterization of thin, aggregative fimbriae from *Salmonella enteritidis. Journal of Bacteriology*, 173, 4773–4781.

Collinson, S. K., Emody, L., Trust, T. J. and Kay, W. W. (1992). Thin aggregative fimbriae from diarrheagenic *Escherichia coli. Journal of Bacteriology*, 174, 4490–4495.

Danese, P. N. and Silhavy, T. J. (1997). The sigma(E) and the Cpx signal transduction systems control the synthesis of periplasmic protein-folding enzymes in *Escherichia coli. Genes and Development*, 11, 1183–1193.

Danese, P. N., Snyder, W. B., Cosma, C. L., Davis, L. J. and Silhavy, T. J. (1995). The Cpx two-component signal transduction pathway of *Escherichia coli* regulates transcription of the gene specifying the stress-inducible periplasmic protease, DegP. *Genes and Development*, 9, 387–398.

Davies, D. G., Parsek, M. R., Pearson, J. P., Iglewski, B. H., Costerton, J. W. and Greenberg, E. P. (1998).The involvement of cell-to-cell signals in the development of a bacterial biofilm. *Science*, 280, 295–298.
De Wulf, P., Kwon, O. and Lin, E. C. (1999). The CpxRA signal transduction system of *Escherichia coli*: growth-related autoactivation and control of unanticipated target operons. *Journal of Bacteriology*, 181, 6772–6778.
Dibb-Fuller, M. P., Allen-Vercoe, E., Thorns, C. J. and Woodward, M. J. (1999). Fimbriae- and flagella-mediated association with and invasion of cultured epithelial cells by *Salmonella enteritidis*. *Microbiology*, 145, 1023–1031.
Doran, J. L., Collinson, S. K., Burian, J., Sarlos, G., Todd, E. C., Munro, C. K., Kay, C. M., Banser, P. A., Peterkin, P. I. and Kay, W. W. (1993). DNA-based diagnostic tests for Salmonella species targeting agfA, the structural gene for thin, aggregative fimbriae. *Journal of Clinical Microbiology*, 31, 2263–2273.
Dorel, C., Vidal, O., Prigent-Combaret, C., Vallet, I. and Lejeune, P. (1999). Involvement of the Cpx signal transduction pathway of *E. coli* in biofilm formation. *FEMS Microbiology Letters*, 178, 169–175.
Gaal, T., Ross, W., Estrem, S. T., Nguyen, L. H., Burgess, R. R. and Gourse, R. L. (2001). Promoter recognition and discrimination by EsigmaS RNA polymerase. *Molecular Microbiology*, 42, 939–954.
Garrett, E. S., Perlegas, D. and Wozniak, D. J. (1999). Negative control of flagellum synthesis in *Pseudomonas aeruginosa* is modulated by the alternative sigma factor AlgT (AlgU). *Journal of Bacteriology*, 181, 7401–7404.
Gerstel, U. and Romling, U. (2001). Oxygen tension and nutrient starvation are major signals that regulate agfD promoter activity and expression of the multicellular morphotype in *Salmonella typhimurium*. *Environmental Microbiology*, 3, 638–648.
Gophna, U., Barlev, M., Seijffers, R., Oelschlager, T. A., Hacker, J. and Ron, E. Z. (2001). Curli fibers mediate internalization of *Escherichia coli* by eukaryotic cells. *Infection and Immunity*, 69, 2659–2665.
Hammar, M., Arnqvist, A., Bian, Z., Olsen, A. and Normark, S. (1995). Expression of two *csg* operons is required for production of fibronectin- and congo red-binding curli polymers in *Escherichia coli* K-12. *Molecular Microbiology*, 18, 661–670.
Hammar, M., Bian, Z. and Normark, S. (1996). Nucleator-dependent intercellular assembly of adhesive curli organelles in *Escherichia coli*. *Proceedings of the National Academy of Sciences of the USA*, 93, 6562–6566.
Harshey, R. M. and Matsuyama, T. (1994). Dimorphic transition in *Escherichia coli* and *Salmonella typhimurium*: surface-induced differentiation into hyperflagellate swarmer cells. *Proceedings of the National Academy of Sciences of the USA*, 91, 8631–8635.
Hecht, G. B. and Newton, A. (1995). Identification of a novel response regulator required for the swarmer-to-stalked-cell transition in *Caulobacter crescentus*. *Journal of Bacteriology*, 177, 6223–6229.
Heldwein, E. E. and Brennan, R. G. (2001). Crystal structure of the transcription activator BmrR bound to DNA and a drug. *Nature*, 409, 378–382.
Herwald, H., Morgelin, M., Olsen, A., Rhen, M., Dahlback, B., Muller-Esterl, W. and Bjorck, L. (1998). Activation of the contact-phase system on bacterial surfaces – a clue to serious complications in infectious diseases. *Nature Medicine*, 4, 298–302.

Heyde, M., Laloi, P. and Portalier, R. (2000). Involvement of carbon source and acetyl phosphate in the external-pH-dependent expression of porin genes in *Escherichia coli*. *Journal of Bacteriology*, 182, 198–202.

Huang, K. J. and Igo, M. M. (1996). Identification of the bases in the ompF regulatory region, which interact with the transcription factor OmpR. *Journal of Molecular Biology*, 262, 615–628.

Hung, D. L., Raivio, T. L., Jones, C. H., Silhavy, T. J. and Hultgren, S. J. (2001). Cpx signaling pathway monitors biogenesis and affects assembly and expression of P pili. *Embo Journal*, 20, 1508–1518.

Jameson, J. E. (1966). Differentiation of Salmonella strains by colonial morphology. *Journal of Pathology and Bacteriology*, 91, 141–148.

Johansson, C., Nilsson, T., Olsen, A. and Wick, M. J. (2001). The influence of curli, a MHC-I-binding bacterial surface structure, on macrophage-T cell interactions. *FEMS Immunology and Medical Microbiology*, 30, 21–29.

La Ragione, R. M., Cooley, W. A. and Woodward, M. J. (2000a). The role of fimbriae and flagella in the adherence of avian strains of *Escherichia coli* O78:K80 to tissue culture cells and tracheal and gut explants. *Journal of Medical Microbiology*, 49, 327–338.

La Ragione, R. M., Sayers, A. R. and Woodward, M. J. (2000b). The role of fimbriae and flagella in the colonization, invasion and persistence of *Escherichia coli* O78:K80 in the day-old-chick model. *Epidemiology and Infection*, 124, 351–363.

Lee, A. K., Detweiler, C. S. and Falkow, S. (2000). OmpR regulates the two-component system SsrA-ssrB in *Salmonella* pathogenicity island 2. *Journal of Bacteriology*, 182, 771–781.

Liljestrom, P., Laamanen, I. and Palva, E. T. (1988). Structure and expression of the ompB operon, the regulatory locus for the outer membrane porin regulon in *Salmonella typhimurium* LT-2. *Journal of Molecular Biology*, 201, 663–673.

Lingelsheim, V. (1913). Frage der Variation der Typhusbacillen und verwandter Gruppen. *Centralblat Bakteriology (Orig)*, 68, 577–582.

Loferer, H., Hammar, M. and Normark, S. (1997). Availability of the fibre subunit CsgA and the nucleator protein CsgB during assembly of fibronectin-binding curli is limited by the intracellular concentration of the novel lipoprotein CsgG. *Molecular Microbiology*, 26, 11–23.

Maurer, J. J., Brown, T. P., Steffens, W. L. and Thayer, S. G. (1998). The occurrence of ambient temperature-regulated adhesins, curli and the temperature-sensitive hemagglutinin tsh among avian *Escherichia coli*. *Avian Diseases*, 42, 106–118.

Olsen, A., Arnqvist, A., Hammar, M., Sukupolvi, S. and Normark, S. (1993). The RpoS sigma factor relieves H-NS-mediated transcriptional repression of csgA, the subunit gene of fibronectin-binding curli in *Escherichia coli*. *Molecular Microbiology*, 7, 523–536.

Olsen, A., Jonsson, A. and Normark, S. (1989). Fibronectin binding mediated by a novel class of surface organelles on *Escherichia coli*. *Nature*, 338, 652–655.

Olsen, A., Wick, M. J., Morgelin, M. and Bjorck, L. (1998). Curli, fibrous surface proteins of *Escherichia coli*, interact with major histocompatibility complex class I molecules. *Infection and Immunity*, 66, 944–949.

Patri, E., Szabo, E., Pal, T. and Emody, L. (2000). Thin aggregative fimbriae on urinary *Escherichia coli* isolates. *Advances in Experimental and Medical Biology*, 485, 219–224.

Pedersen, A. G., Jensen, L. J., Brunak, S., Staerfeldt, H. H. and Ussery, D. W. (2000). A DNA structural atlas for *Escherichia coli*. *Journal of Molecular Biology*, 299, 907–930.

Pei, J. and Grishin, N. V. (2001). GGDEF domain is homologous to adenylyl cyclase. *Proteins*, 42, 210–216.

Pratt, L. A., Hsing, W., Gibson, K. E. and Silhavy, T. J. (1996). From acids to osmZ: multiple factors influence synthesis of the OmpF and OmpC porins in *Escherichia coli*. *Molecular Microbiology*, 20, 911–917.

Prigent-Combaret, C., Brombacher, E., Vidal, O., Ambert, A., Lejeune, P., Landini, P. and Dorel, C. (2001). Complex regulatory network controls initial adhesion and biofilm formation in *Escherichia coli* via regulation of the csgD gene. *Journal of Bacteriology*, 183, 7213–7223.

Prigent-Combaret, C., Prensier, G., Le Thi, T. T., Vidal, O., Lejeune P. and Dorel, C. (2000). Developmental pathway for biofilm formation in curli-producing *Escherichia coli* strains: role of flagella, curli and colanic acid. *Environmental Microbiology*, 2, 450–464.

Prigent-Combaret, C., Vida, O., Dorel, C. and Lejeune, P. (1999). Abiotic surface sensing and biofilm-dependent regulation of gene expression in *Escherichia coli*. *Journal of Bacteriology*, 181, 5993–6002.

Rabsch, W., Andrews, H. L., Kingsley, R. A., Prager, R., Tschape, H., Adams, L. G. and Baumler, A. J.(2002). *Salmonella enterica* serotype Typhimurium and its host-adapted variants. *Infection and Immunity*, 70, 2249–2255.

Raivio, T. L. and Silhavy, T. J. (1997). Transduction of envelope stress in *Escherichia coli* by the Cpx two-component system. *Journal of Bacteriology*, 179, 7724–7733.

Rajashekara, G., Munir, S., Alexeyev, M. F., Halvorson, D. A., Wells, C. L. and Nagaraja, K. V. (2000). Pathogenic role of SEF14, SEF17, and SEF21 fimbriae in *Salmonella enterica* serovar enteritidis infection of chickens. *Applied and Environmental Microbiology*, 66, 1759–1763.

Romling, U. (2002). Molecular biology of cellulose production in bacteria. *Research in Microbiology*, 153, 205–212.

Romling, U., Bian, Z., Hammar, M., Sierralta, W. D. and Normark, S. (1998a). Curli fibers are highly conserved between *Salmonella typhimurium* and *Escherichia coli* with respect to operon structure and regulation. *Journal of Bacteriology*, 180, 722–731.

Romling, U. and Rohde, M. (1999). Flagella modulate the multicellular behavior of *Salmonella typhimurium* on the community level. *FEMS Microbiology Letters*, 180, 91–102.

Romling, U., Rohde, M., Olsen, A., Normark, S. and Reinkoster, J. (2000). AgfD, the checkpoint of multicellular and aggregative behaviour in *Salmonella typhimurium*, regulates at least two independent pathways. *Molecular Microbiology*, 36, 10–23.

Romling, U., Sierralta, W. D., Eriksson, K. and Normark, S. (1998b). Multicellular and aggregative behaviour of *Salmonella typhimurium* strains is controlled by mutations in the agfD promoter. *Molecular Microbiology*, 28, 249–264.

Sakellaris, H., Hannink, N. K., Rajakumar, K., Bulach, D., Hunt, M., Sasakawa, C. and Adler, B. (2000). Curli loci of *Shigella* spp. *Infection and Immunity*, 68, 3780–3783.

Sjobring, U., Pohl, G. and Olsen, A. (1994). Plasminogen, absorbed by *Escherichia coli* expressing curli or by *Salmonella enteritidis* expressing thin aggregative fimbriae, can be activated by simultaneously captured tissue-type plasminogen activator (t-PA). *Molecular Microbiology*, 14, 443–452.

Solano, C., Garcia, B., Valle, J., Berasain, C., Ghigo, J. M., Gamazo, C. and Lasa, I. (2002). Genetic analysis of *Salmonella enteritidis* biofilm formation: critical role of cellulose. *Molecular Microbiology*, 43, 793–808.

Styles, D. K. and Flammer, K. (1991). Congo red binding of *Escherichia coli* isolated from the cloacae of psittacine birds. *Avian Diseases*, 35, 46–48.

Sukupolvi, S., Edelstein, A., Rhen, M., Normark, S. J. and Pfeifer, J. D. (1997a). Development of a murine model of chronic Salmonella infection. *Infection and Immunity*, 65, 838–842.

Sukupolvi, S., Lorenz, R. G., Gordon, J. I., Bian, Z., Pfeifer, J. D., Normark, S. J. and Rhen, M. (1997b). Expression of thin aggregative fimbriae promotes interaction of Salmonella typhimurium SR-11 with mouse small intestinal epithelial cells. *Infection and Immunity*, 65, 5320–5325.

Townsend, S. M., Kramer, N. E., Edwards, R., Baker, S., Hamlin, N., Simmonds, M., Stevens, K., Maloy, S., Parkhill, J., Dougan, G. and Baumler, A. J. (2001). *Salmonella enterica* serovar Typhi possesses a unique repertoire of fimbrial gene sequences. *Infection and Immunity*, 69, 2894–2901.

Uhlich, G. A., Keen, J. E. and Elder, R. O. (2001). Mutations in the csgD promoter associated with variations in curli expression in certain strains of *Escherichia coli* O157:H7. *Applied and Environmental Microbiology*, 67, 2367–2370.

Uhlich, G. A., Keen, J. E. and Elder, R. O. (2002). Variations in the csgD promoter of *Escherichia coli* O157:H7 associated with increased virulence in mice and increased invasion of HEp-2 cells. *Infection and Immunity*, 70, 395–399.

Updegraff, D. M. (1969). Semimicro determination of cellulose in biological materials. *Analytical Biochemistry*, 32, 420–424.

van der Velden, A. W., Baumler, A. J., Tsolis, R. M. and Heffron, F. (1998). Multiple fimbrial adhesins are required for full virulence of *Salmonella typhimurium* in mice. *infection and immunity*, 66, 2803–2808.

Vidal, O., Longin, R., Prigent-Combaret, C., Dorel, C., Hooreman, M. and Lejeune, P. (1998). Isolation of an *Escherichia coli* K-12 mutant strain able to form biofilms on inert surfaces: involvement of a new ompR allele that increases curli expression. *Journal of Bacteriology*, 180, 2442–2449.

Volz, K. (1993). Structural conservation in the CheY superfamily. *Biochemistry*, 32, 11741–11753.

White, A. P., Collinson, S. K., Banser, P. A., Gibson, D. L., Paetzel, M., Strynadka, N. C. and Kay, W. W. (2001). Structure and characterization of AgfB from *Salmonella enteritidis* thin aggregative fimbriae. *Journal of Molecular Biology*, 311, 735–749.

Wong, H. C., Fear, A. L., Calhoon, R. D., Eichinger, G. H., Mayer, R., Amikam, D., Benziman, M., Gelfand, D. H., Meade, J. H., Emerick, A. W., et al. (1990). Genetic organization of the cellulose synthase operon in *Acetobacter xylinum. Proceedings of the National Academy of Sciences of the USA*, 87, 8130–8134.
Zogaj, X., Nimtz, M., Rohde, M., Bokranz, W. and Romling, U. (2001). The multicellular morphotypes of *Salmonella typhimurium* and *Escherichia coli* produce cellulose as the second component of the extracellular matrix. *Molecular Microbiology*, 39, 1452–1463.

CHAPTER TWELVE

Bacterial Growth on Mucosal Surfaces and Biofilms in the Large Bowel

S. Macfarlane and G. T. Macfarlane

1 THE LARGE INTESTINAL MICROBIOTA

It has been estimated that of the 10^{14} cells associated with the human body, approximately 90 per cent are microorganisms, and the vast majority of these organisms are bacteria growing in the large intestine (Savage, 1977). The large bowel is the main area of permanent microbial colonisation of the human gastrointestinal tract; gastric acid kills most oral and environmental microorganisms in the stomach, whereas the rapid passage of digestive materials through the upper gut does not allow time for significant bacterial growth to occur (Macfarlane and Cummings, 1991). However, the rate of movement of intestinal contents slows markedly in the large gut, which facilitates development of rich and diverse bacterial communities (Cummings, 1978; Cummings et al., 1993). The growth and metabolic activities of these microbial populations are influenced to a considerable degree by diet, as well as by the structure and physiology of the colon (Macfarlane et al., 1995).

The large intestine is an open system in the sense that food residues from the small intestine enter at one end and, together with bacterial cell mass, are excreted at the other end. Because of this, the colon is often viewed as being a continuous culture system, although only the caecum and ascending colon really exhibit characteristics of a continuous culture (Cummings et al., 1987; Macfarlane, Macfarlane, and Gibson, 1998). Culturing studies (Moore and Holdeman, 1974; Finegold et al., 1975) show that faecal material contains large numbers of viable bacteria ranging from 10^{11} to 10^{13} per gram (see Table 12.1), but direct microscope counts and molecular analyses of bacterial populations in the gut indicate that considerably more bacteria are present

Table 12.1: Numerically predominant anaerobes and facultative anaerobes isolated from faeces

Genus	Type	Mean Count	Range	Nutrition
Bacteroides	Gram-negative rods	11.3	9.2–13.5	Saccharolytic
Eubacterium	Gram-positive rods	10.7	5.0–13.3	Saccharolytic, some amino acid fermenting species
Bifidobacterium	Gram-positive rods	10.2	4.9–13.4	Saccharolytic
Lactobacillus	Gram-positive rods	9.8	3.3–13.1	Saccharolytic
Clostridium	Gram-positive rods	9.6	3.6–12.5	Saccharolytic, some amino acid fermenting species
Ruminococcus	Gram-positive cocci	10.2	4.6–12.8	Saccharolytic
Peptostreptococcus	Gram-positive cocci	10.1	3.8–12.6	Saccharolytic, some amino acid fermenting species
Peptococcus	Gram-positive cocci	10.0	5.1–12.9	Amino acid fermenters
Propionibacterium	Gram-positive rods	9.4	4.3–12.0	Saccharolytic and lactate fermenters
Actinomyces	Gram-positive rods	9.2	5.7–11.1	Saccharolytic
Enterococcus	Gram-positive cocci	8.9	3.9–12.9	Carbohydrate and amino acid fermenters
Fusobacterium	Gram-negative rods	8.4	5.1–11.0	Amino acid fermenters, but carbohydrate assimilated
Escherichia	Gram-negative rods	8.6	3.9–12.3	Carbohydrate and amino acid fermenting

Note: Cell counts are per gram dry weight faeces and are taken from Finegold, Sutter, and Mathisen (1983).

(Macfarlane and Gibson, 1994; Sharp and Ziemer, 1999; Hopkins, Sharp, and Macfarlane, 2001). For example, many bacterial morphologies can be viewed during microscopic examination of faecal material, but cannot be found in subsequent isolation procedures (Croucher et al., 1983). This is similar to what has been found in studies on mice, where, despite extensive research over many years, it was considered that only about 10 per cent of the bacteria could be characterised (Wilkins, 1981). This might be due to non-viability of large numbers of bacterial cells in faeces or the inability of cultural methods to facilitate their isolation.

1.1 Role of the Gut Microflora in Health and Disease

The large intestine is a dynamic environment in which individual bacteria exist in a great diversity of microhabitats and metabolic niches; yet in healthy adults, the microbiota is relatively stable in composition. Although little is known of the metabolic relationships and interactions that occur between individual groups of bacteria in the large bowel, or of the multicellular organisation of the microbiota, intestinal microorganisms are known to play a major role in health and disease. The normal microbiota affects human physiology in a multiplicity of ways, through, for example, bile acid and steroid transformations, metabolism of xenobiotic substances, vitamin synthesis, mineral absorption, maintenance of colonisation resistance to bacterial pathogens, activation or destruction of genotoxins and mutagens, and modulation of immune system function (Macfarlane and Cummings, 2002). Of particular importance, the colonic microflora functions symbiotically in carbohydrate and protein digestion, where the host provides dietary residues or endogenously produced substrates for the bacteria, which in turn supply metabolic end products such as short chain fatty acids (SCFA), that are required by the body.

2 BACTERIAL GROWTH ON SURFACES IN THE LARGE INTESTINE

The existence of distinct mucosal and lumenal bacterial populations in intestinal ecosystems is well recognised, and, in many animals, a specific microflora has been found growing in association with epithelial surfaces (Breznak and Pankratz, 1977; Wallace et al., 1979; Lee, 1980). There is also evidence for independent mucosal communities in humans (Lee et al., 1971; Croucher et al., 1983), although some authors have reported that the composition of epithelial populations is broadly similar to those that exist in the gut lumen (Nelson and Mata, 1970).

The first steps in bacterial colonisation of surfaces probably occur through the attachment of single cells or small groups of organisms, followed by a non-linear increase in cell mass that could lead to formation of a biofilm. Sessile bacteria growing in biofilms often behave very differently from their non-adherent forms, for example, changes in the nature and efficiency of their metabolism have been reported (McCabe, Mann, and Bowie, 1998), whereas the biofilms exhibit greater resistance to antibiotics and other environmental factors that are inhibitory to planktonic cells (Anwar, Dasgupta, and Costerton, 1990; Van Loosdrecht et al., 1990; Mozes and Rouxhet, 1992). Close spatial relationships between bacterial cells on surfaces are important in metabolic communication between microorganisms in the microbiota and are ecologically significant in that they reduce the latent growth limiting effects on syntrophic populations associated with mass transfer resistance (Conrad, Phelps, and Zeikus, 1985).

2.1 Biofilm Populations in the Gut Lumen

Bacteria colonising surfaces of digestive residues in the large bowel are more directly involved in the key role of breaking down complex insoluble carbohydrates and proteins than non-adherent organisms, giving them an important competitive advantage in the gut, while being important in releasing and cycling nutrients for other groups of microorganisms in the ecosystem (Macfarlane and Macfarlane, 1995).

Bacterial communities forming biofilms in the large intestine might therefore be expected to manifest distinctive biochemical properties when compared to non-adherent populations. This has indeed been shown in the gut lumen, where significant differences in polysaccharidase, glycosidase, protease, and peptidase synthesis were demonstrated in biofilm communities colonising the surfaces of food residues (Macfarlane, McBain, and Macfarlane, 1997). Moreover, subsequent studies showed that SCFA production was also distinct in these biofilms. Acetate was the main fermentation product in biofilm and planktonic populations; however, butyrate formation was always greater in non-adherent communities (Macfarlane and Macfarlane, 2001). No significant differences were found in the bacterial composition of biofilm and non-adherent populations in these experiments. Since butyrate is used as an electron sink by a number of colonic anaerobes, and acetate formation is an indication of carbon limitation and ATP generation, these results suggested that bacteria living in mature biofilms in the gut lumen were growing under more energy limiting conditions than non-adherent species. Reduced butyrate production in biofilm communities may have health implications because it is an essential fuel for the colonic epithelium (Roediger, 1980), particularly in the distal gut. Production of butyrate by bacteria in the large bowel is also important from the viewpoint of preventing colon cancer, since it inhibits DNA synthesis and induces differentiation in human cancer lines, while reducing proliferation of neoplastic cells and the effects of contact-independent growth (Young, 1991).

2.2 Mucosal Populations

Secretory intestinal epithelia in the human large intestine are covered in a mucus coating up to 200 μm in thickness (Pullan et al., 1994), which is important as a substratum and as a source of carbon and energy for bacterial populations growing in association with the underlying epithelium (Savage, 1978; Macfarlane, Cummings, and Macfarlane, 1999). Mucosal bacterial communities in the large bowel are difficult to study in healthy people for ethical and practical reasons, and this has limited their investigation. Consequently, the metabolic and health-related significance of bacteria colonising the large intestinal epithelium is unclear. Nevertheless, some reports suggest that mucosal communities in humans are generally similar to those present in the gut lumen (Nelson and Mata, 1970; Poxton et al., 1997), with bacteroides and fusobacteria predominating, but a wide range of other organisms such as clostridia, eubacteria, and anaerobic Gram-positive cocci have also been

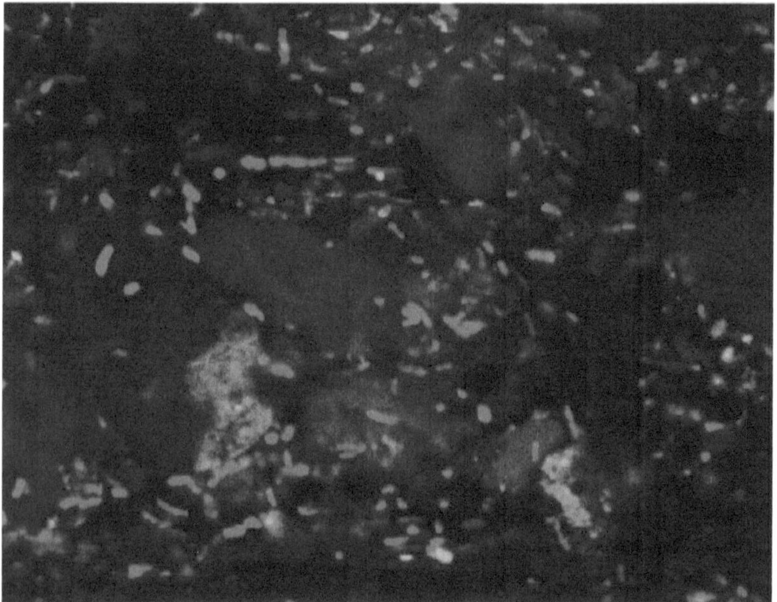

Figure 12.1: Light micrograph of the colonic mucosal surface stained with 16S rRNA oligonucleotide probes targeted against bacteroides (red, cy3), bifidobacteria (blue, cy5), and escherichia (green, fluorescein isothiocyanate). (See colour section.)

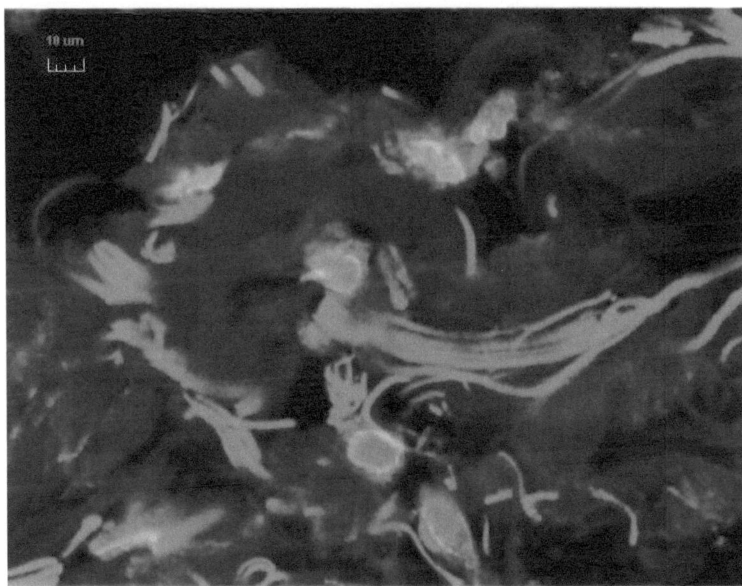

Figure 12.2: Light micrograph of the mucosal surface in tissue taken from the colon and stained with a eubacterial 16S rRNA oligonucleotide probe labelled with FITC. (See colour section.)

reported (Edmiston, Avant, and Wilson, 1982; Croucher et al., 1983). Large numbers of bacteroides are indeed present on the mucus layer, as evidenced by Figure 12.1, which shows a colonic biopsy sample stained with specific 16S rRNA oligonucleotide probes directed against bacteroides, bifidobacteria, and escherichia. In addition, some bacteria growing on the epithelial surface exhibit unusual morphological properties and cannot be seen in, or cultured from, faeces (Lee et al., 1971). Indeed, bacteria with distinct morphological characteristics have been visualised *in situ* on the mucosa, where microscopic analysis of biopsy specimens demonstrates the presence of large spirochaete-like organisms colonising the mucus layer (Figure 12.2). Several reports suggest that a number of bacteria associated with the colonic epithelium are spiral shaped (Takeuchi et al., 1974; Croucher et al., 1983), as shown in Figure 12.3. There is also some evidence which suggests that many of the organisms associated with the intestinal epithelium inhabit the mucus layer rather than the mucosal surface, as indicated in Figure 12.4, which shows bacteria in microcolonies and diffusely dispersed through the mucus in a biopsy sample taken from the proximal large intestine.

Commensal and parasitic species living in close association with host tissues often directly exploit the nutritional potential of the substratum.

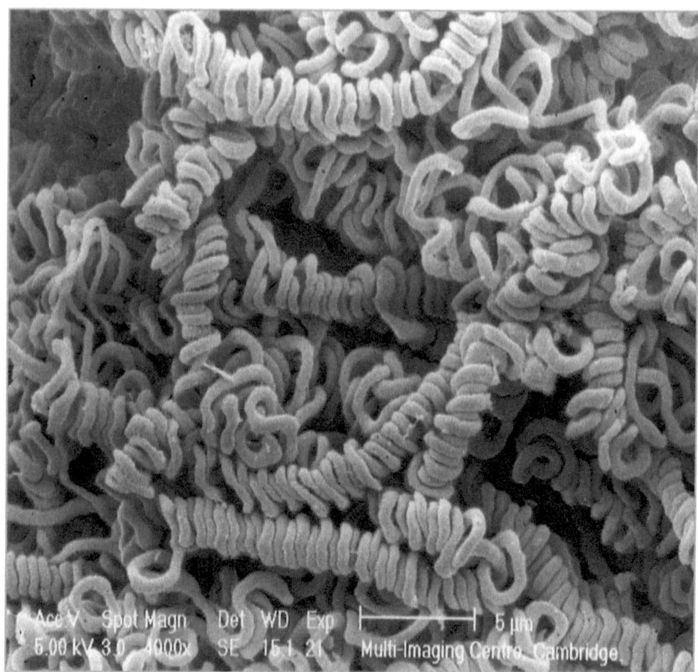

Figure 12.3: Scanning electron micrograph of large spiral bacterial forms associated with the epithelial surface.

Examples include bacterial utilisation of complex host macromolecules such as mucins (Macfarlane and Gibson, 1991), as well as cell matrix constituents such as vitronectin (Dehio et al., 1998) and fibronectin (Patti et al., 1994). Recent developments have demonstrated that some adhesive bacteria are able to recruit a variety of structurally diverse host proteins, adhesive glycoproteins, growth factors, and cytokines by initially binding heparin and functionally similar sulphated polysaccharides to their surfaces, where they act as non-specific, secondary recruiting sites for other host molecules (Duensing, Wing, and van Putten, 1999).

2.3 Colonisation of the Large Intestinal Mucosa

As summarised in Table 12.2, species composition, community structure, and the metabolic activities of bacterial populations growing in association with the colonic mucosa and the mucus layer are determined by a complexity of host, environmental, and microbiological factors. Through the actions of bacterial products with immunomodulatory properties such as endotoxic lipopolysaccharide, lipoteichoic acids (LTA), and peptidoglycans (Standiford,

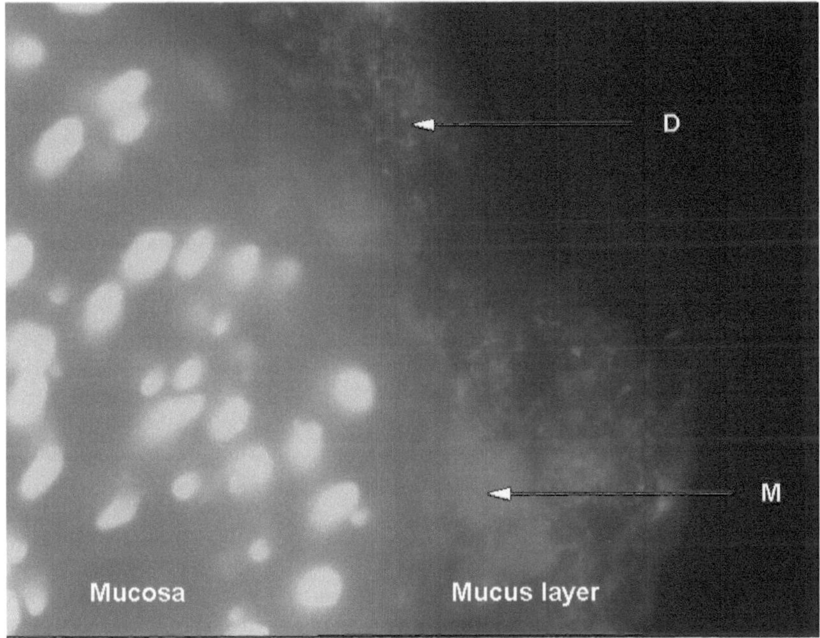

Figure 12.4: Light micrograph of a tranverse section of mucosal tissue taken from the colon and stained with 4,6-diamidino-2-phenylindole. The bacteria can be seen occurring as microcolonies (M) and diffusely (D) in the mucus layer. (See colour section.)

Arenberg, and Danforth, 1994), microorganisms colonising epithelial surfaces in the large bowel affect local and systemic immunity in the host. This involves B and T cells, blood leukocytes, and the intestinal epithelium (Schiffrin et al., 1997). Preservation of immune homeostasis is therefore controlled, to some extent, by direct cellular interactions between mucosal organisms and these effector cells of the immune system. In concordance with this principle, lactobacilli have been demonstrated to bind to circulating peripheral blood CD4 and CD8 T lymphocytes (De Simone et al., 1992), and species that attach to colonic epithelial cells have been shown to be capable of macrophage activation (Perdigon et al., 1993).

Much of what we know about bacterial adherence and colonisation of human gastrointestinal mucosae has come from studies on putatively probiotic lactobacilli and bifidobacteria. These investigations have suggested that adherence of some lactobacilli to intestinal epithelial cells is important in colonisation resistance in the gut. For example, *Lactobacillus acidophilus* has been found to interfere with the binding of several intestinal pathogens, including *Salmonella typhimurium* and *Yersinia pseudotuberculosis*, as well as enterotoxigenic and enteropathogenic *Escherichia coli* (Bernet et al., 1994).

Table 12.2: Factors affecting bacterial colonisation in the large intestine

Host	Environmental	Bacterial
Diet	Amounts and types of substrate in the microbiota	Competition between bacteria for limiting nutrients and adhesion sites on food particles, mucus, and intestinal mucosa Cooperative interactions between microorganisms
Colonic transit time, epithelial cell turnover rates	pH of intestinal contents	Generic and species composition of microbiota
Disease, drugs, antibiotic therapy, rates of mucus production and its chemical composition, and pancreatic and other secretions	Redox potential	Inhibition of allochthonous species by fermentation products including HS^-, SCFA, phenolic compounds, deconjugated bile salts, etc. Bacterial secretion of antagonistic substances
IgA production and innate immunity (defensin secretion, lysozyme production, etc.) at mucosal surface; possible stimulatory interactions of host hormones and neurotransmitters with some Gram-negative species	Geographical residence/cultural factors associated with host	Synergistic effects of bacterial antagonism and local immunity in the mucus layer and on the colonic mucosa

In *L. plantarum*, adherence seems to be mannosespecific and to occur via protease-sensitive structures on the bacterial cell surface (Adlerberth et al., 1996). *In vivo* experiments have demonstrated that probiotic lactobacilli are able to colonise the small and large intestinal epithelial surface temporarily and to supplant other bacterial species (Johansson et al., 1993). In these experiments, nineteen strains of lactobacilli (each 5×10^6 mL^{-1}) were fed to volunteers. The organisms were found to be persistent in the gut, where high numbers of adherent *L. plantarum, L. reuteri, L. agilis,* and *L. casei* were recovered from jejunal biopsies nearly 2 weeks after probiotic feeding was discontinued.

Studies on interactions between bifidobacteria and human colonic cell lines (Sato, Mochizuki, and Homma, 1982) showed that these organisms are also enteroadherent. Subsequent work demonstrated that in a similar way to lactobacilli, bifidobacteria inhibited the attachment and invasion of Caco-2 cells by pathogenic *E. coli* (diffusely adhering, enteropathogenic, enterotoxigenic), *S. typhimurium*, and *Y. pseudotuberculosis* (Bernet et al., 1993). However, bifidobacterial adhesive properties appear to vary markedly within the genus, with marked species and strain variations being evident (Crociani et al., 1995).

2.4 Bacterial Antagonism

Some bacteria isolated from the large intestine are able to secrete low molecular mass antagonistic substances that inhibit the growth of other intestinal microorganisms. In lactobacilli, bacteriocin formation appears to be growth associated in some species and to be dependent on carbon availability (Lejeune et al., 1998). Whereas bacteriocin formation has been widely studied in lactic acid bacteria (Dodd and Gasson, 1994; De Vuyst and Vandamme, 1994), there is increasing interest in bifidobacterial antagonism. Although antimicrobial secretory products of bifidobacteria have not been studied extensively, they seem to be active against a wide range of organisms when compared to conventional bacteriocins. For example, *Bifidobacterium infantis* NCFB 2255 and *B. breve* NCFB 2258 secrete two different types of antimicrobial substance: one mainly affects Gram-positive bacteria, whereas the other is inhibitory towards Gram-negative organisms (O'Riordan, Condon, and Fitzgerald, 1995). Gibson and Wang (1994) reported that eight different bifidobacterial species secreted inhibitory substances that were antagonistic to a range of Gram-positive and Gram-negative pathogens, including listeria, salmonella, shigella, vibrio, and campylobacter. However, this property seems to be variable in bifidobacteria, since only one out of thirteen test strains investigated by Meghrous et al. (1990) formed a protease-sensitive inhibitory substance. This bacteriocin had broad spectrum activity against other bifidobacteria, streptococci, lactobacilli, and clostridia, though it was not active against Gram-negative species such as klebsiella, pseudomonas, escherichia, or proteus.

Virtually all of the studies made on bacteriocin production have focussed on the use of pure cultures in the laboratory; consequently, the effectiveness of these substances in the hostile environment of the large intestine is uncertain. Is the formation of inhibitory substances likely to be an important determinant of bacterial growth in the large bowel? The colon contains large amounts of proteases and peptidases (Gibson et al., 1989) that could destroy or inactivate some bacteriocins. Furthermore, bacteriocins could be adsorbed onto the surfaces of food residues and non-sensitive bacterial species, thereby negating their antimicrobial potential. However, studies on an enterocin secreted by a ruminal strain of *Enterococcus faecium* have shown that it retained antimicrobial activity for long periods in rumen fluid (Laukova and Czikkova, 1998), whereas biofilms in the gut lumen and on the mucosal surface may physically protect bacteriocins and enhance their inhibitory effects in localised microenvironments.

2.5 Mucosal Populations in Ulcerative Colitis

Many diseases that occur in the large gut are of unknown aetiology, but microorganisms have been implicated either as causative agents or maintenance factors in a number of bowel disorders. A number of bacterial species perturb normal gut homeostasis and evoke an acute inflammatory response in the host. In most, though not all, cases, the principal organisms involved are adherent or invasive to the gut epithelium and include enterotoxigenic strains of *Escherichia coli*, as well as species belonging to the genera *Yersinia, Shigella, Salmonella, Campylobacter*, and *Aeromonas* (Cohen and Giannella, 1991; Macfarlane and Gibson, 1995). The clinical effects of these bacteria are usually acute rather than chronic, and their pathogenicity and host responses have been well studied. However, the role of bacteria in other, more chronic forms of gut disease is less clear. Antibiotic-associated colitis, inflammatory bowel disease (IBD), and large bowel cancer are all thought to have an aetiology connected in some way with the activities of the microbiota. The inflammatory response of ulcerative colitis (UC) is primarily located in the colonic mucosa and submucosa. The distal colon is always affected, with the condition expressing itself in acute attacks followed by periods of symptom-free remission. Interestingly, the disease frequently appears first in the rectum, progressing up towards the proximal bowel. Bacterial involvement has been proposed in both the initiation and the maintenance stages of UC (Hill, 1986). A variety of species, including *Streptococcus mobilis*, shigella, and, fusobacteria, have received attention as being aetiologic agents in UC (Onderdonk, 1983), largely because these organisms either are able to penetrate the gut epithelium or cause a similar range of disease symptoms in experimental animals. Evidence also points towards strains of *E. coli* isolated from the colitic bowel having increased adhesive properties (Chadwick, 1991), although this may be an adaptation to the disease state in the host. Bradley, Wyatt, and Bayliss (1987) observed a higher than normal proportion of facultative anaerobes and clostridia in IBD patients; however, other workers have reported that antimicrobial agents specifically active against obligate anaerobes prevented ulceration in guinea pigs (Onderdonk and Bartlett, 1979), as well as increased antibody production against strictly anaerobic species (Monteiro et al., 1971). In general, however, evidence for a specific transmissible agent in UC is weak, since antibody production is usually low and the majority of bacteria that have been implicated by various workers are not found in all patients with the disease.

Despite this, there is a good case for bacteria growing on the gut wall to have a fundamental role in IBD, either as a result of members of the normal commensal microflora eliciting an inappropriate response by the innate immune system or by pathogenic organisms proliferating on the epithelial surface and invading the underlying mucosa. Alternatively, mucosa-associated bacteria may be involved through colonisation resistance, where non-pathogenic commensal species occupy adhesion sites on the mucosa and prevent the establishment of harmful microorganisms. This was recently demonstrated when non-pathogenic *E. coli* were successfully used to treat UC patients (Rembacken et al., 1999).

Because bacteria growing on the mucosal surface in the large intestine exist in close proximity to host tissues, they interact to a greater extent with the immune and neuroendocrine systems than their lumenal counterparts. As discussed earlier, very few investigations have been made on the bacteria that inhabit the human colonic mucosa. There are a number of reasons for this apparent incongruity: first, faeces and material taken directly from the lumen of the bowel are readily available for study, whereas in healthy people there are considerable practical and ethical problems in obtaining fresh biopsy tissue. Although this can be circumvented, to some degree, by using fresh tissue obtained at autopsy (Cummings et al., 1987), the material must be acquired within a few hours of death. In the normal course of events, operative specimens are also of questionable value from the microbiologist's viewpoint, as in the vast majority of cases the patients will have been pretreated with drugs and antibiotics and starved, whereas the bowel itself will have been cleansed before surgery.

Not having the problems associated with obtaining untreated tissue samples from the normal colon, rectal biopsies are useful for studying epithelial populations in the gut, as they are routinely available from gastroenterology outpatients clinics, the patients do not need to be treated before the tissues are removed, and rectal inflammatory manifestations are easily recognisable to the physician. Other benefits associated with the use of rectal biopsies are that, for most of the time, the rectum is empty and the mucosa is clean and uncontaminated with faecal material.

Microbiological analysis of the rectal mucosa has shown that it harbours a rich and diverse microbiota. Figure 12.5 demonstrates that bacteria often exist in microcolonies on the rectal mucosa and that these organisms are

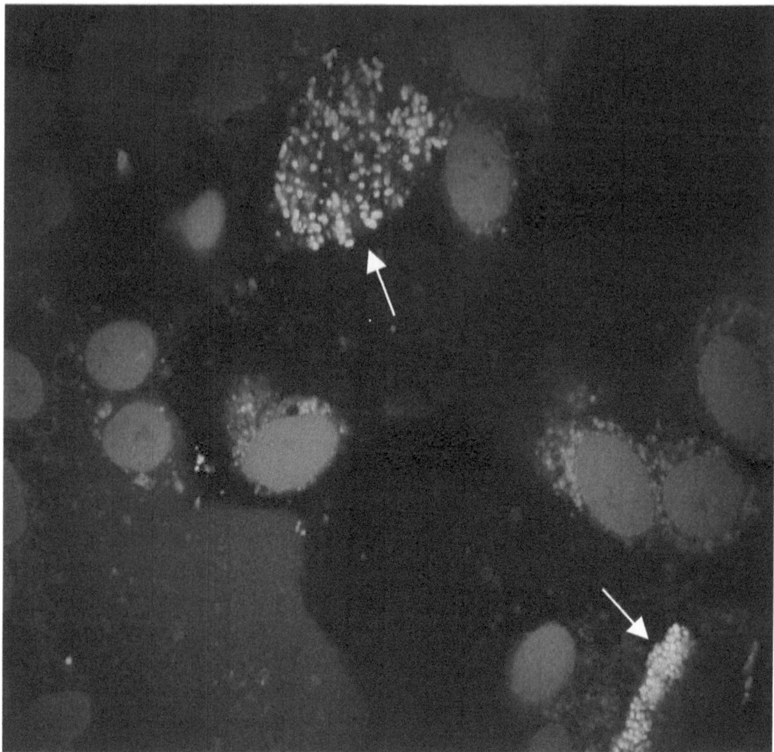

Figure 12.5: Light micrograph showing a live/dead stain of two bacterial microcolonies (arrows) on the surface of the rectal mucosa. Yellow cells are live, and red bacteria are dead. (See colour section.)

actively growing on the epithelial surface. The confocal image in Figure 12.6 indicates that the small bacterial colony which appears at the bottom of Figure 12.5 is mainly composed of live cells at the mucosal surface, whereas on the side of the gut lumen, the majority of bacteria are dead. Table 12.3 shows that although facultative anaerobes are present in comparatively high numbers on rectal epithelia, strictly anaerobic species predominate by a ratio of about 5:1 in UC and 10:1 in healthy persons. In numerical terms, the composition of the rectal microbflora contrasts markedly with microbiotas in the gut lumen, where facultatively anaerobic bacteria are outnumbered 1,000-fold by strictly anaerobic species. These measurements are in broad agreement with results obtained from colonic tissue at autopsy (Croucher et al., 1983), but differ from the data obtained by Poxton et al. (1997), where strict anaerobes on the mucosal surface were 10- to 100-fold higher than facultative species.

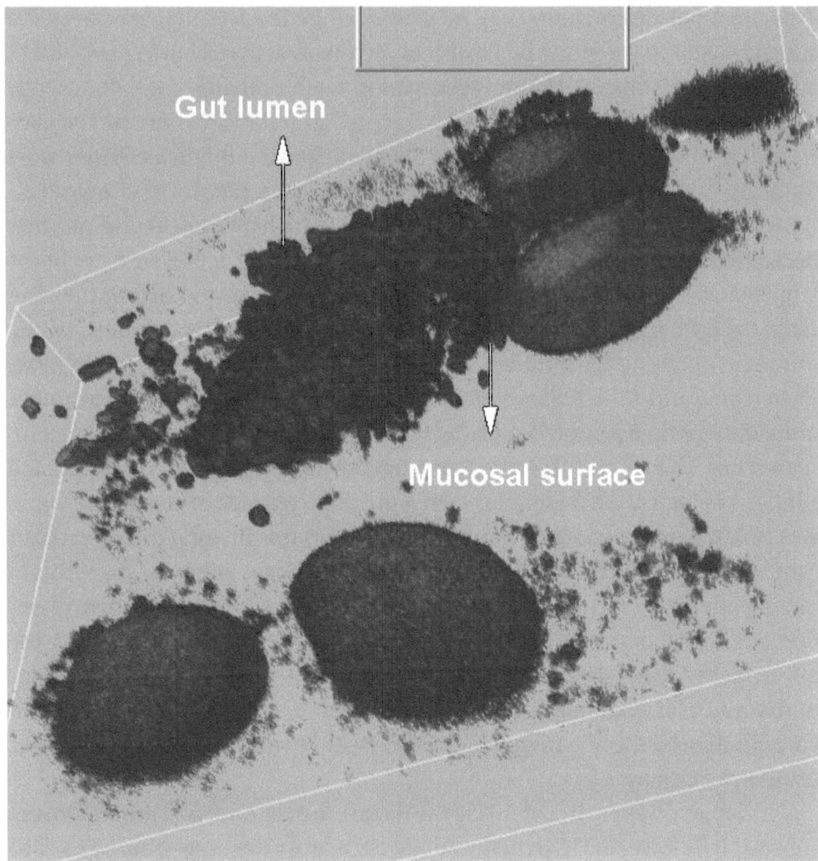

Figure 12.6: Confocal image of the small bacterial microcolony in Figure 12.5. Green bacteria are living cells, and red bacteria are dead. (See colour section.)

Table 12.3: Enumeration and identification of major groups of bacteria in rectal biopsies obtained from UC patients and healthy subjects

	Cell Count ($\log_{10}$ cm^{-2})	
Bacterial Group	**UC (n = 9)**	**Healthy (n = 11)**
Bacteroides	5.2 ± 0.7	5.2 ± 1.4
Bifidobacteria	3.6 ± 0.7	4.6 ± 0.8
Clostridia	3.4 ± 1.2	2.9 ± 1.1
Lactobacilli	4.6 ± 0.2	3.7 ± 1.3
Gram-positive cocci	4.2 ± 1.4	4.4 ± 0.7
Gram-negative cocci	4.6 ± 0.5	3.3 ± 1.7
Gram-negative facultative anaerobes	4.7 ± 1.3	4.2 ± 1.3

This apparent discrepancy may be explained by the fact that several of the subjects used in the study by Poxton and co-workers were taking antibiotics and all of the subjects in the investigation were prepared for colonoscopy 48 hours before biopsy samples were taken using co-dansthrusate and Picolax.

Species belonging to the genera *Bacteroides* and *Bifidobacterium* were found to be the most ubiquitous anaerobes in our studies. Other investigations involving the use of colonic and rectal biopsies have also determined that bacteroides are the major anaerobes associated with mucosal surfaces in the large bowel (Poxton et al., 1997). These experiments showed that *Bacteroides vulgatus* and *B. fragilis* were the predominant organisms, whereas of the fourteen different bacteroides species isolated in our work, *B. vulgatus, B. stercoris,* and *B. thetaiotaomicron* were the most prevalent. Interestingly, individuals who harboured *B. vulgatus* did not exhibit carriage of *B. thetaiotaomicron.* The reverse was also true, suggesting that these species occupied similar ecological or metabolic niches in their respective hosts. *B. fragilis* was isolated from only one (colitic) individual in this investigation, yet when Namavar et al. (1989) studied lumenal and mucosal adherent bacteroides in ten patients with colon cancer, using biopsies obtained at operative resection and faecal material, they observed that although *B. fragilis* counts were low in faeces, these organisms acccounted for 42 per cent of total bacteroides in the mucosal population. However, apart from suffering from severe and chronic illness, these patients had been treated with antibiotics and other drugs prior to surgery.

Bifidobacteria heavily colonised the rectal mucosa, and their numbers were substantially lower in UC patients (Table 12.3), possibly suggesting a link with the disease. A number of different bifidobacterial species were isolated from rectal biopsies in our investigations, and marked qualitative, as well as quantitative, differences were observed in mucosal carriage of these organisms. For example, of the seven different species detected, only *Bifidobacterium angulatum* and *B. bifidum* were found in both subject groups. Similar observations were made with respect to Gram-positive cocci such as peptostreptococcus and enterococcus, as well as lactobacilli, and certain strictly anaerobic Gram-negative cocci (veillonellas), which were only found in the colitic subjects. These results show that carriage of certain groups of bacteria on the rectal mucosa is different in UC patients and healthy individuals. However, whether these differences can be attributed to cause or effect in relation to disease aetiology remains to be established.

Mucosal bacterial populations in the human large bowel are members of complex multi-species consortia. In other environments, such as the oral cavity, these surface-dwelling communities are often highly evolved assemblages, where partner recognition appears to be very specific during the formative stages of co-aggregation and ecosystem development (Kolenbrander, 1989). Biofilm communities frequently exhibit highly coordinated multicellular behaviour within and between species, and many biofilm properties are dependent on local cell population densities, as, for example, in quorum sensing transcriptional activation in Gram-negative bacteria (Salmond et al., 1995). Future studies on mucosal communities in the human large bowel need to address the mechanisms whereby these microbiotas develop and to determine in ecological, physiological, and biochemical terms the processes whereby individual groups of microorganisms interact with each other and the host.

3 *IN VITRO* MODELLING OF BIOFILMS FORMED BY COLONIC MICROORGANISMS

Hydrogen, which is formed as a byproduct during the fermentation of carbohydrates and amino acids, is an important metabolic waste product in the human large intestine. It serves as an electron donor in many bacterial processes in the gut, allowing growth of a variety of syntrophic species (Macfarlane and Gibson, 1996), including acetogenic bacteria, methanogens, and dissimilatory sulphate-reducing bacteria (SRB). SRB belonging to the genus *Desulfovibrio* predominate in the large bowel, and, although these organisms are nutritionally versatile, H_2 is a major electron donor (Gibson, Cummings, and Macfarlane, 1988; Gibson, 1990). SRB use sulphate as a terminal electron acceptor in dissimilatory metabolism, reducing the anion to H_2S, a potent cellular toxin. Because of this, SRB are believed to be involved in UC (Gibson, Cummings, and Macfarlane, 1991; Cummings and Macfarlane, 2001).

Because of their H_2 scavenging abilities, SRB can affect the outcome of the fermentation process in the large gut (Macfarlane, Gibson, and Cummings, 1992). This is clearly demonstrated in Figure 12.7, which shows results from a study in which human gut contents, obtained at autopsy, were incubated with either sulphate, to stimulate SRB activities, or molybate, a competitive inhibitor of dissimilatory sulphate metabolism. Compared to the control, acetate and propionate were stimulated by sulphate, whereas butyrate formation was reduced and lactate was not detected. Conversely, when molybdate inhibited SRB metabolism, SCFA production declined markedly, whereas pyruvate and lactate, which are fermentation intermediates, accumulated in the cultures.

To further investigate this phenomenon, chemostat studies were undertaken to determine how *D. desulfuricans* interacted with simplified

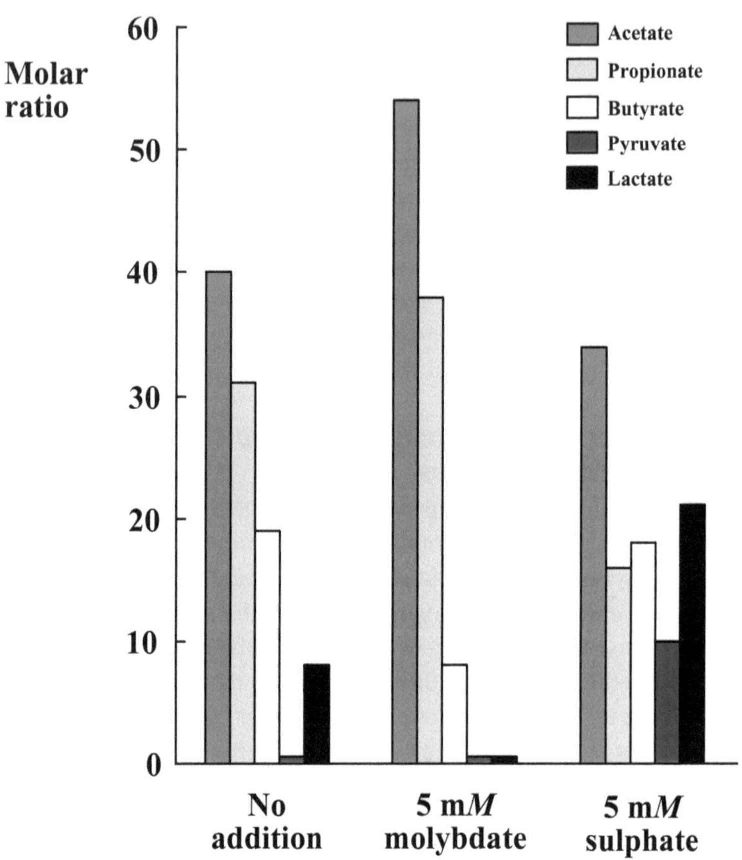

Figure 12.7: Effect of intestinal sulphate-reducing bacteria on fermentation product formation in incubations of human intestinal contents.

microbiotas based on defined populations of intestinal microorganisms (Newton et al., 1998). The continuous cultures used in these experiments were not intended to simulate the colon, but were designed to investigate physiological and ecological interactions between specific groups of microorganisms under controlled conditions. The bacteria used in these studies and the effects of introducing *D. desulfuricans* are shown in Table 12.4. The results show that stable and reproducible multi-species communities comprising common intestinal bacteria could be maintained for extended periods in chemostat culture. *B. longum, B. adolescentis,* and *Bacteroides thetaiotaomicron* were numerically predominant in these continuous cultures. However, when *D. desulfuricans* was added, planktonic *Bifidobacterium longum,*

Table 12.4: Effect of *Desulfovibrio desulfuricans* on the establishment and species composition of biofilms in defined continuous cultures of colonic bacteria

Organism	Planktonic without *Desulfovibrio*	With Added *Desulfovibrio*	
		Planktonic	Wall Biofilm
D. desulfuricans	Not applicable	8.4	9.2
Bacteroides vulgatus	8.7	8.7	9.0
B. thetaiotaomicron	9.3	9.4	9.5
Escherichia coli	8.7	8.0	8.5
Enterococcus faecalis	9.1	8.5	8.6
E. faecium	8.1	8.1	8.4
Clostridium innocuum	8.7	8.5	Not detected
C. perfringens	9.2	8.2	8.7
C. butyricum	8.5	9.2	7.2
Bifidobacterium adolescentis	9.6	9.2	7.0
B. pseudolongum	9.3	9.4	Not detected
B. longum	9.8	8.6	8.4
B. infantis	8.3	8.6	7.5

Note: Bacterial cell counts are $\log_{10}$ per mL of culture fluid or wall biofilm.

B. pseudolongum, and *Clostridium perfringens* decreased tenfold. Many other populations declined to a lesser extent, including *E. coli*, *Enterococcus faecalis*, and *B. adolescentis*, but both bacteroides were relatively unaffected, while numbers of *C. butyricum* increased.

An extensive multispecies biofilm developed rapidly on the chemostat wall after introduction of the SRB. The fact that wall biofilms only formed in the presence of the SRB suggests that this was due to the extracellular polysaccharides that are synthesised by these organisms (Beech et al., 2001). *D. desulfuricans*, *Bacteroides thetaiotaomicron*, and *B. vulgatus* predominated in the biofilms, although other planktonic species were also present in significant numbers. However, with the exception of *Bifidobacterium longum*, bifidobacteria poorly colonised the biofilm, while *B. pseudolongum* and *C. innocuum* were important planktonic species that were not detected at any time.

In addition to their effects on bacterial species composition and structure in the chemostat, SRB profoundly affected metabolic processes and carbon flow in the community (Table 12.5). This was evident by the reduction in total SCFA formed and the increased production of more oxidised fermentation products such as acetate, which is a characteristic of other habitats in which SRB occur (Gibson, 1990). Reduction in the branched chain fatty acids isobutyrate, isovalerate/2-methylbutyrate, and isocaproate showed that *D. desulfuricans* also affected dissimilatory amino acid metabolism (Macfarlane, Cummings,

Table 12.5: Effect of *Desulfovibrio desulfuricans* on production of fermentation products by defined populations of other intestinal bacteria in continuous culture

Fermentation Acid (m*M*)	Chemostat without *Desulfovibrio*	Chemostat with *Desulfovibrio*
Acetate	32.4 (43)	46.2 (71)
Propionate	23.4 (31)	12.4 (19)
Butyrate	19.2 (26)	6.4 (10)
Branched chain fatty acids	8.0	Trace
Total	83.0	65.0
Lactate	13.5	Trace

Note: Values in parenthesis are molar ratios (acetate:propionate:butyrate).

and Allison, 1986). Lactate, an important fermentation product formed by enterococci and bifidobacteria, was not detected, which indicated that this fermentation intermediate was being used as an electron donor by *D. desulfuricans.* However, it is clear that the desulfovibrios were also scavenging H_2, since acetate concentrations increased while, despite the presence of

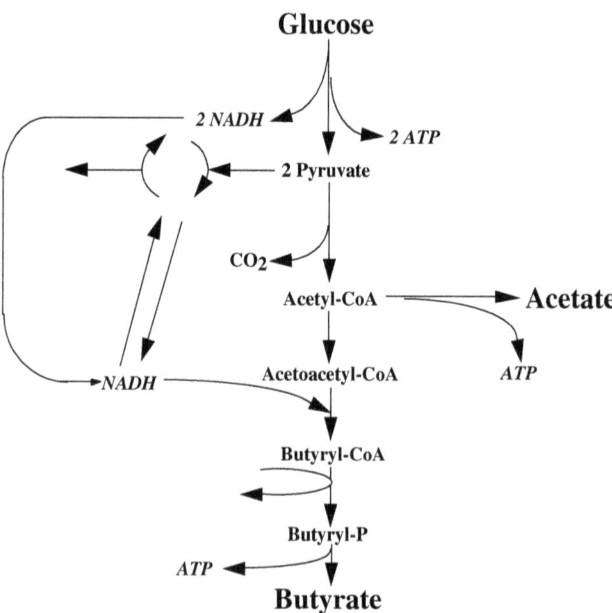

Figure 12.8: Pathway of butyrate formation in saccharolytic clostridia showing interactions between energy (ATP) generation, reducing power (NADH, FdH_2) consumption, and electron sink product (H_2, butyrate, lactate) formation.

substantial numbers of *C. butyricum*, butyrate formation was reduced three-fold. This shows that reduction of pH_2 by *D. desulfuricans* enabled the clostridium to dispose of excess reducing equivalents via ferredoxin: NADH oxidoreductase and hydrogenase (Macfarlane, 1991), thereby diverting carbon flow from butyrate, which is used as an electron sink, towards acetate and enabling the bacterium to produce extra ATP (see Figure 12.8). This explains why numbers of *C. butyricum* increased in the SRB culture.

These experiments highlighted the occurrence of wide-ranging interactions between *D. desulfuricans* and saccharolytic and amino acid fermenting bacteria isolated from the large intestine. The extent to which this was due to biofilm creation by the SRB is unclear. However, through their ecological and physiological effects on butyrate production, the occurrence and activities of SRB in the large bowel are likely to be of considerable metabolic significance to the host.

REFERENCES

Adlerberth, I., Ahrne, S., Johansson, M.-L., Molin, G., Hanson, L. A. and Wold, A. E. (1996). A mannose-specific adherence mechanism in *Lactobacillus plantarum* conferring binding to the human colonic cell line HT-29. *Applied and Environmental Microbiology*, 62, 2244–2251.

Anwar, H., Dasgupta, M. K. and Costerton, J. W. (1990). Testing the susceptibility of bacteria in biofilms to antibacterial agents. *Antimicrobial Agents and Chemotherapy*, 34, 2043–2046.

Beech, I. B., Paiva, M., Caus, M. and Coutinho, C. M. L. (2001). Enzymatic activity within biofilms of sulphate-reducing bacteria. In *Biofilm Community Interactions: Chance or Necessity?* eds. P. Gilbert, D. Allison, M. Brading, J. Verran and J. Walker, pp. 231–239. Cardiff: Bioline.

Bernet, M.-F., Brassart, D., Neeser, J.-R. and Servin, A. L. (1993). Adhesion of human bifidobacterial strains to cultured human intestinal epithelial cells and inhibition of enteropathogen-cell interactions. *Applied and Environmental Microbiology*, 59, 4121–4128.

Bernet, M.-F., Brassart, D., Neeser, J.-R. and Servin, A. L. (1994). *Lactobacillus acidophilus* LA 1 binds to cultured human intestinal cell lines and inhibits cell attachment and cell invasion by enterovirulent bacteria. *Gut*, 35, 483–489.

Bradley, H. K., Wyatt, G. M. and Bayliss, C. E. (1987). Instability in the faecal flora of a patient suffering from food-related irritable bowel syndrome. *Medical Microbiology*, 23, 29–32.

Breznak, J. A. and Pankratz, H. S. (1977). In situ morphology of the gut microbiota of wood eating termites [*Reticulitennes flaviceps* Kollar and *Coptotermes formosanus* Shiraki]. *Applied and Environmental Microbiology*, 33, 406–426.

Chadwick, V. S. (1991). Etiology of chronic ulcerative colitis and Crohn's disease. In *The Large Intestine: Physiology, Pathophysiology and Disease*, eds. S. F. Phillips, J. H. Pemberton and R. G. Shorter, pp. 445–463. New York: Raven Press.

Cohen, M. B. and Giannella, R. A. (1991). Bacterial infections: pathophysiology, clinical features and treatment. In *The Large Intestine: Physiology, Pathophysiology and Disease*, eds. S. F. Phillips, J. H. Pemberton and R. G. Shorter, pp. 395–428. New York: Raven Press.

Conrad, R., Phelps, T. J. and Zeikus, J. G. (1985). Gas metabolism evidence in support of the juxtaposition of hydrogen-producing and methanogenic bacteria in sewage sludge and lake sediments. *Applied and Environmental Microbiology*, 50, 595–601.

Crociani, J., Grill J.-P., Huppert, M. and Ballongue, J. (1995). Adhesion of different bifidobacteria strains to human enterocyte-like Caco-2 cells and comparison with in vivo study. *Letters in Applied Microbiology*, 21, 146–148.

Croucher, S. C., Houston, A. P., Bayliss, C. E. and Turner, R. J. (1983). Bacterial populations associated with different regions of the human colon wall. *Applied and Environmental Microbiology*, 45, 1025–1033.

Cummings, J. H. (1978). Diet and transit through the gut. *Journal of Plant Foods*, 3, 83–95.

Cummings, J. H., Bingham, S. A., Heaton, K. W. and Eastwood, M. A. (1993). Fecal weight, colon cancer risk and dietary intake of non-starch polysaccharides (dietary fiber). *Gastroenterology*, 103, 1783–1789.

Cummings, J. H. and Macfarlane, G. T. (2001). Is there a role for microorganisms? In *Questions and Uncertainties in Inflammatory Bowel Disease*, eds. D. P. Jewell, N. Mortensen and B. F. Warren, pp. 42–51. Oxford: Blackwell Science.

Cummings, J. H., Pomare, E. W., Branch, W. J., Naylor, C. P. E. and Macfarlane, G. T. (1987). Short chain fatty acids in human large intestine, portal, hepatic and venous blood. *Gut*, 28, 1221–1227.

Dehio, M., Gomez-Duarte, O. G., Dehio, C. and Meyer, T. F. (1998). Vitronectin-dependent invasion of epithelial cells by *Neisseria gonorrhoea* involves alpha(v) integrin receptors. *FEBS Letters*, 424, 84–88.

De Simone, C., Ciardi, A., Grassi, A., Lambert Gardini, S., Tzantzoglou, S., Trinchieri, V., Moretti, S. and Jirillo, E. (1992). Effect of *Bifidobacterium bifidum* and *Lactobacillus acidophilus* on gut mucosa and peripheral blood B lymphocytes. *Immunopharmacology and Immunotoxicology*, 14, 331–340.

De Vuyst, L. and Vandamme, E. J. (1994). Antimicrobial potential of lactic acid bacteria. In *Bacteriocins of Lactic Acid Bacteria*, eds. L. De Vuyst and E. J. Vandamme, pp. 91–142. Glasgow: Blackie Academic and Professional.

Dodd, H. M. and Gasson, M. J. (1994). Bacteriocins of lactic acid bacteria. In *Genetics and Biotechnology of Lactic Acid Bacteria*, eds. M. J. Gasson and W. M. de Vos, pp. 211–251. Glasgow: Blackie Academic and Professional.

Duensing, T. D., Wing, J. S. and van Putten, J. P. M. (1999). Sulfated polysaccharide-directed recruitment of mammalian host proteins: a novel strategy in microbial pathogenisis. *Infection and Immunity*, 67, 4463–4468.

Edmiston, C. E., Jr., Avant, G. R. and Wilson, F. A. (1982). Anaerobic bacterial populations on normal and diseased human biopsy tissue obtained at colonoscopy. *Applied Environmental Microbiology*, 43, 1173–1181.

Finegold, S. M., Flora, D. J., Attlebury, H. R. and Sutter, L. V. (1975). Fecal bacteriology of colonic polyp patients and control patients. *Cancer Research*, 35, 3407–3417.

Finegold, S. M., Sutter, V. L. and Mathisen, G. E. (1983). Normal indigenous intestinal flora. In *Human Intestinal Microflora in Health and Disease*, ed. D. J. Hentges, pp. 3–31. London: Academic Press.

Gibson, G. R. (1990). Physiology and ecology of the sulphate-reducing bacteria. *Journal of Applied Bacteriology*, 69, 769–797.

Gibson, G. R., Cummings, J. H. and Macfarlane, G. T. (1988). Competition for hydrogen between sulphate-reducing bacteria and methanogenic bacteria from the human large intestine. *Journal of Applied Bacteriology*, 65, 241–247.

Gibson, G. R., Cummings, J. H. and Macfarlane, G. T. (1991). Growth and activities of sulphate-reducing bacteria in gut contents from healthy subjects and patients with ulcerative colitis. *FEMS Microbiology Ecology*, 86, 103–112.

Gibson, G. R. and Wang, X. (1994). Regulatory effects of bifidobacteria on the growth of other colonic bacteria. *Journal of Applied Bacteriology*, 77, 412–420.

Gibson, S. A. W., McFarlan, C., Hay, S. and Macfarlane, G. T. (1989). Significance of the microflora in proteolysis in the colon. *Applied and Environmental Microbiology*, 55, 679–683.

Hill, M. J. (1986). The possible role of bacteria in inflammatory bowel disease. *Current Concepts in Gastroenterology*, 3, 10–14.

Hopkins, M. J., Sharp, R. and Macfarlane, G. T. (2001). Age and disease-related changes in intestinal bacterial populations assessed by cell culture, 16S rRNA abundance and community cellular fatty acid profiles. *Gut*, 48, 198–205.

Johansson, M.-L., Molin, G., Jeppsson, B., Nobaek, B., Ahrne, S. and Bengmark, S. (1993). Administration of different *Lactobacillus* strains in fermented oatmeal soup: in vivo colonization of human intestinal mucosa and effect on the indigenous flora. *Applied and Environmental Microbiology*, 59, 15–20.

Kolenbrander, P. E. (1989). Surface recognition among oral bacteria: multigeneric coaggregations and their mediators. *Critical Reviews in Microbiology*, 17, 137–159.

Laukova, A. and Czikkova, S. (1998). Inhibition effect of enterocin CCM 4231 in the rumen fluid environment. *Letters in Applied Microbiology*, 26, 215–218.

Lee, A. (1980). Normal flora of animal intestinal surfaces. In *Adsorption of Microorganisms to Surfaces*, eds. G. Bitton and K. C. Marshall, pp. 145–174. New York: John Wiley & Sons.

Lee, F. D., Kraszewski, A., Gordon, J., Howie, J. G. R., McSeveney, D. and Harland, W. A. (1971). Intestinal spirochaetosis. *Gut*, 12, 126–133.

Lejeune, R., Callewaert, R., Crabbe, K. and De Vuyst, L. (1998). Modelling the growth and bacteriocin production by *Lactobacillus amylovorus* DCE 471 in batch cultivation. *Journal of Applied Microbiology*, 84, 159–168.

Macfarlane, G. T. (1991). Fermentation reactions in the large intestine. In *Short-Chain Fatty Acids: Metabolism and Clinical Importance*, eds. J. H. Cummings, J. L. Rombeau and T. Sakata, pp. 5–10. Columbus: Ross Laboratories Press.

Macfarlane, G. T. and Cummings, J. H. (1991). The colonic flora, fermentation and large bowel digestive function. In *The Large Intestine: Physiology, Pathophysiology and Disease*, eds. S. F. Phillips, J. H. Pemberton and R. G. Shorter, pp. 51–92. New York: Raven Press.

Macfarlane, G. T. and Cummings, J. H. (2002). Diet and the metabolism of intestinal bacteria. In *Food Allergy and Intolerance*, 2nd ed. (eds. Brostoff, J., Challacombe, S. J., Kniker, W. T.). W.B. Saunders Company Ltd., London, pp. 321–343.

Macfarlane, G. T., Cummings, J. H. and Allison, C. (1986). Protein degradation by human intestinal bacteria. *Journal of General Microbiology*, 132, 1647–1656.

Macfarlane, G. T. and Gibson, G. R. (1991). Formation of glycoprotein degrading enzymes by *Bacteroides fragilis. FEMS Microbiology Letters*, 77, 289–294.

Macfarlane, G. T. and Gibson, G. R. (1994). Metabolic activities of the normal colonic flora. In *Human Health: The Contribution of Microorganisms*, ed. S. A. W. Gibson, pp. 17–52. London: Springer-Verlag.

Macfarlane, G. T. and Gibson, G. R. (1995). Bacterial infections and diarrhea. In *Human Colonic Bacteria: Role in Nutrition, Physiology and Pathology*, eds. G. R. Gibson and G. T. Macfarlane, pp. 201–226. Boca Raton: CRC Press.

Macfarlane, G. T. and Gibson, G. R. (1996). Carbohydrate fermentation, energy transduction and gas metabolism in the human large intestine. In *Ecology and Physiology of Gastrointestinal Microbes Vol. 1: Gastrointestinal Fermentations and Ecosystems*, eds. R. I. Mackie and B. A. White, pp. 269–318. New York: Chapman & Hall.

Macfarlane, G. T., Gibson, G. R. and Cummings, J. H. (1992). Comparison of fermentation reactions in different regions of the human colon. *Journal of Applied Bacteriology*, 72, 57–64.

Macfarlane, G. T., Gibson, G. R., Drasar, B. S. and Cummings, J. H. (1995). Metabolic significance of the colonic microflora. In *Gastrointestinal and Oesophageal Physiology*, ed. R. Whitehead, pp. 249–274. Edinburgh: Churchill Livingstone.

Macfarlane, G. T. and Macfarlane, S. (1995). Human intestinal biofilm communities. In *The Life and Death of Biofilm*, eds. J. Wimpenny, P. Handley, P. Gilbert and H. Lappin-Scott, pp. 83–89. Cardiff: Bioline.

Macfarlane, G. T., Macfarlane, S. and Gibson, G. R. (1998). Use of a three-stage compound continuous culture system to investigate bacterial growth and metabolism in the human colonic microbiota. *Microbial Ecology*, 35, 180–187.

Macfarlane, S., Cummings, J. H. and Macfarlane, G. T. (1999). Bacterial colonisation of surfaces in the large intestine. In *Colonic Microflora, Nutrition and Health*, eds. G. R. Gibson and M. Roberfroid, pp. 71–87. London: Chapman & Hall.

Macfarlane, S. and Macfarlane, G. T. (2001). Community structure and interactions in the large intestine. In *Biofilm Community Interactions: Chance or Necessity?* eds. P. Gilbert, D. Allison, M. Brading, J. Verran and J. Walker, pp. 83–96. Cardiff: Bioline.

Macfarlane, S., McBain, A. J. and Macfarlane, G. T. (1997). Consequences of biofilm and sessile growth in the large intestine. *Advances in Dental Research*, 11, 59–68.

McCabe, K., Mann, M. D. and Bowie, M. D. (1998). D-Lactate production and [^{14}C] succinic acid uptake by adherent and nonadherent *Escherichia coli. Infection and Immunity*, 66, 907–911.

Meghrous, J., Euloge, P., Junelles, A. M., Ballongue, J. and Petitdemange, H. (1990). Screening of *Bifidobacterium* strains for bacteriocin production. *Biotechnology Letters*, 12, 575–580.

Moore, W. E. C. and Holdeman, L. V. (1974). Human fecal flora. The normal flora of 20 Japanese-Hawaiians. *Applied Environmental Microbiology*, 27, 961–979.

Monteiro, E., Fossey, J., Shiner, M., Drasar, B. S. and Allison, A. C. (1971). Antibacterial antibodies in rectal and colonic mucosa in ulcerative colitis. *Lancet*, 6(1), 249–251.

Mozes, N. and Rouxhet, P. G. (1992). Influence of surfaces on microbial activity. In *Biofilms-Science and Technology*, eds. L. F. Melo, T. R. Bott and B. Capdeville, pp. 125–136. Dordrecht: Kluwer Academic Publishers.

Namavar, F., Theunissen, E. B., Verweij-Van Vught, A. M., Peerbooms, P. G., Bal, M., Hoitsma, H. F. and Maclaren, D. M. (1989). Epidemiology of the *Bacteroides fragilis* group in the colonic flora in 10 patients with colonic cancer. *Journal of Medical Microbiology*, 29, 171–176.

Nelson, D. P. and Mata, L. J. (1970). Bacterial flora associated with the human gastrointestinal mucosa. *Gastroenterology*, 58, 56–61.

Newton, D. F., Cummings, J. H., Macfarlane, S. and Macfarlane, G. T. (1998). Growth of a human intestinal *Desulfovibrio desulfuricans* in continuous cultures containing defined populations of saccharolytic and amino acid fermenting bacteria. *Journal of Applied Microbiology*, 85, 372–380.

Onderdonk, A. B. (1983). Role of the intestinal microflora in ulcerative colitis. In *Human Intestinal Microflora in Health and Disease*, ed. D. J. Hentges, pp. 481–493. London: Academic Press.

Onderdonk, A. B. and Bartlett, M. D. (1979). Bacteriological studies of experimental ulcerative colitis. *American Journal of Clinical Nutrition*, 32, 258–265.

O'Riordan, K. C., Condon, S. and Fitzgerald, G. F. (1995). Bacterial interference by *Bifidobacterium* species and a comparative analysis of genomic profiles from strains of this genus. *Proceedings of the Lactic Acid Bacteria Conference, Cork, Ireland*, p. 207.

Patti, J. M., Allen, B. L., McGavin, M. J. and Hook, M. (1994). MSCRAMM-mediated adherence of microorganisms to host tissues. *Annual Review in Microbiology*, 48, 585–617.

Perdigon, G., Medici, M., Bibas Bonet de Jorrat, M. E., de Budeguer, M. V. and Pesce de Ruiz Holgado, A. (1993). Immunomodulating effects of lactic acid bacteria on mucosal and tumoral immunity. *International Journal of Immunotherapy*, IX, 29–52.

Poxton, I. R., Brown, R., Sawyerr, A. and Ferguson, A. (1997). Mucosa-associated bacterial flora of the human colon. *Journal of Medical Microbiology*, 46, 85–91.

Pullan, R. D., Thomas, G. A. O., Rhodes, M., Newcombe, R. G., Williams, G. T., Allen, A. and Rhodes, J. (1994). Thickness of adherent mucus gel on colonic mucosa in humans and its relevance to colitis. *Gut*, 35, 353–359.

Rembacken, B. J., Snelling, A. M., Hawkey, P. M., Chalmers, D. M. and Axon, A. T. R. (1999). Non-pathogenic *Escherichia coli* versus mesalzine for the treatment of ulcerative colitis: a randomised trial. *Lancet*, 354, 635–639.

Roediger, W. E. W. (1980). Role of anaerobic bacteria in the metabolic welfare of the colonic mucosa of man. *Gut*, 21, 793–798.

Salmond, G. P. C., Bycroft, B. W., Stewart, G. S. A. B. and Williams, P. (1995). The bacterial 'enigma': cracking the code of cell–cell communication. *Molecular Microbiology*, 16, 615–624.

Sato, J., Mochizuki, K. and Homma, N. (1982). Affinity of the *Bifidobacterium* to intestinal mucosal epithelial cells. *Bifidobacteria Microflora,* 1, 51–54.

Savage, D. C. (1977). Microbial ecology of the gastrointestinal tract. *Annual Review in Microbiology,* 31, 107–133.

Savage, D. C. (1978). Factors involved in colonization of the gut epithelial surface. *American Journal of Clinical Nutrition,* 31, S131–S135.

Schiffrin, E. J., Brassart, D., Servin, A. L., Rochat, F. and Donnet-Hughes, A. (1997). Immune modulation of blood leukocytes in humans by lactic acid bacteria: criteria for strain selection. *American Journal of Clinical Nutrition,* 66, 15S–20S.

Sharp, R. and Ziemer, C. J. (1999). Application of taxonomy and systematics to molecular techniques in intestinal microbiology. In *Colonic Microflora, Nutrition and Health,* eds. G. R. Gibson and M. Roberfroid, pp. 167–190. London: Chapman & Hall.

Standiford, T. K., Arenberg, D. A. and Danforth, J. M. (1994). Lipoteichoic acid induces secretion of interleukin-8 from human blood monocytes: a cellular and molecular analysis. *Infection and Immunity,* 62, 119–25.

Takeuchi, A., Jervis, H. R., Nakagawa, H. and Robinson, D. M. (1974). Spiral-shaped organisms on the surface colonic epithelium of the monkey and man. *American Journal of Clinical Nutrition,* 27, 1287–1296.

Van Loosdrecht, M. C. M., Lyklema, J., Norde, W. and Zehnder, A. J. B. (1990). Influence of interfaces on microbial activity. *Microbiological Reviews,* 54, 75–87.

Wallace, R. J., Cheng, K.-J., Dinsdale, D. and Orskov, E. R. (1979). An independent microbial flora of the epithelium and its role in the microbiology of the rumen. *Nature,* 279, 424–426.

Wilkins, T. D. (1981). Microbiological considerations in interpretation of data obtained from experimental animals. *Banbury Report,* 7, 3–9.

Young, G. P. (1991). Butyrate and the molecular biology of the large bowel. In *Short Chain Fatty Acids: Metabolism and Clinical Importance,* eds. J. H. Cummings, J. L. Rombeau and T. Sakata, pp. 39–44. Columbus: Ross Laboratories Press.

CHAPTER THIRTEEN

Pseudomonas aeruginosa Biofilms in Lung Infections

Kimberly K. Jefferson and Gerald B. Pier

1 INTRODUCTION

In 1962, Nobel Laureate Frank MacFarlane Burnet wrote 'one can think of the middle of the twentieth century as the end of one of the most important social revolutions in history, the virtual elimination of the infectious diseases as a significant factor in social life' (Burnet, 1962). Indeed, advances in medical technology have come forth at an impressive rate. However, pathogens have adapted to and even taken advantage of the altered host environments created by modern therapies and medical devices, and new forms of microbial infections continue to challenge and confound modern medical technology. Development of microbial resistance to antibiotics and biocide disinfectants represent well-known examples of microbial adaptation. The increasing frequency with which indwelling medical devices, including intravenous catheters, replacement heart valves, prosthetic hip and knee joints, and artificial hearts, are being used represents a tremendous advancement in modern medicine, but it has also afforded certain microbes with a new opportunity to breach the primary host defences and initiate a focal point of infection. These devices penetrate protective dermal or mucosal layers and provide many pathogenic organisms with a surface on which they can attach and grow. The mode of growth that occurs, that of a community of interconnected and communicating cells known as a biofilm, allows a high density cluster of cells in various physiologic states to coexist. The net result for patients and clinicians is that the adaptation of the bacterial community to the biofilm mode of growth promotes bacterial survival, resistance to antibiotics, and persistence of infection.

Another interesting environment that has emerged in the past 50 years due to modern medical treatment is that of the cystic fibrosis (CF) lung. In the first half of the twentieth century, CF patients died very early, often from malnutrition and acute infection. Improved diagnosis and therapy led to an increased median survival of CF patients to over 31 years by 2000. As the survival of CF patients has increased, the unique environment in their lungs has provided a new niche for chronic infection. The predominant pathogen to emerge was *Pseudomonas aeruginosa*, accounting for over 80 per cent of the mortality due to chronic infection and respiratory failure. Although there are many interesting facets to the pathology of *P. aeruginosa* infection in CF patients, the most challenging component of this disease is the inability to eradicate the infecting organisms once they firmly establish infection. This disease provides the premiere example of how the biofilm mode of growth is critical to understanding and proper management of infectious disease problems. Thus, the growth of *P. aeruginosa* as a biofilm in the lungs of CF patients is one of the most studied infectious disease pathologies because of the many opportunities it provides for understanding biofilm-based infections. What is now clear is that multiple bacterial factors and host components contribute to the formation of a biofilm in the CF lung, and manipulation and control of many of these factors may be required for effective therapy of CF lung disease.

2 CF

CF is inherited in an autosomal recessive fashion and is caused by a mutation in the gene encoding a chloride ion channel, the CF transmembrane conductance regulator (CFTR). The most common mutation is a 3-nucleotide deletion that results in the loss of a phenylalanine residue at position 508 (ΔF508). The ΔF508 mutation impedes proper folding and glycosylation of the CFTR protein, which affects its trafficking from the Golgi and results in its degradation within the cytoplasm (Riordan et al., 1989; Cheng et al., 1990). Between 4 and 5 per cent of Caucasians are heterozygous carriers of alleles comprising either ΔF508 or some other loss-of-function mutation within the *CFTR* gene. This high prevalence and maintenance of a mutation that, when present in the homozygous state, is lethal prior to reproductive years seems counterintuitive. Recent evidence suggests that a possible basis for the perpetuation of *CFTR* mutations in the gene pool may be enhanced resistance of heterozygous carriers to certain infectious diseases. One proposal is based on the finding that CFTR is a receptor for pathogenic bacteria such as *Salmonella enterica* serovar Typhi and that the reduced level of CFTR present on the gastrointestinal epithelial cells of individuals with a non-functional copy of the *CFTR* gene confers increased resistance to typhoid fever (Pier et al., 1998). In addition, heterozygocity for the *CFTR* gene appears to endow gastrointestinal epithelial cells with resistance to toxins such as cholera toxin or perhaps the related heat-labile toxin of *Escherichia coli* (Gabriel et al., 1994). As carriage of a single mutated *CFTR* gene is clinically silent, heterozygous individuals may have historically had a significant survival advantage.

The major consequences of a lack of functional CFTR are altered conductance of chloride ions in and out of cells and altered regulation of other ion channels such as the epithelial sodium channel (Kunzelmann et al., 2000). Consequently, certain tissues produce dehydrated and abnormally viscous secretions, which tend to clog secretory ducts such as the pancreas and affect the function of other secretory tissues such as the lungs and sweat glands. Most CF males are infertile because the vas deferens fails to develop normally, and it has been shown that mutations in the *CFTR* gene, including mutations that do not result in other clinical manifestations of CF, are the most common cause of congenital bilateral absence of the vas deferens (Chillon et al., 1995).

The emergence of chronic *P. aeruginosa* lung infection as the primary cause of morbidity and ultimately death in CF patients has occurred along with improvements in disease therapy. Historically, mortality was associated with pancreatic insufficiency and typically took the lives of its victims before the end of their first year (Harper, 1930; Andersen, 1938). Implementation of dietary control essentially eliminated pancreatic insufficiency as a cause of death in CF patients, although most patients still have nutritional deficiencies. Nutritional supplementation would keep afflicted individuals alive past their first year, but they generally developed *Staphylococcus aureus* pneumonia and typically died in infancy (Gilligan, 1991). With the advent of appropriate antibiotics, staphylococcal pneumonia became manageable and was no longer a significant cause of mortality. *S. aureus* and *Haemophilus influenzae* cause some morbidity in CF patients, but respiratory infections caused by these pathogens can usually be controlled with antibiotic therapy (Pedersen, 1992). By age 8, however, over 50 per cent of CF patients become infected with *P. aeruginosa*, and carriage increases to 80 per cent by 15–20 years of age. Eventually, the infecting strains of *P. aeruginosa* become refractory to antibiotic therapy and remain a chronic and insidious inhabitant of the lungs of the vast majority of CF patients until they succumb to respiratory failure, usually by the age of 30–35.

3 EXPLANATION FOR THE HIGH PREVALENCE OF *P. AERUGINOSA* INFECTION IN CF

P. aeruginosa is a common environmental organism that can be isolated from soil and water. Despite its pervasiveness in the environment, however, it seldom infects healthy individuals (Costerton, 1984). Deficiencies in host defences such as burn wounds, chemotherapy-induced neutropenia, and acquired immunodeficiency syndrome can lead to infections with *P. aeruginosa* (Kielhofner et al., 1992; Lyczak, Cannon, and Pier, 2000). This does not, however, explain the association of *P. aeruginosa* with CF patients who are otherwise immunocompetent. A number of explanations for the predisposition of the CF lung to infection with *P. aeruginosa* have been proposed.

In the normal lung, inhaled bacteria such as *P. aeruginosa* are typically removed by the mucociliary escalator which pushes the organisms, entrapped in respiratory mucus, up the trachea so that they are diverted to the oesophagus, swallowed, and ultimately destroyed within the acidic juices of the stomach (Govan and Deretic, 1996). The absence of CFTR on the surface of lung epithelial cells in the CF patient and the resultant abnormal chloride ion conductance leads to a relative dehydration of their pulmonary secretions. The dehydrated lung secretions are abnormally viscous and counteract the function of the mucociliary escalator. This deficiency may partially account for the predisposition of the CF lung to bacterial infections in general, but does not explain the extremely high affinity of *P. aeruginosa* for the CF lung.

Another explanation that has been proposed for the association of *P. aeruginosa* with the CF lung is enhanced adherence of *P. aeruginosa* to the CF lung epithelium. The lack of functional CFTR purportedly leads to an altered glycosylation status of a number of membrane glycoproteins, possibly due to a defect in the regulation of pH in intracellular compartments (Scanlin and Glick, 1999). The effect of CFTR mutations on the pH of the trans-Golgi network was reported by one group to result in defective acidification and, more recently, by another group to result in hyperacidification (Barasch et al., 1991; Scanlin and Glick, 1999). Both groups claim, however, that the defect decreases sialylation of glycoproteins on the cell surface and increases the presence of oligosaccharides such as asialo GM1. Krivan, Ginsburg, and Roberts, (1988) initially found that *P. aeruginosa* binds specifically to asialo GM1, but also reported a large number of other pathogens, including many that do not infect CF patients, such as *Streptococcus pneumoniae*, similarly bound to this oligosaccharide. Numerous other studies shed doubt on the asialo GM1 finding, but increased binding of *P. aeruginosa* to CF lung epithelial cells due to alterations of either cell surface receptors or defects in removal of *P. aeruginosa* from the CF epithelial cell remains a viable hypothesis (Schroeder, Zaidi, and Pier, 2001). Overall, the binding of *P. aeruginosa* to CF cells is only augmented by 10–15 per cent when compared with non-CF cells, making it unlikely to be a prominent factor in the high infection rate of CF patients, particularly when considering that infection of the lungs of healthy individuals with *P. aeruginosa* is almost unheard of (Hoiby, 1982; Pier, 2000).

Structural changes resulting in lung disease develop in patients once they become colonised with *P. aeruginosa* (Pedersen, 1992). Pulmonary damage in CF patients is due to both chronic infection and the ensuing inflammation. It is not yet clear if lung damage is also due to infections with pathogens such as *Staphylococcus aureus* and *H. influenzae* or viral pathogens such as respiratory syncitial virus or influenza virus. Some have argued that the CF lung inherently produces cytokines that lead to inflammation and pulmonary damage prior to detectable infection, thus predisposing patients to pseudomonal infection. *P. aeruginosa* is well known to infect non-CF lungs that have been damaged by chronic inflammation, making it an important complication associated with bronchiectasis (Ip and Lam, 1996). In summary, *P. aeruginosa* appears to be especially well adapted for survival in lungs that have been structurally damaged by chronic inflammation.

The epithelium of certain tissues may rid themselves of adherent and epithelial-cell-ingested bacteria via desquamation. It is likely that mammalian hosts have evolved mechanisms to detect bacteria that have penetrated mucus layers and bound to or invaded epithelial cells. This would represent a host defence against organisms that need to establish themselves on the cell surface to initiate infection or hide inside the cell as a means of evading host defence. As an example of this defence mechanism, epithelial cells of the bladder that contain internalised or surface-bound *E. coli* slough off and undergo an apoptotic death. They are then removed from the bladder during urination (Mulvey et al., 2000). The epithelial cells of the lung appear to utilise a similar mode of defence. Initial contact between normal pulmonary epithelial cells and *P. aeruginosa* results in an increase in the amount of surface CFTR (Pier, Grout, and Zaidi, 1997). This increase is rapid and likely results from the mobilisation of intracellular CFTR stores. CFTR binds specifically to *P. aeruginosa* lipopolysaccharide (LPS). Only LPS that bears a complete outer core oligosaccharide will bind to CFTR, but it does not appear that the O polysaccharide side chains that give rise to smooth LPS are directly involved in binding (Pier, 2000). The binding of CFTR to the complete core oligosaccharide triggers endocytosis of the bacteria (Pier et al., 1996). Following endocytosis of *P. aeruginosa*, the apoptotic pathway is triggered within infected host cells, resulting in exfoliation. Pulmonary epithelial cells within the CF lung apparently still bind *P. aeruginosa* to an extent, but due to their lack of membrane-bound CFTR, they are incapable of internalising the pathogen and eradicating it via cellular desquamation.

A combination of these and other factors probably underlies the predilection of *P. aeruginosa* for the CF lung. It does not, however, explain how *P. aeruginosa* chronically persists in the CF lung despite intact host acquired immune defences and antibiotic therapy. It is seldom contested that one of the most important factors that contributes to the survival of *P. aeruginosa* in the lung is the tendency of the organism to grow as a biofilm, but the molecular and cellular components critical to biofilm formation and persistence in the lung of CF patients are only now being defined.

4 PSEUDOMONAL BIOFILMS

One of the problems in elucidating the genotypic and phenotypic characteristics of biofilms is that the definition of a biofilm is inexact. Even when just considering the contribution of bacterial factors to biofilm formation, these structures recall the famous quote of former U.S. Supreme Court Justice Potter Stewart about pornography: unable to define 'hard core' pornography precisely, he once wrote 'but I know it when I see it'. Similarly, the structural motifs and architecture of a bacterial biofilm may be obvious to those with expertise in these matters, but the specific motifs and features themselves are not clearly defined. For *P. aeruginosa*, the problem is particularly vexing in that there is a biofilm architecture defined mostly by *in vitro* observations of cells adhering to inert surfaces and by *in vivo* observations of *P. aeruginosa* growing in the lung of a CF patient. These may appear similar in many aspects, but whether the *in vitro* studies serve as an appropriate model for the *in vivo* situation is simply not known. Critically important host factors that contribute to the overall composition of airway secretions in the lungs of infected CF patients and also impact the molecular and cellular characteristics of the *in vivo* biofilm are missing from the current *in vitro* models of biofilm formation.

In culture, *P. aeruginosa* is often grown in nutrient-rich broth, where the organisms tend to grow in a non-adherent, free-swimming, planktonic state. In the environment, where the cells come in contact with solid surfaces, the bacteria are likely exposed to nutrient deprivation and other unfavourable conditions, making the planktonic mode of growth less desirable. In these more natural conditions, *P. aeruginosa* is induced to express the biofilm mode of growth. Biofilm formation is a multistep process that is initiated by the adherence of bacteria to a substrate and continued by the formation of microcolonies. The microcolonies develop over a period of days, and the mature biofilm is characterised by a thick pillar-like growth formation 'vascularised' with aquatic channels that allow nutrients and oxygen to flow and bathe cells within the biofilm. Biofilm formation confers a number of advantages upon *P. aeruginosa*. It enables sequestration of the organism to an area that is relatively rich in nutrients or otherwise favourable for growth and promotes resistance to phagocytosis (Costerton, 1984). In addition, the biofilm cells are more resistant to physical and nutritional changes in the environment, to adverse conditions, and to antimicrobial agents such as disinfectants and antibiotics.

The formation of *P. aeruginosa* biofilms *in vitro* is not simply due to a random conglomeration of bacterial cells, but rather appears to be a deliberate and structured process involving an advanced system of communication between bacterial cells (Costerton, Stewart, and Greenberg, 1999). The communication system is referred to as quorum sensing, and it directs the bacteria to increase or decrease expression of certain genes so that each bacterial cell can assume its appropriate functional role in the elaboration of biofilm. Quorum sensing in *P. aeruginosa* is regulated by the *lasR-lasI* and *rhlR-rhlI* systems (Figure 13.1) (Pesci and Iglewski, 1997; Pesci et al., 1997; Davies et al., 1998). LasI is involved in the synthesis of an extracellular 'pheromone', *N*-(3-oxododecanoyl)-L-homoserine lactone (3-O-C_{12}-HSL), which diffuses into adjacent bacterial cells and cooperates with LasR to induce expression of a number of virulence genes, including *rhlI*. The *rhlI* gene product produces another extracellular signal, *N*-butyryl-L-homoserine lactone (C_4-HSL), which regulates expression of additional virulence factors. The effects of 3-O-C_{12}-HSL and C_4-HSL on gene expression are concentration dependent; consequently, their activity is directly related to cell density. 3-O-C_{12}-HSL is not required for the initial substrate adherence, but appears to be necessary for the later stages of biofilm formation. C_4-HSL is secreted by developed biofilms and has been used as an indicator that *P. aeruginosa* cells are in the biofilm mode of growth, but it does not appear that C_4-HSL is required for biofilm formation (Davies et al., 1998; Singh et al., 2000). The quorum sensing system is intimately involved in the careful coordination of the multistep process of biofilm formation.

Initial attachment of *P. aeruginosa* to a surface to initiate biofilm formation depends upon flagella-based motility, which presumably allows the bacterium

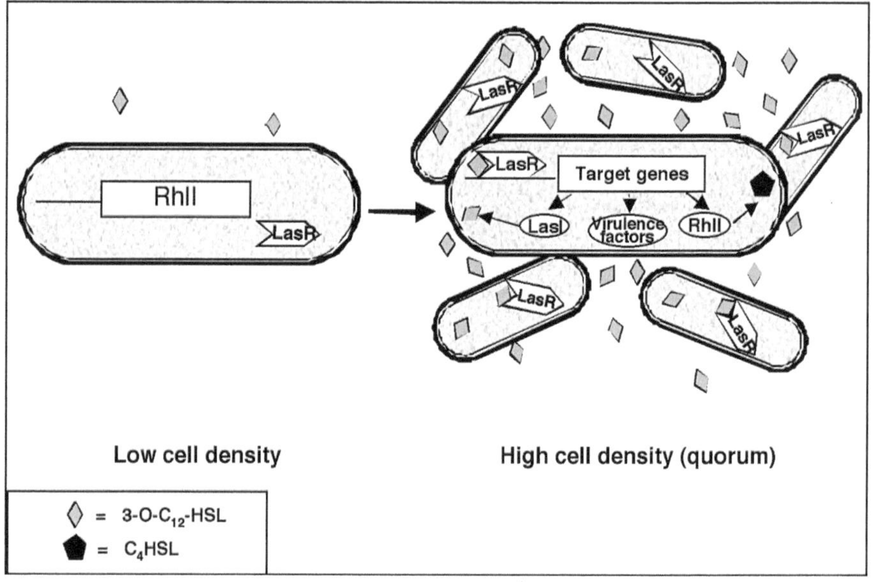

Figure 13.1: *P. aeruginosa* constitutively secretes 3-O-C_{12}-HSL. At low cell density, the concentration of 3-O-C_{12}-HSL is too low to activate the transcription factor LasR. At high cell density, 3-O-C_{12}-HSL accumulates and cooperates with LasR to induce the expression of a number of virulence genes, including *lasI* and *rhlI*. RhlI then catalyses the synthesis of C_4-HSL.

to localise to a suitable substrate and overcome any hydrophobic forces otherwise preventing bacterial contact with a surface (O'Toole and Kolter, 1998). Twitching motility, which is dependent upon the type IV pilus, is required for microcolony formation (O'Toole and Kolter, 1998). Twitching probably allows the cells to swarm together and initiate the formation of the quorum necessary for LasI-induced activation of the LasR transcription factor. LasR decreases the expression of flagella and type IV pili, thereby stabilising the biofilm and also regulating transcription of a number of other genes (Whiteley et al., 2001). The roles of many of these LasR-regulated genes in biofilm maturation have not yet been identified. Interestingly, a recent comparison of gene expression of planktonic and biofilm *P. aeruginosa* cells using microarray technology showed only a 1 per cent difference in expression between cells growing in these two states; about 0.5 per cent of genes showed increased expression in biofilms compared to planktonic cells, and about 0.5 per cent showed decreased expression (Whiteley et al., 2001). Thus, whatever the effects of the quorum sensing genes are, the effect is limited to about 60 out of the roughly 6,000 genes in the *P. aeruginosa* chromosome.

In vivo, in the CF lung, determinations of the biofilm mode of growth are limited to microscopic determinations from explanted or autopsy tissues. Microcolony formation by *P. aeruginosa* in the CF lung was initially observed two decades ago (Costerton et al., 1983), and more conclusive evidence that the biofilm mode of growth is actually utilised in this situation was provided only very recently (Singh et al., 2000). In addition to the microscopic observations of microcolony formation, Singh et al. (2000) noted that *in vitro,* certain strains of *P. aeruginosa* expressed different ratios of 3-O-C_{12}-HSL and C_4-HSL depending on whether they were assuming a planktonic or a biofilm mode of growth. They determined the relative amounts of these signalling molecules in sputum samples from chronically infected CF patients and found that most of the strains made more C_4-HSL than 3-O-C_{12}-HSL, a ratio indicative of the biofilm rather than the planktonic growth state.

Even before the pathologic evidence suggesting that *P. aeruginosa* grows as a biofilm in CF lungs, it was observed that sputum cultures of CF patients often yielded *P. aeruginosa* isolates that exhibited a very mucoid phenotype *in vitro* (Doggett, Harrison, and Wallis, 1964; Doggett et al., 1966). The over-produced mucoid exopolysaccharide (MEP), also called alginate, is a high molecular weight, *O*-acetylated α-1-4 linked random polymer of mannuronic and guluronic acids similar in structure to seaweed alginate, which lacks the *O*-acetyl group. MEP encases both individual and small clusters of *P. aeruginosa* cells, imparts a mucoid phenotype on colonies (Figure 13.2), and likely affords additional protection to the bacterial cells within a biofilm.

MEP is often associated with biofilms in the CF lung, but is not required for classic biofilm formation, as it has been defined by *in vitro* studies. Moreover, MEP is generally not present at high levels in environmental biofilms formed by non-mucoid *P. aeruginosa* strains or in biofilms formed *in vitro* by isolates from non-CF patients (Lam et al., 1980). Epidemiologic studies on the pathophysiology of CF lung disease clearly show that it is the emergence of the mucoid phenotype of *P. aeruginosa* that is associated with increased rates of decline in lung function (Pedersen et al., 1992; Demko, Byard, and Davis, 1995). Thus, the contribution of the MEP component of biofilm formation and the quorum sensing component is unclear. Even though MEP is not required for *in vitro* biofilm formation, MEP production does augment the thickness of biofilms, possibly by acting as an intercellular adhesin or a 'cement' and promoting formation of the mushroom- or pillar-like structures associated with mature biofilms (Nivens et al., 2001). *In vivo,* MEP increases the tolerance of biofilms to antibiotics and to host immune effectors, particularly antibody, complement, and phagocytes. Probably as a result of the tenacity imparted

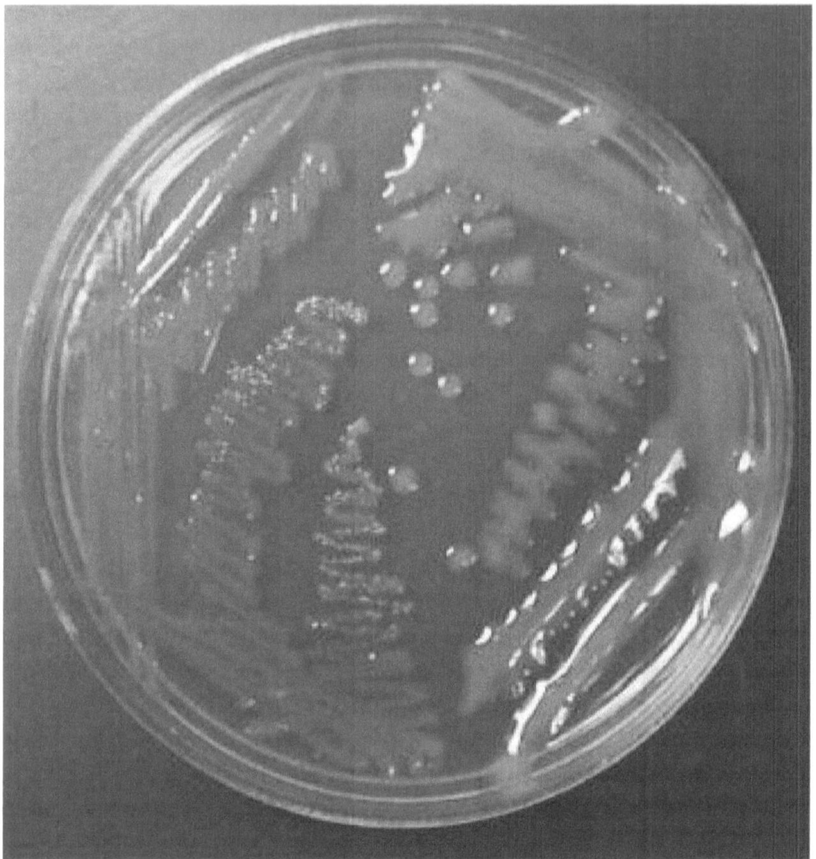

Figure 13.2: CF patients are initially colonised by non-mucoid *P. aeruginosa* (left half of plate). Eventually, the mucoid phenotype emerges (right half of plate). Once established in the CF lung, mucoid *P. aeruginosa* is very difficult, if not impossible, to eradicate. (See colour section.)

on biofilms by MEP, the presence of mucoid *P. aeruginosa* in the lung is prognostic of a poor clinical outcome, and most patients, once colonised, will be chronically infected until they succumb to respiratory failure.

Quorum sensing, although required for biofilm formation, is not required for MEP production, and deletion of the gene that encodes LasI does not inhibit production of MEP (Davies et al., 1998). The transcription factor responsible for inducing the MEP biosynthetic operon does not require a cofactor such as LasI. MEP becomes constitutively expressed when the repressor that normally inhibits transcription of genes involved in its biosynthesis is no longer produced. The MucA protein interacts with another factor responsible

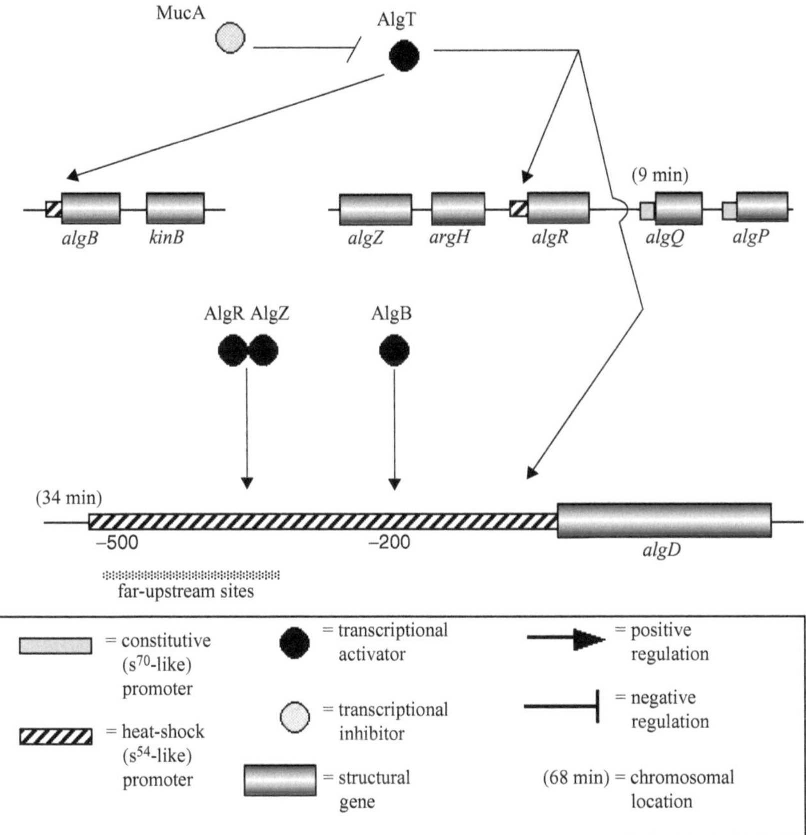

Figure 13.3: MucA normally represses synthesis of MEP. In the absence of MucA, AlgT induces transcription of *algD*, either directly or through transcriptional activation of *algR*, *algZ*, and *algB*. AlgD synthesises the precursor for MEP.

for activation of the MEP biosynthetic operon, AlgU (also called AlgT in some publications and in Figure 13.3). AlgU is an alternative sigma factor with homology to the *E. coli* extreme heat shock sigma factor σ^E (RpoE), and MucA acts as its anti-sigma factor by binding to AlgU and preventing interaction with its target DNA (Govan and Deretic, 1996). Thus, a loss-of-function mutation within the *mucA* gene is responsible for conversion of *P. aeruginosa* from the non-mucoid to the mucoid phenotype (Martin et al., 1993; Boucher et al., 1997). In the absence of MucA, AlgU becomes available to promote transcription of *algD*, directly and indirectly through activation of the *algB* and *algR* promoters (Figure 13.3) (Lyczak et al., 2000). The *algD* gene is part of the MEP biosynthetic gene cluster and encodes the enzyme for GDP-mannose dehydrogenase, which catalyses the first step in the synthesis of the precursor for MEP monosaccharide constituents (Govan and Deretic, 1996).

Even though the genes responsible for this phenotypic switch in morphology have been characterised, the precise reason for the pervasive assumption of the mucoid phenotype by *P. aeruginosa* in the CF lung is unknown. One theory is that hydrogen peroxide and similar stress factors, which are present in the CF lung and possibly associated with chronic neutrophilic inflammation, increase the frequency of mutations within *mucA* (Mathee et al., 1999). Another possibility is that the bacteria are under strong selective pressure due to immune responses elicited during infection of the CF lung and that variants overproducing MEP are better able to survive the onslaught of host defences. *P. aeruginosa* induces a potent antibody response that is ineffective at clearing the mucoid microbial cells, but probably contributes to lung damage by promoting inflammation. Overall, both MEP-dependent and MEP-independent biofilm formation may be key factors in the successful adaptation of mucoid *P. aeruginosa* to the CF lung, since this phenotype promotes resistance to host defences.

5 BIOFILMS AND *P. AERUGINOSA* EVASION OF IMMUNE RESPONSES

There are no formal studies evaluating how the quorum sensing aspect of biofilm formation leads to *P. aeruginosa* resistance to host immune effectors. A number of studies have been carried out addressing the role of MEP in *P. aeruginosa* evasion of host defences, and the growth of *P. aeruginosa* in an MEP-encased biofilm appears to be a key factor in bacterial resistance to opsonic antibodies. Morbidity and mortality in CF patients do not generally result from bacterial virulence factors themselves, and, in fact, the production of virulence factors actually appears to be decreased among mucoid isolates of *P. aeruginosa* (Pedersen, 1992; Deretic, Schurr, and Yu, 1995). Chronic inflammation, rather than the effects of bacterial virulence factors, is generally regarded as the primary cause of tissue damage and subsequent pulmonary insufficiency. *P. aeruginosa* antigens provoke excessive antibody production, leading to the formation of immune complexes that exacerbate inflammation. In essence, *P. aeruginosa* thwarts the immune system by secreting MEP and growing as a biofilm, which evokes an ineffective immune response (Pedersen et al., 1992).

Opsonising antibodies against MEP have been a major focus of study, as these antibodies have been associated with naturally acquired resistance to mucoid *P. aeruginosa* infection (Pier et al., 1987). MEP-specific opsonic antibodies deposit opsonins derived from the complement pathway onto the outer surface of the MEP-encased cells within the pseudomonal biofilm, where the opsonins are available for interaction with complement receptors on phagocytes (Meluleni et al., 1995). This process promotes phagocytosis and killing of the mucoid *P. aeruginosa* cells. However, only a small percentage of CF patients (<5 per cent) actually produce opsonising antibodies to MEP, and the reason for their ability to do this is unknown (Pier et al., 1987). Most CF, along with non-CF, individuals produce only non-opsonising antibodies against MEP (Pier et al., 1987). Non-opsonising, MEP-specific antibodies fail to deposit complement component C3 on the bacterial surface, which is required for efficient phagocytic killing (Meluleni et al., 1995). CF patients do produce opsonising antibodies against other *P. aeruginosa* surface antigens, and although these antibodies can promote phagocytosis of planktonic cells, they cannot promote phagocytosis of the bacteria growing in a biofilm (Pier, Grout, and Desjardins, 1991). Opsonising antibodies that are specific for surface antigens other than MEP and are effective at mediating killing of planktonic cells do trigger C3 deposition onto the bacterial surface, but phagocytosis still fails to occur when the organisms are growing in a biofilm (Meluleni et al., 1995). Most cell surface antigens are buried below the thick layer of MEP, and the antibodies and complement bound to these antigens appear to be unable to bind to receptors on the surfaces of phagocytes (Meluleni et al., 1995). Thus, the MEP-dependent biofilm form of growth interferes with effective immunity to *P. aeruginosa* infection in the CF lung.

Why do CF patients, who are otherwise immunocompetent, fail to produce opsonic, MEP-specific antibodies? One means for this immune evasion seems to relate to the natural production of non-opsonising antibodies that have been found in all humans examined. These ineffective antibodies appear to give rise to a T cell-mediated killing of B cells that otherwise would produce opsonising antibodies to MEP (Pier et al., 1993). When an individual is exposed to MEP, the preexisting non-opsonic antibodies induce formation of immune complexes. These immune complexes bind simultaneously to Fc receptors on activated, cytotoxic T cells and to membrane-bound, MEP-specific opsonic antibody on B cells, effecting T cell-mediated killing of the B cell. For unknown reasons, T cells do not appear to induce apoptosis in B cells that produce non-opsonising MEP antibody. The mechanism by which such B cells escape T cell-mediated killing may be due to a greater density of the epitopes within MEP that these cells recognise, which would sufficiently activate the B cells and make them resistant to the regulatory effects of T cells.

It is not simply the physical encasement of *P. aeruginosa* in an MEP-biofilm, but also the chemical composition of the MEP that is critical for resistance to host defences. Acetylation of MEP constitutes a second mechanism utilised by *P. aeruginosa* to avoid phagocytosis. Three genes within the MEP biosynthetic operon, *algF*, *algI*, and *agJ*, catalyse the transfer of acetate substituents to hydroxyl groups on mannuronic acid residues of MEP (Franklin and Ohman, 1993, 1996). These acetate residues prevent covalent linkage of complement components C3b and C4b to the hydroxyl groups. AlgJ-deficient *P. aeruginosa* produce non-acetylated MEP and are very sensitive to opsonophagocytosis (Pier et al., 2001). Thus, acetylation of MEP augments the resistance of MEP-encased biofilms to the immune system.

Elastase and other proteases may also play a role in protecting *P. aeruginosa* from phagocytic clearance. Both neutrophil and *P. aeruginosa* elastase cleaves complement from the bacterial surface and complement receptors from the surface of phagocytes (Doring et al., 1986; Tosi, Zakem, and Berger, 1990). Elastase could presumably, therefore, prevent complement-mediated phagocytosis, although the relative importance of this phenomenon with respect to immune evasion has not been determined.

In addition to inhibiting phagocytosis, MEP has other properties that allow it to stifle immune effectors, such as the ability to sequester hypochlorite and reactive oxygen intermediates and to inhibit chemotaxis and oxidative burst of neutrophils (Learn, Brestel, and Seetharama, 1987; Simpson, Smith, and Dean, 1989). Together, these immune-evasion techniques facilitate the establishment of a chronic infection by mucoid *P. aeruginosa.* In addition to its ability to dodge the immune system, the organism is selected out to become unresponsive to antibiotic therapy, making resolution of the infection nearly impossible.

6 ANTIBIOTIC TREATMENT, PATHOPHYSIOLOGY, AND BIOFILMS

The CF lung is initially colonised with non-mucoid *P. aeruginosa,* which probably causes intermittent infections. The decision to treat CF patients with antibiotics at this stage varies highly from treatment centre to treatment centre and is the subject of much controversy. Some clinical investigators, particularly those from Denmark, claim that routine treatment following initial isolation of *P. aeruginosa* from the respiratory tract increases life expectancy and overall outcome (Hoiby and Koch, 2000). Other CF centres do not use such an approach and claim that there are no good data to support its practice. One trial comparing routine administration of antibiotics with antibiotic therapy given only in response to symptomatic episodes claimed that there was no difference in the clinical course in response to these approaches, but the study had some serious limitations (Elborn et al., 2000; Marshall and Liou, 2000).

The United Kingdom Cystic Fibrosis Trust has a routine recommendation for immediate use of both oral and nebulised antibiotics upon initial isolation of *P. aeruginosa* from CF patients, whereas in the United States there is no such consensus. Routine administration of nebulised antibiotics to chronically colonised patients does improve numerous aspects of disease pathology in the short term, but it is not clear yet if these aggressive interventions will have an impact on the overall quality of life or life expectancy. As with any prolonged antibiotic treatment, the development of resistance and selection for other organisms that can cause significant pathology has to be closely monitored. Nonetheless, as more data are accumulated and better clinical trials are conducted, it should be feasible to determine if early antibiotic intervention is effective in maintaining pulmonary function in CF and whether this treatment might also effectively inhibit the development of biofilms.

The major biologic and clinical challenges in antibiotic treatment of established *P. aeruginosa* in the CF lung stem from the need to find ways to treat the mucoid biofilm mode of growth in a beneficial manner. Antimicrobial agents such as tobramycin generally retain their efficacy against the planktonic form of *P. aeruginosa*, although resistance can occur (Jensen et al., 1987; Ciofu et al., 1994). When growing as a biofilm, however, *P. aeruginosa* is highly resistant to most antibiotics, and eradication of the biofilm with antimicrobial chemotherapy is very rarely possible. To treat the biofilm organisms, many clinicians use aerosolised antibiotics, which result in much higher local levels of the drug. Therefore, antibiotics can often provide patients with some relief from acute respiratory symptoms.

That high-level antibiotic resistance is due to the biofilm mode of growth is exemplified by the fact that sputum samples, when cultured to yield the planktonic mode of growth, give rise to cells that are sensitive to certain antibiotics to which cells growing as a biofilm are resistant. The mechanism of antibiotic resistance is not well understood and is certainly multifactorial. It is possible that MEP and other factors that contribute to the biofilm mode of growth form a physical barrier against the antibiotic either by excluding its diffusion or by binding to the antibiotic and restricting its access to the periphery of the biofilm (Nichols et al., 1988). Antimicrobial resistance may also result from the reduced rate of bacterial growth within the biofilm (Hoiby et al., 2001). The bottom line, however, is that during chronic infection, *P. aeruginosa* becomes more and more resistant to antibiotics, and many CF patients die due to untreatable, multiresistant *P. aeruginosa*. Although exact molecular mechanisms leading to this state are not fully understood, there is a general consensus that the biofilm mode of growth is a key component to the development of multiresistant *P. aeruginosa* in CF lungs.

7 ALTERNATIVE THERAPY FOR *PSEUDOMONAS* BIOFILM LUNG INFECTIONS

The issue of eventual ineffectiveness of antibiotics in pseudomonal lung infections in CF has lead to the evaluation of alternative therapies. Such alternative therapies fall into two general categories. The first type of approach targets inflammation, and the second targets *P. aeruginosa* itself. However, these two approaches may be interrelated. It is now clear that host factors contribute to the overall formation of the *P. aeruginosa* biofilm in the CF lung. In particular, DNA and actin filaments derived from neutrophils that enter the CF lung in response to inflammatory signals are found in copious amounts in CF sputum and contribute to the thickness of airway secretions (Sheils et al., 1996). Thus, in a chronically infected CF lung, the molecular components of the biofilm are also modified by host factors.

Overall, inflammation subsequent to chronic lung infection appears to be the major cause of decline in respiratory function in infected CF patients. Pseudomonal biofilms, especially mucoid biofilms, are extremely difficult, if not impossible, to eradicate with antibiotics, so several trials have been performed to try to reduce the associated inflammation instead. Patients treated with the antiinflammatory agents ibuprofen and budenoside had a slower rate of respiratory decline than patients given a placebo, suggesting some promise for this approach (Konstan et al., 1995; Bisgaard et al., 1997). Inhaled steroids have not shown a clear benefit in regard to reducing inflammation, although a recent safety study suggested inhaled beclomethasone diproprionate may lead to reduced markers of inflammation (Kennedy, 2001; Wojtczak et al., 2001). Oral corticosteroids have also been tried, but the side effects on growth and physical development of patients were unacceptable, thus preventing their long-term use. The use of high-dose ibuprofen appears to be effective in patients with mild lung disease in regard to reducing the annual rate of decline in pulmonary function, but has not been widely adapted, mostly due to concerns about gastrointestinal bleeding, long-term consequences, and their efficacy in children with FEV1 (forced expiratory volume in 1 second) of <60 per cent. Other antiinflammatory approaches, although promising, require further study before they are put into common therapeutic practice.

Bronchopulmonary secretions in CF patients are extremely viscous due to relative dehydration and to the presence of DNA from dead neutrophils. This thick mucus results in suboptimal performance of the mucociliary escalator and impedes clearance of inhaled bacteria. Recombinant human deoxyribonuclease (rhDNase), an enzyme that degrades DNA, reduces the viscosity of CF sputum and has been shown to decrease the consequences of lung infections (Hodson, 1995). Although rhDNase is frequently used in CF patients, its value over the long term is just beginning to be revealed. Several studies suggest a benefit from use of rhDNAse, and, although the therapy is expensive, its use may be associated with reduced overall costs of care (Ollendorf et al., 2000; Furuya et al., 2001; Suri et al., 2001). Since actin filaments also derived from dead neutrophils can bind DNA and inhibit the activity of rhDNase, forms of the enzyme more active on actin-bound DNA are now being developed (Zahm et al., 2001). In addition, gelsolin, an actin-degrading enzyme, has shown efficacy at reducing the viscosity of CF sputum in *in vitro* studies and has potential as a future mucolytic agent (Vasconcellos et al., 1994).

Some other potential approaches to treatment of *P. aeruginosa* biofilms remain quite speculative, but nonetheless intriguing. Planktonic cells are more sensitive to antibiotics than biofilm cells. Therefore, inducing the release of planktonic cells from biofilms and effectively increasing the time of exposure to antibiotics to destroy planktonic cells before they can form new microcolonies in different locations is another approach. Such an approach might use an enzyme called alginase, which degrades the MEP (alginate), enhances release of bacterial cells from the biofilm, and thus reduces the density of cells in the biofilm until it is extirpated (Linker and Evans, 1984). However, the only alginase enzymes identified to date are of microbial origin, which would make them immunogenic if used in CF patients and likely ineffective. Nonetheless, in a rabbit model of right-sided endocarditis due to mucoid *P. aeruginosa*, alginase had a positive therapeutic effect (Bayer et al., 1992).

Administering long-term antibiotic therapy or encapsulating antibiotics within a liposome to increase the overall exposure of the infecting cells to the drug is another consideration, but this method may not be as successful at eradication of *P. aeruginosa* from the CF lung as is early treatment (Omri et al., 1994; Beaulac et al., 1996; Ramsey et al., 1999).

An innovative and possibly beneficial, but highly speculative potential therapy is based on the increased sensitivity of biofilm cells to the effects of antibiotics when the biofilm is subjected to a mild electric current (Jass and Lappin-Scott, 1996). It is not understood how the electric current increases antibiotic sensitivity of the biofilm. The effect may be due to an increase in the metabolic rate of cells within the biofilm or to a charge-induced increase in penetration of the antibiotic within the biofilm. Alternatively, electrons may bind to oxygen and generate oxygen radicals, which act in concert with the antibiotic (Hoiby et al., 2001). This technique has not yet been used *in vivo*, so its efficacy remains to be determined, as does a means to introduce the electric field into the infected lung.

Overexpression of natural antimicrobial agents which are normally present in the lungs, such as lysozyme and antimicrobial peptides, is an additional approach to prevent or treat biofilm organisms (Akinbi et al., 2000). Presumably, the small size of these agents could allow them to penetrate into biofilms. One drawback is that because many of these agents are highly cationic, the presence of strong anions such as MEP and DNA may prevent these peptides from accessing the *P. aeruginosa* cells. Non-charged antimicrobial peptides may be a more realistic possibility. Again, however, this approach requires further study to support its usefulness as an antipseudomonal therapy.

8 CONCLUSIONS

As chronic *P. aeruginosa* infection in the CF lung is considered to be the hallmark of non-device-related biofilm infections, the recent improvements in identifying and characterising the molecular components and physical properties of these structures has advanced our insight into the pathophysiology of this disease. Although the concept of a biofilm has helped to define the process and to guide thinking about new ways to attack the infecting organisms and improve clinical outcomes for CF patients, we are still highly limited in our overall understanding. This is because the bacterial virulence factors that combine to form a biofilm in the CF lung, notably, the quorum-sensing regulated factors and MEP, and host factors such as DNA and actin all interact in poorly understood ways to contribute to microbial persistence, biofilm formation, sputum viscosity, inhibition of host defence, and exacerbation of inflammation. Attacking individual components, such as the use of antibiotics to limit bacterial growth, antiinflammatory drugs, and rhDNase, can all provide some relief and improvement in clinical condition. Ultimately, some means of preventing the establishment of the *P. aeruginosa* biofilm must be developed and used very early in the lives of CF patients in order to make the major strides needed to improve the quality of life and to increase the life expectancy in CF.

REFERENCES

Akinbi, H. T., Epaud, R., Bhatt, H. and Weaver, T. E. (2000). Bacterial killing is enhanced by expression of lysozyme in the lungs of transgenic mice. *Journal of Immunology*, 165, 5760–5766.

Andersen, D. D. (1938). Cystic fibrosis of the pancreas and its relation to celiac disease. *American Journal of Diseases of Children*, 56, 344–399.

Barasch, J., Kiss, B., Prince, A., Saiman, L., Gruenert, D. and al-Awqati, Q. (1991). Defective acidification of intracellular organelles in cystic fibrosis. *Nature*, 352, 70–73.

Bayer, A. S., Park, S., Ramos, M. C., Nast, C. C., Eftekhar, F. and Schiller, N. L. (1992).

Effects of alginase on the natural history and antibiotic therapy of experimental endocarditis caused by mucoid *Pseudomonas aeruginosa*. *Infection and Immunity*, 60, 3979–3985.

Beaulac, C., Clement-Major, S., Hawari, J. and Lagace, J. (1996). Eradication of mucoid *Pseudomonas aeruginosa* with fluid liposome-encapsulated tobramycin in an animal model of chronic pulmonary infection. *Antimicrobial Agents and Chemotherapy*, 40, 665–669.

Bisgaard, H., Pedersen, S. S., Nielsen, K. G., Skov, M., Laursen, E. M., Kronborg, G., Reimert, C. M., Hoiby, N. and Koch, C. (1997). Controlled trial of inhaled budesonide in patients with cystic fibrosis and chronic bronchopulmonary *Pseudomonas aeruginosa* infection. *American Journal of Respiratory and Critical Care Medicine*, 156, 1190–1196.

Boucher, J. C., Yu, H., Mudd, M. H. and Deretic, V. (1997). Mucoid *Pseudomonas aeruginosa* in cystic fibrosis: characterization of *muc* mutations in clinical isolates and analysis of clearance in a mouse model of respiratory infection. *Infection and Immunity*, 65, 3838–3846.

Burnet, M. (1962). *Natural History of Infectious Disease*. Cambridge: Cambridge University Press.

Cheng, S. H., Gregory, R. J., Marshall, J., Paul, S., Souza, D. W., White, G. A., O'Riordan, C. R. and Smith, A. E. (1990). Defective intracellular transport and processing of CFTR is the molecular basis of most cystic fibrosis. *Cell*, 63, 827–834.

Chillon, M., Casals, T., Mercier, B., Bassas, L., Lissens, W., Silber, S., Romey, M. C., Ruiz-Romero, J., Verlingue, C. and Claustres, M. (1995). Mutations in the cystic fibrosis gene in patients with congenital absence of the vas deferens. *New England Journal of Medicine*, 332, 1475–1480.

Ciofu, O., Giwercman, B., Pedersen, S. S. and Hoiby, N. (1994). Development of antibiotic resistance in *Pseudomonas aeruginosa* during two decades of antipseudomonal treatment at the Danish CF Centre. *Apmis*, 102, 674–680.

Costerton, J. W. (1984). The etiology and persistence of cryptic bacterial infections: a hypothesis. *Reviews of Infectious Diseases*, 6, S608–S616.

Costerton, J. W., Lam, J., Lam, K. and Chan, R. (1983). The role of the microcolony mode of growth in the pathogenesis of *Pseudomonas aeruginosa* infections. *Reviews of Infectious Diseases*, 5, S867–S873.

Costerton, J. W., Stewart, P. S. and Greenberg, E. P. (1999). Bacterial biofilms: a common cause of persistent infections. *Science*, 284, 1318–1322.

Davies, D. G., Parsek, M. R., Pearson, J. P., Iglewski, B. H., Costerton, J. W. and Greenberg, E. P. (1998). The involvement of cell-to-cell signals in the development of a bacterial biofilm. *Science*, 280, 295–298.

Demko, C. A., Byard, P. J. and Davis, P. B. (1995). Gender differences in cystic fibrosis: *Pseudomonas aeruginosa* infection. *Journal of Clinical Epidemiology*, 48, 1041–1049.

Deretic, V., Schurr, M. J. and Yu, H. (1995). *Pseudomonas aeruginosa*, mucoidy and the chronic infection phenotype in cystic fibrosis. *Trends in Microbiology*, 3, 351–356.

Doggett, R. G., Harrison, G. M., Stillwell, R. N. and Wallis, E. S. (1966). An atypical *Pseudomonas aeruginosa* associated with cystic fibrosis of the pancreas. *Journal of Paediatrics*, 68, 215–221.

Doggett, R. G., Harrison, G. M. and Wallis, E. S. (1964). Comparison of some properties of *Pseudomonas aeruginosa* isolated from infections in persons with and without cystic fibrosis. *Journal of Bacteriology*, 87, 427–431.
Doring, G., Goldstein, W., Botzenhart, K., Kharazmi, A., Schiotz, P. O., Hoiby, N. and Dasgupta, M. (1986). Elastase from polymorphonuclear leucocytes: a regulatory enzyme in immune complex disease. *Clinical and Experimental Immunology*, 64, 597–605.
Elborn, J. S., Prescott, R. J., Stack, B. H., Goodchild, M. C., Bates, J., Pantin, C., Ali, N., Shale, D. J. and Crane, M. (2000). Elective versus symptomatic antibiotic treatment in cystic fibrosis patients with chronic *Pseudomonas* infection of the lungs. *Thorax*, 55, 355–358.
Franklin, M. J. and Ohman, D. E. (1993). Identification of *algF* in the alginate biosynthetic gene cluster of *Pseudomonas aeruginosa* which is required for alginate acetylation. *Journal of Bacteriology*, 175, 5057–5065.
Franklin, M. J. and Ohman, D. E. (1996). Identification of *algI* and *algJ* in the *Pseudomonas aeruginosa* alginate biosynthetic gene cluster which are required for alginate O acetylation. *Journal of Bacteriology*, 178, 2186–2195.
Furuya, M. E., Lezana-Fernandez, J. L., Vargas, M. H., Hernandez-Sierra, J. F. and Ramirez-Figueroa, J. L. (2001). Efficacy of human recombinant DNase in pediatric patients with cystic fibrosis. *Archives of Medical Research*, 32, 32–34.
Gabriel, S. E., Brigman, K. N., Koller, B. H., Boucher, R. C. and Stutts, M. J. (1994). Cystic fibrosis heterozygote resistance to cholera toxin in the cystic fibrosis mouse model. *Science*, 266, 107–109.
Gilligan, P. H. (1991). Microbiology of airway disease in patients with cystic fibrosis. *Clinical Microbiology Reviews*, 4, 35–51.
Govan, J. R. and Deretic, V. (1996). Microbial pathogenesis in cystic fibrosis: mucoid *Pseudomonas aeruginosa* and *Burkholderia cepacia*. *Microbiological Reviews*, 60, 539–574.
Harper, M. H. (1930). Two cases of congenital pancreatic steattorhea with infantilism. *Medical Journal of Australia*, 2, 663–666.
Hodson, M. E. (1995). Clinical studies of rhDNase in moderately and severely affected patients with cystic fibrosis – an overview. *Respiration*, 62, 29–32.
Hoiby, N. (1982). Microbiology of lung infections in cystic fibrosis. *Acta Paediatrica Scandanavica Supplement*, 301, 33–54.
Hoiby, N. and Koch, C. (2000). Maintenance treatment of chronic *Pseudomonas aeruginosa* infection in cystic fibrosis. *Thorax*, 55, 349–350.
Hoiby, N., Krogh Johansen, H., Moser, C., Song, Z., Ciofu, O. and Kharazmi, A. (2001). *Pseudomonas aeruginosa* and the *in vitro* and *in vivo* biofilm mode of growth. *Microbes and Infection*, 3, 23–35.
Ip, M. S. and Lam, W. K. (1996). Bronchiectasis and related disorders. *Respirology*, 1, 107–114.
Jass, J. and Lappin-Scott, H. M. (1996). The efficacy of antibiotics enhanced by electrical currents against *Pseudomonas aeruginosa* biofilms. *Journal of Antimicrobial Chemotherapy*, 38, 987–1000.
Jensen, T., Pedersen, S. S., Nielsen, C. H., Hoiby, N. and Koch, C. (1987). The efficacy

and safety of ciprofloxacin and ofloxacin in chronic *Pseudomonas aeruginosa* infection in cystic fibrosis. *Journal of Antimicrobial Chemotherapy*, 20, 585–594.

Kennedy, M. J. (2001). Inflammation and cystic fibrosis pulmonary disease. *Pharmacotherapy*, 21, 593–603.

Kielhofner, M., Atmar, R. L., Hamill, R. J. and Musher, D. M. (1992). Life-threatening *Pseudomonas aeruginosa* infections in patients with human immunodeficiency virus infection. *Clinical Infectious Diseases*, 14, 403–411.

Konstan, M. W., Byard, P. J., Hoppel, C. L. and Davis, P. B. (1995). Effect of high-dose ibuprofen in patients with cystic fibrosis. *New England Journal of Medicine*, 332, 848–854.

Krivan, H. C., Ginsburg, V. and Roberts, D. D. (1988). *Pseudomonas aeruginosa* and *Pseudomonas cepacia* isolated from cystic fibrosis patients bind specifically to gangliotetraosylceramide (asialo GM1) and gangliotriaosylceramide (asialo GM2). *Archives of Biochemistry and Biophysics*, 260, 493–496.

Kunzelmann, K., Schreiber, R., Nitschke, R. and Mall, M. (2000). Control of epithelial Na^+ conductance by the cystic fibrosis transmembrane conductance regulator. *Pflugers Archiv-European Journal of Physiology*, 440, 191–201.

Lam, J., Chan, R., Lam, K. and Costerton, J. W. (1980). Production of mucoid microcolonies by *Pseudomonas aeruginosa* within infected lungs in cystic fibrosis. *Infection and Immunity*, 28, 546–556.

Learn, D. B., Brestel, E. P. and Seetharama, S. (1987). Hypochlorite scavenging by *Pseudomonas aeruginosa* alginate. *Infection and Immunity*, 55, 1813–1818.

Linker, A. and Evans, L. R. (1984). Isolation and characterization of an alginase from mucoid strains of *Pseudomonas aeruginosa*. *Journal of Bacteriology*, 159, 958–964.

Lyczak, J. B., Cannon, C. L. and Pier, G. B. (2000). Establishment of *Pseudomonas aeruginosa* infection: lessons from a versatile opportunist. *Microbes and Infection*, 2, 1051–1060.

Marshall, B. C. and Liou, T. G. (2000). Elusiveness of ideal approach to *Pseudomonas aeruginosa* infection complicating cystic fibrosis. *Lancet*, 356, 613–614.

Martin, D. W., Schurr, M. J., Mudd, M. H., Govan, J. R., Holloway, B. W. and Deretic, V. (1993). Mechanism of conversion to mucoidy in *Pseudomonas aeruginosa* infecting cystic fibrosis patients. *Proceedings of the National Academy of Sciences of the USA*, 90, 8377–8381.

Mathee, K., Ciofu, O., Sternberg, C., Lindum, P. W., Campbell, J. I., Jensen, P., Johnsen, A. H., Givskov, M., Ohman, D. E., Molin, S., Hoiby, N. and Kharazmi, A. (1999). Mucoid conversion of *Pseudomonas aeruginosa* by hydrogen peroxide: a mechanism for virulence activation in the cystic fibrosis lung. *Microbiology*, 145, 1349–1357.

Meluleni, G. J., Grout, M., Evans, D. J. and Pier, G. B. (1995). Mucoid *Pseudomonas aeruginosa* growing in a biofilm *in vitro* are killed by opsonic antibodies to the mucoid exopolysaccharide capsule but not by antibodies produced during chronic lung infection in cystic fibrosis patients. *Journal of Immunology*, 155, 2029–2038.

Mulvey, M. A., Schilling, J. D., Martinez, J. J. and Hultgren, S. J. (2000). Bad bugs and beleaguered bladders: interplay between uropathogenic *Escherichia coli* and innate host defenses. *Proceedings of the National Academy of Sciences of the USA*, 97, 8829–8835.

Nichols, W. W., Dorrington, S. M., Slack, M. P. and Walmsley, H. L. (1988). Inhibition of tobramycin diffusion by binding to alginate. *Antimicrobial Agents and Chemotherapy*, 32, 518–523.

Nivens, D. E., Ohman, D. E., Williams, J. and Franklin, M. J. (2001). Role of alginate and its O-acetylation in formation of *Pseudomonas aeruginosa* microcolonies and biofilms. *Journal of Bacteriology*, 183, 1047–1057.

O'Toole, G. A. and Kolter, R. (1998). Flagellar and twitching motility are necessary for *Pseudomonas aeruginosa* biofilm development. *Molecular Microbiology*, 30, 295–304.

Ollendorf, D. A., McGarry, L. J., Watrous, M. L. and Oster, G. (2000). Use of rhDNase therapy and costs of respiratory-related care in patients with cystic fibrosis. *Annals of Pharmacotherapy*, 34, 304–308.

Omri, A., Beaulac, C., Bouhajib, M., Montplaisir, S., Sharkawi, M. and Lagace, J. (1994). Pulmonary retention of free and liposome-encapsulated tobramycin after intratracheal administration in uninfected rats and rats infected with *Pseudomonas aeruginosa*. *Antimicrobial Agents and Chemotherapy*, 38, 1090–1095.

Pedersen, S. S. (1992). Lung infection with alginate-producing, mucoid *Pseudomonas aeruginosa* in cystic fibrosis. *Apmis Supplementum*, 1–79.

Pedersen, S. S., Hoiby, N., Espersen, F. and Koch, C. (1992). Role of alginate in infection with mucoid *Pseudomonas aeruginosa* in cystic fibrosis. *Thorax*, 47, 6–13.

Pesci, E. C. and Iglewski, B. H. (1997). The chain of command in *Pseudomonas* quorum sensing. *Trends in Microbiology*, 5, 132–134.

Pesci, E. C., Pearson, J. P., Seed, P. C. and Iglewski, B. H. (1997). Regulation of las and rhl quorum sensing in *Pseudomonas aeruginosa*. *Journal of Bacteriology*, 179, 3127–3132.

Pier, G. B. (2000). Role of the cystic fibrosis transmembrane conductance regulator in innate immunity to *Pseudomonas aeruginosa* infections. *Proceedings of the National Academy of Sciences of the USA*, 97, 8822–8828.

Pier, G. B., Coleman, F., Grout, M., Franklin, M. and Ohman, D. E. (2001). Role of alginate O-acetylation in resistance of mucoid *Pseudomonas aeruginosa* to opsonic phagocytosis. *Infection and Immunity*, 69, 1895–1901.

Pier, G. B., Grout, M. and Desjardins, D. (1991). Complement deposition by antibodies to *Pseudomonas aeruginosa* mucoid exopolysaccharide (MEP) and by non-MEP specific opsonins. *Journal of Immunology*, 147, 1869–1876.

Pier, G. B., Grout, M. and Zaidi, T. S. (1997). Cystic fibrosis transmembrane conductance regulator is an epithelial cell receptor for clearance of *Pseudomonas aeruginosa* from the lung. *Proceedings of the National Academy of Sciences of the USA*, 94, 12088–12093.

Pier, G. B., Grout, M., Zaidi, T. S. and Goldberg, J. B. (1996). How mutant CFTR may contribute to *Pseudomonas aeruginosa* infection in cystic fibrosis. *American Journal of Respiratory and Critical Care Medicine*, 154, S175–S182.

Pier, G. B., Grout, M., Zaidi, T., Meluleni, G., Mueschenborn, S. S., Banting, G., Ratcliff, R., Evans, M. J. and Colledge, W. H. (1998). *Salmonella typhi* uses CFTR to enter intestinal epithelial cells. *Nature*, 393, 79–82.

Pier, G. B., Saunders, J. M., Ames, P., Edwards, M. S., Auerbach, H., Goldfarb, J., Speert, D. P. and Hurwitch, S. (1987). Opsonophagocytic killing antibody to *Pseudomonas aeruginosa* mucoid exopolysaccharide in older noncolonized patients with cystic fibrosis. *New England Journal of Medicine*, 317, 793–798.

Pier, G. B., Takeda, S., Grout, M. and Markham, R. B. (1993). Immune complexes from immunized mice and infected cystic fibrosis patients mediate murine and human T cell killing of hybridomas producing protective, opsonic antibody to *Pseudomonas aeruginosa. Journal of Clinical Investigation*, 91, 1079–1087.

Ramsey, B. W., Pepe, M. S., Quan, J. M., Otto, K. L., Montgomery, A. B., Williams-Warren, J., Vasiljev, K. M., Borowitz, D., Bowman, C. M., Marshall, B. C., Marshall, S. and Smith, A. L. (1999). Intermittent administration of inhaled tobramycin in patients with cystic fibrosis. Cystic Fibrosis Inhaled Tobramycin Study Group. *New England Journal of Medicine*, 340, 23–30.

Riordan, J. R., Rommens, J. M., Kerem, B., Alon, N., Rozmahel, R., Grzelczak, Z., Zielenski, J., Lok, S., Plavsic, N. and Chou, J. L. (1989). Identification of the cystic fibrosis gene: cloning and characterization of complementary DNA. *Science*, 245, 1066–1073.

Scanlin, T. F. and Glick, M. C. (1999). Terminal glycosylation in cystic fibrosis. *Biochimica et Biophysica Acta*, 1455, 241–253.

Schroeder, T. H., Zaidi, T. and Pier, G. B. (2001). Lack of adherence of clinical isolates of *Pseudomonas aeruginosa* to asialo-GM(1) on epithelial cells. *Infection and Immunity*, 69, 719–729.

Sheils, C. A., Kas, J., Travassos, W., Allen, P. G., Janmey, P. A., Wohl, M. E. and Stossel, T. P. (1996). Actin filaments mediate DNA fiber formation in chronic inflammatory airway disease. *American Journal of Pathology*, 148, 919–927.

Simpson, J. A., Smith, S. E. and Dean, R. T. (1989). Scavenging by alginate of free radicals released by macrophages. *Free Radical Biology and Medicine*, 6, 347–353.

Singh, P. K., Schaefer, A. L., Parsek, M. R., Moninger, T. O., Welsh, M. J. and Greenberg, E. P. (2000). Quorum-sensing signals indicate that cystic fibrosis lungs are infected with bacterial biofilms. *Nature*, 407, 762–764.

Suri, R., Metcalfe, C., Lees, B., Grieve, R., Flather, M., Normand, C., Thompson, S., Bush, A. and Wallis, C. (2001). Comparison of hypertonic saline and alternate-day or daily recombinant human deoxyribonuclease in children with cystic fibrosis: a randomised trial. *Lancet*, 358, 1316–1321.

Tosi, M. F., Zakem, H. and Berger, M. (1990). Neutrophil elastase cleaves C3bi on opsonized pseudomonas as well as CR1 on neutrophils to create a functionally important opsonin receptor mismatch. *Journal of Clinical Investigation*, 86, 300–308.

Vasconcellos, C. A., Allen, P. G., Wohl, M. E., Drazen, J. M., Janmey, P. A. and Stossel, T. P. (1994). Reduction in viscosity of cystic fibrosis sputum *in vitro* by gelsolin. *Science*, 263, 969–971.

Whiteley, M., Bangera, M. G., Bumgarner, R. E., Parsek, M. R., Teitzel, G. M., Lory, S. and Greenberg, E. P. (2001). Gene expression in *Pseudomonas aeruginosa* biofilms. *Nature*, 413, 860–864.

Wojtczak, H. A., Kerby, G. S., Wagener, J. S., Copenhaver, S. C., Gotlin, R. W., Riches, D. W. and Accurso, F. J. (2001). Beclomethasone diproprionate reduced airway inflammation without adrenal suppression in young children with cystic fibrosis: a pilot study. *Pediatric Pulmonology*, 32, 293–302.

Zahm, J. M., Debordeaux, C., Maurer, C., Hubert, D., Dusser, D., Bonnet, N., Lazarus, R. A. and Puchelle, E. (2001). Improved activity of an actin-resistant DNase I variant on the cystic fibrosis airway secretions. *American Journal of Respiratory Critical Care Medicine*, 163, 1153–1157.

Index

For EU product safety concerns, contact us at Calle de José Abascal, 56–1°, 28003 Madrid, Spain or eugpsr@cambridge.org.

www.ingramcontent.com/pod-product-compliance
Ingram Content Group UK Ltd.
Pitfield, Milton Keynes, MK11 3LW, UK
UKHW042208080726
473066UK00007B/307

* 9 7 8 1 1 0 7 4 0 3 4 5 1 *